Advances in Factor Analysis and Structural Equation Models

Advances in Factor Analysis and Structural Equation Models

Karl G. Jöreskog
Dag Sörbom

University of Uppsala
Uppsala, Sweden

with an Introduction by William W. Cooley

Edited by Jay Magidson

Abt Books
Cambridge, Massachusetts

Library of Congress Catalog Card Number 79-52433

Printed in the United States of America.

ISBN: 0-89011-535-4

Contents

Preface

> . . . if there be any relation among objects which it imports us to know perfectly, it is that of cause and effect. On this are founded all our reasoning concerning matter of fact or existence. By means of it alone we attain any assurance concerning objects which are removed from the present testimony of our memory and senses. The only immediate utility of all sciences is to teach us how to control and regulate future events by their causes. Our thoughts and enquiries are, therefore, every moment, employed about this relation.
>
> David Hume, 1777

In the 1950s and 1960s some important articles urged social scientists to reconsider their aversion to making causal interpretations of nonexperimental data.[1] These led to an upsurge of interest in path analysis, first introduced by Sewall Wright in 1918, developed further by Wright over the next fourteen years, and rediscovered by sociologists in the 1960s. The United States at this time had established its first wave of Great Society legislation and critics requested that serious attention be given to the problems in evaluating the reforms.[2]

These two developments led the Social Science Research Council (SSRC) in 1968 to recognize the urgent need for further methodological developments in causal modeling, especially advances that would improve the design and evaluation of social interventions – that is, programs designed to produce changes in social conditions, social processes, or individual behavior in areas such as education, welfare, housing, and health.

A conference was organized by SSRC in early 1970 and another later that year on structural equation (causal) models with participation by leading investigators in sociology, economics, political science, psychology, and other areas of the social sciences. Two major textbooks on structural equation models were a direct result of the conference.[3] These and other books have had a major impact on academicians throughout the 1970s. Yet the evaluation of social programs is just now beginning to use some of these new methodological approaches.

One useful approach is Karl Jöreskog's general Linear Structural Relations (LISREL) model and recent extensions of it by him and Dag Sörbom. The articles in this book develop the basic ideas of causal modeling and illustrate them for a wide range of applications in the social sciences.

Let us consider the general problem of making causal interpretations from nonexperimental data, especially when little is known a priori about the causal mechanisms that generated the data (the typical case in the social sciences?). Suppose we run a regression of y, a measure of self-esteem of high school students, on x, a set of variables that may be related to self-esteem. It is not uncommon in this situation, especially if the predictor variables are correlated with each other, to find the signs of one or more of the significant regression coefficients to be negative although we suspected that they would be positive. Does this mean that the associated predictor variables have a negative "effect" on self-esteem? Not necessarily.

Arthur Goldberger points out that regression analysis is a legitimate method for generating unbiased predictions for the dependent variable.[4] Whether the predictors are in fact causally prior to, causally unrelated to, or causally determined by the dependent variable, the estimated regression coefficients generate unbiased predictions based on the given values of the predictors. However, to interpret these coefficients as structural measures of causal effect, certain requirements must be met that are typically not met in most applications of regression analysis in the social sciences. When they are met, the regression model transforms into a structural equation (causal) model and the predictor variables become explanatory variables, or explanatory factors.

These conditions for enabling one to infer causal interpretations from nonexperimental data are not found in traditional textbooks on statistics or data analysis.[5] Their absence is not surprising, since there still is no agreement among philosophers on the meaning of causality or on what a complete set of such requirements should be.

Another way to view the situation is in terms of model misspecification. Causal modeling requires researchers to think through all the conceptual implications associated with a given specification. A causal model is said to be misspecified (and inadequate) if the assumptions are incorrect and do not allow the desired causal interpretation of the parameters. The question now becomes, When is a model misspecified?

One common source of misspecification derives from the omission of a major causative factor that is correlated with at least one predictor variable included in the model. In this case the predictor variable serves as a proxy for the omitted factor, and its regression coefficient reflects the effect of the omitted factor weighted by the correlation between the variable and the omitted factor. If the variable is originally hypothesized to represent some factor other than the omitted factor, the result will be a spurious effect. On the other hand, if the variable was mistakenly preconceived to perfectly measure the omitted causative factor, the coefficient will be biased due to unspecified measurement error. In either case the model is misspecified. The algebraic proof is straightforward.[6]

Causal modeling in the social sciences is especially challenging and complex because the causes of a given outcome are typically unknown or not susceptible to direct observation. Moreover, the observed patterns among the variables may be consistent with a variety of causal explanations. Thus it is often desirable to formulate and test a variety of models. Some models may be ruled out on grounds of common sense. (For example, although educational aspirations may differ by race, aspirations cannot cause race.) Other models might be rejected only if they yield inferences inconsistent (within statistical limits) with the pattern of relationships actually observed.

Causal modeling can be used to state theory more exactly, to test the theory more precisely, and to yield a more thorough understanding of the data. Costner notes that it involves a new mode of thinking for many researchers:

> The pervasive influence of causal models, path analysis, and structural equation models in sociological methodology is evidenced not simply by the increasing use of such modes of analysis in empirical sociological studies but also by the ramifications of this mode of thinking in a variety of methodological issues. During the relatively brief period that sociologists have focused attention on these methods, the mode of thinking that they entail has permeated our conceptions of theory construction, measurement problems, and data analysis. Guided by this mode of thinking, we seem to be in the process of redesigning our approaches to multivariate analysis, reconceptualizing earlier notions of reliability and validity, reevaluating the relative merits of standardized and unstandardized coefficients, rethinking the inferences from experimentation, reinterpreting the utility of factor analysis, revising some of our earlier conceptions of macrosociological knowledge, and reassessing the data needs of sociology. This relatively rapid shift in our way of thinking about the methodology of social research has immersed us in a vocabulary that is still relatively new to many sociologists: *identification problems, recursive systems, specification error, correlated errors, structural parameters,* and numerous other terms that all but the relatively recent Ph.D.'s (and even some of those) missed in their graduate training.[7]

Jöreskog's LISREL model meets many of the needs of the causal modeler. It is sufficiently general to handle latent factors, measurement errors, and reciprocal causation. The LISREL model consists of two parts: *the measurement model* and *the structural equation model.* The measurement model specifies how the latent factors are measured in terms of the observed variables and is used to describe the measurement properties (reliabilities and validities) of the observed variables. The structural equation model specifies the causal relationships among the factors and is used to describe the causal effects and the amount of unexplained variance.

The methods in this book do not provide final answers to the questions of model specification. What they offer the researcher is the flexibility to formulate and test a variety of causal models and to guide the analysis toward more adequate explanations of the relationships embedded in the data. In addition to being a valuable reference, this book is useful as a text or supplementary text for graduate or advanced undergraduate students in the social sciences.

I wish to thank Karl Jöreskog and Dag Sörbom for their support and encouragement in assembling this book. I am also grateful to William Cooley for allowing his state-of-the-art address on methodology in educational research to be reprinted here. Although his address was written specifically for educational researchers, I believe that all social scientists will benefit from it, since we all face similar methodological concerns. I also wish to thank the editors and publishers of the journals and books in which the articles originally appeared for their permission to reprint. The journals and books are *Applications of Statistics* (North Holland Publishing Co.), *British Journal of Mathematical Psychology,* vol. 2 (W. H. Freeman & Co.), *Educational Researcher, Latent Variables in Socioeconomic Models* (North Holland Publishing Co.), and *Psychometrika.*

Jay Magidson

Cambridge, Massachusetts
April 1979

[1] For example, see H. A. Simon, "Spurious Correlation: A Causal Interpretation," *Journal of the American Statistical Association* 49 (1954):467-479; J. W. Tukey, "Causation, Regression and Path Analysis," in *Statistics and Mathematics in Biology,* ed. O. Kempthorne et al. (Ames, Iowa: Iowa State University, 1954), pp. 35-66; H. M. Blalock, Jr., *Causal Inferences in Nonexperimental Research* (Chapel Hill, N.C.: University of North Carolina, 1964); and O. D. Duncan, "Path Analysis: Sociological Examples," *American Journal of Sociology* 72 (1966): 1-16.

[2] See D. T. Campbell, "Reforms as Experiments," *American Psychologist* 24, no. 4 (1969):409-429, reprinted with corrections in *Readings in Evaluation Research,* ed. F. G. Caro (New York: Russell Sage Foundation, 1971), pp. 233-261.

[3] A. S. Goldberger and O. D. Duncan, *Structural Equation Models in the Social Sciences* (New York: Seminar Press, 1973); and O. D. Duncan, *Introduction to Structural Equation Models* (New York: Academic Press, 1975).

[4] Arthur S. Goldberger, "Structural Equation Models: An Overview," in *Structural Equation Models in the Social Sciences,* ed. Goldberger and Duncan, p. 2.

[5] For a recent attempt to state the conditions, see chapter 4 in David Kenny, *Correlation and Causality* (New York: Wiley-Interscience, forthcoming). See also chapter 1 in Bagozzi, R. P., *Causal Models in Marketing* (New York: John Wiley and Sons, forthcoming).

[6] Goldberger, "Structural Equation Models: An Overview," pp. 3-6.

[7] Herbert Costner, *Sociological Methodology, 1973-1974* (San Francisco: Jossey-Bass, 1974), pp. ix-xiv.

Acknowledgments

We wish to thank Jay Magidson for his great help in preparing this volume for publication. He wrote the preface and the introductions to the three sections and did whatever editorial work needed to be done. We also wish to thank Professor William Cooley for allowing his 1978 state-of-the-art address to the American Educational Research Association to be used as a general introduction. Thanks are also due to Lotten Gellin for typing chapter 1 and the addendum to chapter 2.

Karl G. Jöreskog
Dag Sörbom

Uppsala, Sweden
April 1979

Introduction

Structural Equations and Explanatory Observational Studies[1]

William W. Cooley

The task assigned to me for today was to describe the state-of-the-art in the design and analysis of research studies involving relationships among variables. That seemed impossible to do in 25 minutes, so a necessary first step was to decide on a particular focus.

The focus of these remarks is on what the statistical literature refers to as observational studies. More specifically, studies that inquire into the learning and development of human beings in the natural environments in which these processes occur. These studies are multivariate and longitudinal, recording the variances and covariances of events and individual differences as they occur and unfold. I will particularly focus on observational studies that are intended to be explanatory. That is, I am talking about studies of such phenomena as program effectiveness, in which an attempt is made to explain why the students in some programs achieve more than the students in others. This is in contrast to descriptive studies like National Assessment, which attempt to *describe* such things as the levels of achievement obtained in various sectors of American schooling. As Jöreskog (1976) put it:

> . . . a descriptive study involves the collection of data, tabulation and graphical presentation of statistical facts, whereas, an explanatory study also involves the specification of causal hypotheses, estimation of causal models and testing of the validity of these models and hypotheses. (p. 1)

In 1956, Wold observed that the methodology for explanatory observational studies was a vastly underdeveloped field of inquiry, particularly among statisticians. He observed that most of the work that was being done in the area was being done by people in fields such as economics, genetics, and sociology. Although there have been some important improvements in the methodology for such studies since Wold's review, the development and application of this methodology has received far too little attention in educational research.

There are many reasons why this is so. One important factor has been the dominance of experimental paradigms in educational research. Many of us have been taught to believe that the only possible way of supporting causal inferences is through the use of true experimental designs. If randomization is not possible, then we are to conduct a quasi-experiment, realizing that we will suffer some loss of internal validity. However, it is important to remember that most of what is known about people and the universe has not been based on experimentation, but on observation. If we are to further our understanding of educational processes, and do so in a way that will allow us to improve those processes. it is essential that we give more attention to developing methods for conducting explanatory observational studies.

At times you may think that I am coming out against the use of experiments. I want to assure you that this is not the case. Experimentation is clearly a powerful scientific procedure. But we must recognize that experimental paradigms are not the only ones available, and that explanatory observational studies have the potential to make major contributions to our understanding of educational processes.

It is also important to point out that I am talking about *quantitative* observational studies. Admittedly, there has been only limited success in quantifying even those things we all "know" to be true (for instance, that school differences make a difference). However, I strongly believe that quantitative approaches are a necessary ingredient in a total effort toward achieving an understanding of educational processes (Cooley, Leinhardt, & McGrail, 1977).

A major problem with observational studies has been that they are generally defined negatively. For example, reference is made to the fact that randomization is absent, or treatment is *not* controlled, or one is applying *non*-experimental designs, etc. It usually sounds apologetic. What I want to do today is focus on the major requirements for sound observational studies, requirements that, if met, can yield consistent, convincing, useful explanations of educational phenomena. I shall discuss what I consider to be the most critical features of such studies under the following three topics:

(1) the sampling framework, which affects the generalizability of the observed relationships;
(2) the theoretical model, which describes the hypothesized causal structure of the variables under consideration; and
(3) the statistical procedure, which is used to analyze the network of observed relationships for the purpose of establishing the plausibility of the theoretical model and estimating its parameters.

The Sampling Framework

First, let us consider the question of population and its relevance to observational studies. There seem to be "two branches" in the literature on observational studies. One branch, exemplified by the writings of Cochran (1965) and others (e.g., Lord, 1960; Kenny, 1975; Overall & Woodward, 1975), deals with multi-population observational studies, in which different samples are administered different treatments, but random assignment of subjects to treatments is absent. I consider this branch as part of the literature on quasi-experimental design.

Others (e.g., Wold, 1956; Wiley & Hornik, 1973) consider observational studies that involve a single population. In such studies, one defines a population, draws a sample from it, and studies the relationships among variables measured on that sample. For random samples from a population, the errors in estimating relationships for the population are a function of the sample size, the reliability with which the variables can be measured, and the degree to which measures can be obtained for the entire sample. Important contributions of methodologists have been to show how to take the sampling error, the measurement error, and the missing data bias into account in estimating such relationships.

The estimated relationships may certainly include variables that reflect differences in educational treatments, but in single population studies the treatments are observed as they are currently occurring in the population sampled. The question of whether one is dealing with a single population or with two or more populations is really a question of how the sample was drawn, and whether there was any planned intervention following the drawing of the sample. As Thorndike (1942) pointed out over three decades ago, comparing two samples that have been given two treatments without random assignment to each treatment is best thought of as a comparison of two populations. Under these conditions, there are many possible differences between the two populations other than the imposed treatment differences that may be producing differences in the dependent variable.

A prime example of a multi-population study is the evaluation of the National Follow Through Program (FT), a quasi-experiment that has attempted to examine the effectiveness of a variety of approaches to primary grade instruction for compensatory education. This study of planned variation was a quasi-*experiment* because it intervened in the sample selected by introducing one of 20 possible innovative programs into each school selected for participation. It was a quasi-experiment, because the new program that was used in each school was self-selected by that school, not randomly assigned to it.

Intervention following sample selection changes the situation from a single population observational study (with the observed relationships generalizable to the population sampled) to a multi-population quasi-experiment. FT involves contrasts among 20 different treatment groups, each of which has its own controls, and each of which is best thought of as coming from a separate population. (See, for example, DiCostanzo & Eichelberger, 1977, for further discussion of this problem.)

In contrast to FT is the Sustaining Effects Study (SES) of compensatory education (Hoepfner, Wellisch, & Zagorski, 1977), which is currently being conducted by the System Development Corporation. This is an excellent example of a single population observational study. They have drawn a representative sample of 250 of the nation's elementary schools in a manner that allows them to estimate population relationships between such variables as school practices and student achievement.

It may be useful to compare some important aspects of FT and SES for the purpose of illustration. Both were intended to provide information to guide federal policy in the area of compensatory education. Both involved a large number of school sites. But the FT design did not draw a representative sample of schools, nor did it randomly assign schools to treatments. Thus, it suffers from loss of both external validity (i.e., being able to generalize any observed relationships to some population of interest) and internal validity (i.e., being able to attribute student outcome differences to differences in the imposed treatments).

As SES attempts to identify effective practices for improving the development of basic skills, it will not be handicapped because it did not insure a wide variation of school practices, as the FT interventions did. That variety occurs "naturally," as can be seen in the comparison of practices found in the FT "experimental" and "control" schools. Across all participating schools, the classroom practices that were found to be related to student learning were similarly distributed in the experimental and control schools (see Cooley & Lohnes, 1976, pp. 175-180). In addition, enthusiasts of the experimental paradigm in educational research should be considerably sobered by the wide variations in program-relevant practices among classrooms within any one of the 20 FT treatments (Stallings, 1975; Abt, 1977; House et al., 1977).

The Follow Through program has been an extremely important and useful enterprise. Much has been learned from it. But the contrast between SES and FT is most valuable because it illustrates a tendency in education to move prematurely into intervention-type, quasi-experiments before the relationships are established among the major variables currently operating in a particular causal network. Confident descriptions of such networks *prior* to the implementation of quasi-experiments are necessary if we hope to be able to generalize from such manipulations. Let us now turn to the need for theoretical models and return to this important point later.

The Need For Theoretical Models

So far, I have argued that *if* one has a well-defined population, draws a representative sample from that population, measures variables on that sample, and does not impose treatments on any portion of that sample, it is a relatively straightforward task to estimate the relationships among those variables in that population. Statistical science has developed and is continuing to improve the procedures for taking the sampling error, the measurement error, and the missing data into account in such estimations. But if we also wish to provide *causal* explanations in observational studies, it is also necessary to consider specification error. Here the researcher must turn from statistical considerations to subject matter considerations, where the subject matter is what is already known about the phenomenon under investigation.

Specification error occurs when a study is designed with an inadequate theoretical model. In this context, a theoretical model is a proposed causal structure, and it generally includes both established and hypothesized causal relationships among the variables under investigation. For example, specification error occurs when a variable that directly affects a dependent variable in a causal network is not included in the set of independent variables, but is correlated with one or more of those independent variables.

It may be useful to illustrate this specification problem by referring to a favorite example of Palmer Johnson's, whom we honor annually at AERA with the Palmer O. Johnson Memorial Award. Professor Johnson was not exactly a humorous lecturer, but I recall his chortling a bit when he told about the almost perfect correlation that has existed over time between the number of Protestant ministers in New England, and the barrels of rum imported into Boston Harbor. He always felt that the strong relationship between those two variables should be a very sobering illustration for any graduate student who might be tempted to make causal inferences from *mere* correlational coefficients. My point is that anyone who would base a causal attribution on that two variable relationship is suffering from specification error. The simplest causal model for explaining that spurious correlation between rum and ministers incorporates an important third variable, the increasing size of the population of New England over that time period.

Current examples of specification error can be found in most evaluations of educational programs. For example, to evaluate a new reading program, a large school district utilized a non-randomized comparison group design to contrast their new program with the Houghton-Mifflin program, which was in more general use in the district (Webster, 1975). As is the fashion today, they used the familiar tic-tac-toe tables of *O*'s and *X*'s to describe their design:

$$O_1 X_1 O_2$$

$$O_1 X_2 O_2$$

where X_1 was the old reading program, X_2 the new, and O_1 and O_2 represented reading pretest and posttest, respectively.

To examine whether or not this study suffered from specification error, one must ask whether there were other factors than the pretest and the gross differences between the two reading treatments that might have affected posttest differences. One such factor might have been the degree to which the students receiving the two treatments had the same opportunity to learn the knowledge and skills that were sampled in the posttest. Opportunity, in this context, is both a function of how time was allocated, and of the overlap between the curricula actually covered during the school year and the knowledges and skills sampled in the posttest. These two opportunity variables can be estimated, and they *are* very important determinants of school outcomes.[2] When they are not incorporated into the design, there is the danger of attributing instructional effectiveness to specific programs or ways of teaching, when it is really a matter of how time was allocated, or whether the curriculum content was a good fit to the particular achievement test that was selected. (Notice that specification error is as relevant a concern in quasi-experiments as it is in observational studies.)

There are no statistical methods for guarding against specification error in explanatory observational studies. As Wold (1956) pointed out so well, this question of whether we have included the major explanatory variables in an analysis is not a statistical problem, but a question of subject-matter. Fortunately, procedures have been developed to assist the researcher in organizing the relevant subject matter into the kind of explicit theoretical model required in explanatory observational studies. Excellent recent introductions to such procedures can be found in Duncan's (1975) *Introduction to Structural Equation Models,* and Heise's (1975) book on *Causal Analysis.*

To illustrate the role of subject matter and structural equations in the development of a theoretical model, it is helpful to consider briefly a particular subject matter. I have selected the development of the ability to read. To keep it simple, here are just three fairly safe generalizations from what is currently known about the development of this ability:

(1) reading performance varies considerably among children of any given grade or chronological age;

(2) a child's reading performance at one point in time is highly dependent upon reading performance at some prior point in time; and

(3) a child who spends more time learning to read will become a better reader than if that child spends less time learning to read.

Using these generalizations, a theoretical model that attempts to explain the development of the ability to read might begin with a simple, three-variable network, as shown in Figure 1. This schematic may be expressed as a structural equation in Duncan's (1975) notation:

$$z = p_{zy}\, y + p_{zx}\, x + v$$

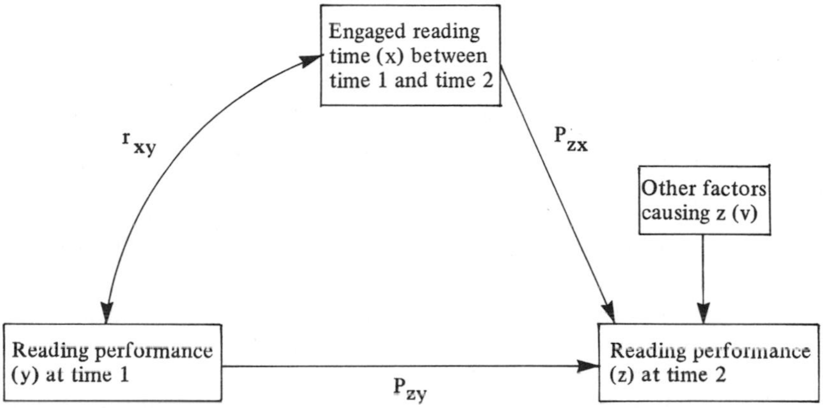

Figure 1. A really simple causal diagram for explaining performance on a reading test.

In the language of structural models, x and y are causally prior to z. The disturbance term v represents the other factors that affect reading performance (z), which are not specified in this model. The variables x and y are exogenous in this model (i.e., their values are not explained by the model), but x and y may be intercorrelated (thus the curved, double-headed arrow), but are assumed to be uncorrelated with v. The structural coefficient (p_{zx}) indicates the amount of change in z that can be expected from a change in x for a given y. Expressing these variables in standard score form,[3] the algebra of structural equations reveals that:

$$p_{zy} = \frac{\rho_{xy} - \rho_{zx}\,\rho_{xy}}{1 - \rho^{2}_{xy}}$$

and

$$p_{zx} = \frac{\rho_{zy} - \rho_{zy}\,\rho_{xy}}{1 - \rho^{2}_{xy}}$$

where the ρ's are the corresponding population correlations. Thus, for a given population, the structural coefficients for x and y are simply a function of the correlations among these three variables.

For example, if the population of third graders, $\rho_{zy} = .60$, $\rho_{zx} = .30$ and $\rho_{xy} = .20$, then the structural equation becomes:

$$z = .56\,y + .19\,x + v$$

This suggests that for a given value of y, an increase of one standard deviation in reading time (x) will result in a .19 standard deviation increase in reading performance, assuming, among other things, that the model is correctly specified and that these correlations for the third grade population are known. Let me illustrate the importance of the latter assumption first. Those of you familiar with third grade reading research may agree that the correlation between fall pretest and spring posttest is about .60, and the other two correlations are probably between .2 and .3. But is time more highly correlated with y or z? It makes a big difference. For example, if we reverse the .20 and .30, the structural coefficient for time goes from an interesting .19 to a dismal .02. This difference shows the importance of having good estimates of the population relationships.

Returning to the first assumption, which is that Figure 1 represents an adequate model of reading achievement, we encounter even greater problems. Although you may be convinced that a child has a better chance of becoming a better reader in reading class than while playing baseball, you may also believe that for a child with a given pretest (y), the effect of engaged time (x) surely depends on whether the time is spent trying to read the *Encyclopedia Britannica* or in one-to-one interaction with a skilled teacher using a highly effective blending algorithm to teach word decoding.

The point is that the structural model must be consistent with what is currently known about the phenomenon being modeled. For example, a given model is inadequately specified if an examination of the relevant subject matter reveals an additional variable that is known to be related functionally to reading achievement, *and* is correlated with one of the other variables already in the model. This is because the algebra of structural equations used in this model assumes that the disturbance variable v is uncorrelated with x and y.

This latter assumption is an important point for it lends optimism to the search for finite models and points to the need to develop networks of models. For example, if research on primary grades reading suggests the importance of a motivating classroom environment in learning to read, such variables need not necessarily be part of this particular model, if it is reasonable to assume (or it can be shown) that motivating variables are relevant because they operate through other variables in the model. To illustrate further, motivating variables may help to explain variance in actual time spent in learning to read (x), or even the correlation between x and y, but they need not be included in this particular model unless it is reasonable to assume that motivators have a direct effect on z (or an effect through some other variable not in the model but correlated with it). Of course, some degree of specification error will always be with us, but we are more likely to have a cumulatively improved understanding of educational processes if we are guided by more and more adequately specified models, and we will continue to perform *non*-cumulative studies if we are *not* guided by causal models.

Another way of looking at this specification problem is from the perspective of "strong inference." Platt (1964) emphasized the importance of being able to deal with alternative hypotheses (i.e., alternative explanations) for the phenomena we are trying to explain. What this means for observational studies is that the variables included represent possible alternatives for explaining the variance in the dependent variable. If, for example, a model for explaining reading performance does not include variables reflecting differences in home environment, the model will not be convincing unless other studies have shown that home environment is not functionally related to reading performance (which is unlikely), or that home environment operates through prior values of the dependent variable (i.e., through y in Figure 1). The latter assumption may be tenable if home environment does not change significantly between measures of y and z. (Another hypothetical possibility is that home environment does have a direct effect but is considered a component of the disturbance term v. That possibility is untenable because of the known relationship between y and home environment.)

Platt was also kind enough to suggest where strong models come from. They come from scientists who are thoroughly familiar with their subject matter. Strong models come from scientists who are also willing and able to spend a portion of each day in quiet, analytical thought, thinking through and anticipating possible challenges to their models as currently defined, and designing new studies or reanalysis of available data to meet those challenges.

To further illustrate this point, I vividly remember the time I heard that Paul Lazarsfeld had edited a book that included a section on multivariate analysis, and how amused I was to discover that it dealt with the analysis of multidimensional contingency tables (Lazarsfeld, Pasanella, & Rosenberg, 1972). Here I was doing elaborate, cross-lagged, multiple-partial canonical correlations involving dozens of variables, and that eminent sociologist was still messing around with chi square tables! What I did not appreciate was that his little analyses were generally more informative than my elaborate ones, because he had the "right" variables. He knew his subject matter. He was aware of the major alternative explanations that had to be guarded against and took that into account when he decided upon the four or five variables that were crucial to include. His work represented the state of the art in model building, while my work represented the state of the art in number crunching.

Paul Lazarsfeld was also very important in helping me think about the qualitative-quantitative distinction. His first big job was to get me out of the qualitative *vs.* quantitative trap. His work is living proof that it is not an either/or proposition, it is a matter of degree of emphasis. Every research study is inevitably both qualitative and quantitative to some degree. But even more important, he recognized the importance of the interaction between qualitative and quantitative "types" in working toward an improved understanding of educational processes. We have to talk to each other and read each others work, not shout at each other. Certainly those of us who emphasize quantitive approaches can benefit greatly from those who provide rich descriptions of what happens in classrooms and schools as we work toward more convincing causal models of the processes of schooling.

Statistical Procedures

This brings us to the statistical procedures that can implement the theoretical structural model. The purpose of the statistical procedures is to assist in establishing the plausibility of the theoretical model and to estimate the degree to which the various explanatory variables seem to be influencing the dependent variables. Various forms of linear regression are generally used for this purpose. It is certainly well-known that regression analysis and analysis of covariance are formally the same statistical procedure. They are both just variations of the general linear model. What is important is not the shift from covariance analysis to regression analysis, but rather a shift to causal modeling.

But it is also important to note that the approach I am advocating also includes the shift to thinking about educational treatments as multi-dimensional domains, and not as distinct, discrete, homogeneous treatments worthy of consideration as levels in a typical analysis-of-covariance contrast. The treatment dimensions reflect the various ways in which a program can be implemented by different teachers, as well as other important ways in which teachers and programs vary.

Of course, it is not enough to suggest that the statistical procedure be linear regression. There are many variations available (step-wise, commonality, ridge, least squares, maximum likelihood, standardized, raw, etc.), and a few well-known problems to consider, such as multi-collinearity among the explanatory variables, the appropriate metric to be used to represent them, sensible levels of data aggregation, and how to deal with their errorful measurement.

In considering possible statistical models, it may be useful to distinguish between exploratory and more "confirmatory" stages in the investigation of a phenomenon. Exploratory approaches are particularly useful in current studies of the effects of educational programs because of the primitive state of relevant theoretical models. Needed at this stage are statistical procedures that allow us to see the relative usefulness of different predictors or sets of predictors, as well as what confounding is occurring among the independent variables, and what differences there are among different possible models for the data. At the exploratory stage, the data analysis should suggest ways in which the theoretical model might be modified, how the measures might be combined or separated, or which variables can safely be deleted from the model. As Lohnes and I argue elsewhere (Lohnes & Cooley, 1978), commonality analysis has been found to be useful for such exploratory data analysis. In a recent, excellent review of regression analysis, Jocking (1976) points out the possible usefulness of commonality analysis in the comparison of possible models, and "as a means of identifying the roles of the variables, both alone and in relation to other variables" (p. 8).

As more convincing models are specified, the algebra of structural equations is useful in the consideration of the appropriateness of different statistical procedures. For example, as you may have noticed, the structural coefficients for Figure 1 can be estimated as linear regression coefficients in good old multiple correlation. But it is also important to notice that I did not start with the notion that I should plug my data into a multiple regression program. The algebra of structural equation modeling showed that for the particular simple causal model of Figure 1, multiple correlation was a sensible statistical procedure for estimating the structural coefficients.

One promising set of statistical procedures for estimating the unknown structural coefficients in a set of linear structural equations has been built into a computer program called LISREL (Jöreskog & van Thillo, 1973). One advantage of LISREL is that its architects recognized the fact that the variables of primary interest are often not directly measurable, but must be estimated from measures of multiple indicators of the unmeasured, hypothetical variables (or latent factors). Another important advantage of LISREL is that it incorporates procedures for estimating the reliabilities of the variables, and good estimates of reliabilities are essential if we hope to correct for the attenuation inherent in research involving errorful measures.

The State-of-the-Art

Having said all this, what is "the state-of-the-art" in the design and analysis of research studies involving relationships among variables? My sense of it is that, thanks to geneticists, economists, sociologists, and others, a very exciting "art form" exists that educational research has largely ignored. The few educational applications that have been tried or suggested (for example, Anderson & Evans, 1974; Rao, Morton, Elston & Yee, 1977; Werts & Linn, 1970; Magidson, 1977; Werts & Hilton, 1977; McDonald & Elias, 1976; Wiley & Harnischfeger, 1973) illustrate its potential, as well as the primitive state of the causal models that have been developed, to date, for guiding such efforts.

We must clearly produce more convincing causal models. The challenge is considerable, but the efforts will surely be more productive than continuing to conduct meaningless quasi-experiments, or averaging all of the t-tests they may have produced. As educational research is guided by increasingly valid models of educational phenomena, and the causal networks that are currently operating are better understood, we may eventually draw close to that "complete covariate" that Cronbach and his associates (1976) have shown is necessary for interpretable quasi-experiments. It may then be sensible "to push entities into locations in the predictor space where they are not naturally found" (Lohnes & Cooley, 1978), and see what results.

References

Abt Associates Inc. *Education as experimentation: A planned variation model. Volume IVA.* Cambridge, Mass.: Author, 1977.

Anderson, J. G., & Evans, F. B. Causal models in educational research: Recursive models. *American Educational Research Journal,* 1974, *11,* 1, 29-39.

Cochran, W. G. The planning of observational studies of human populations. *Royal Statistical Society Journal,* 1965, series A, *128,* 234-266.

Cooley, W. W., & Leinhardt, G. *The Instructional Dimensions Study.* Pittsburgh: University of Pittsburgh Learning Research and Development Center, 1978.

Cooley, W. W., Leinhardt, G., & McGrail, J. How to identify effective teaching. *Anthropology and Education Quarterly,* 1977, *8,* 2, 119-126.

Cooley, W. W., & Lohnes, P. R. *Evaluation research in education.* New York: Irvington Publishers, 1976.

Cronbach, L. J., Rogosa, D. R., Floden, R. E., & Price, G. G. *Analysis of covariance: Angel of salvation, or temptress and deluder?* Stanford, Calif.: Stanford University, 1976.

DiCostanzo, J. L., & Eichelberger, R. T. *Design, analysis, and reporting considerations when ANCOVA-type techniques are used in evaluation settings.* Paper presented at the annual meeting of the American Educational Research Association, New York, April 1977.

Duncan, O. D. *Introduction to structural equation models.* New York: Academic Press, 1975.

Heise, D. R. *Causal Analysis.* New York: John Wiley & Sons, 1975.

Hocking, R. R. The analysis and selection of variables in linear regression. *Biometrics,* 1976, *32,* 1, 2-49.

Hoepfner, R., Zagorski, H., & Wellisch, J. B. *The sample for the sustaining effects study and projections of its characteristics to the national population.* Santa Monica, Calif.: System Development Corporation, 1977.

House, E. R., Glass, G. V., McLean, L. D., & Walker, D. F. *No simple answer: Critique of the "Follow Through" evaluation.* Urbana, Ill.: University of Illinois, Center for Instructional Research and Curriculum Evaluation, 1977.

Jöreskog, K. G. *Causal models in the social sciences: The need for methodological research.* (Research report 76-8). Uppsala, Sweden: University of Uppsala, 1976.

Jöreskog, K. G., & van Thillo, M. *LISREL: A general computer program for estimating a linear structural equation system involving multiple indicators of unmeasured variables.* (Research report 73-5). Uppsala, Sweden: University of Uppsala, 1973.

Kenny, D. A. A quasi-experimental approach to assessing treatment effects in nonequivalent control group design. *Psychological Bulletin,* 1975, *82,* 345-362.

Lazarsfeld, P. F., Pasanella, A. K., & Rosenberg, M. *Continuities in the language of social research.* New York: The Free Press, 1972.

Lohnes, P. R., & Cooley, W. W. *Regarding criticisms of commonality analysis.* Paper presented at the annual meeting of the American Educational Research Association, Toronto, Canada, March 1978.

Lord, F. M. Large-scale covariance analysis when the control variable is fallible. *Journal of the American Statistical Association,* 1960, *55,* 307-321.

Magidson, J. Toward a causal model approach for adjusting for preexisting differences in the nonequivalent control group situation: A general alternative to ANCOVA. *Evaluation Quarterly,* 1977, *1,* 3, 399-420.

McDonald, F. J., & Elias, P. *Beginning teacher evaluation study, phase II. Executive summary report, 1973-74.* Princeton, N. J.: Educational Testing Service, 1976.

Overall, J. E., & Woodward, J. A. *Non-random assignment and the analysis of covariance.* Galveston, Texas: University of Texas, Psychometric Laboratory, 1975.

Platt, J. R. Strong inference. *Science,* 1964, *146,* 347-353.

Rao, D. C., Morton, N. E., Elston, R. C., & Yee, S. Causal analysis of academic performance, *Behavior Genetics,* 1977, *7,* 2, 147-159.

Ross, L., & Cronbach, L. J. Review of *Handbook of evaluation research* by M. Guttentag & E. L. Struening (Eds.). *Educational Researcher,* 1976, *5,* 10, 9-19.

Stallings, J. Implementation and child effects of teaching practices in Follow Through classrooms. *Monographs of the Society for Research in Child Development,* 1975, *40* (7-8, Serial No. 163).

Thorndike, R. L. Regression fallacies in the matched groups experiment. *Psychometrika,* 1942, *7,* 85-102.

Webster, W. J. *Abstracts of research and evaluation report 1974-75.* Dallas, Texas: Dallas Independent School District, 1975.

Werts, C. E., & Linn, R. L. Path analysis: Psychological examples. *Psychological Bulletin,* 1970, *74,* 3, 193-212.

Werts, C. E., & Hilton, R. L. Intellectual status and intellectual growth, again. *American Educational Research Journal,* 1977, *14,* 2, 137-146.

Wiley, D. E., & Harnischfeger, A. *Post hoc, ergo propter hoc: Problems in the attribution of change.* (Report No. 7). Chicago: University of Chicago, 1973.

Wiley, D. E., & Hornik, R. *Measurement error and the analysis of panel data.* (Report No. 5). Chicago: University of Chicago, 1973.

Wold, H. Causal inferences from observational data: A review of ends and means. *Journal of the Royal Statistical Society,* Series A, 1956, *119,* 351-390. Reprinted in M. C. Wittrock, & D. E. Wiley (Eds.), *The evaluation of instruction: Issues and problems.* New York: Holt, Rinehardt and Winston, 1970.

Notes

[1]One does not normally dedicate a paper, but I would like to dedicate this one to Paul Lazarsfeld. The Learning Research and Development Center benefited greatly from his presence during the final years of his very distinguished career.

[2]Cooley and Leinhardt (1978) have demonstrated this in the Instructional Dimensions Study.

[3]Not necessary, but desirable for ease of illustration.

Part I
Advances in Factor Analysis

Introduction to Part I

Theories in the social sciences generally relate hypothetical constructs, such as intelligence, anxiety, or self-esteem, to each other. For example, the theory "poverty is a cause of crime" relates the construct "poverty" to another construct "crime." How to measure or operationalize the constructs is a natural question that arises after the theory takes form. The factor analysis model can often be useful at this stage of model building.

Factor analysis is one of the only statistical procedures that involves the relationship between observed variables (measurements) and the underlying latent factors. As such it has been often used by social scientists as an exploratory technique – to help identify the constructs that underlie a set of observed variables. However, it can also be used to test whether a set of items designed to measure certain construct(s) do, in fact, reveal the hypothesized factor structure.

The history of factor analysis offers insight into its original intended use. The pioneer factor analyst Charles Spearman was interested in measuring the construct "general intelligence."[1] He proposed a model to explicate how various measurements of mental abilities could share a common component associated with the general intelligence factor and also have unique components representing both the measurement error and irrelevant component associated with the measurements.

Spearman's original confirmatory model has since been extended to multiple factors, each factor being measured by two or more variables. In many applications of multiple factor analysis the factors are typically assumed to be uncorrelated (orthogonal) with each other. Although this assumption results in some algebraic simplifications, the more general approach that allows for correlated (oblique) factors is more realistic. Jöreskog's approach to factor analysis postulates oblique factors and allows the researcher to play a more active role in defining these factors than do more traditional approaches.

In chapter 1 Jöreskog presents the basic ideas of factor analysis using terms familiar to social scientists – correlations, partial correlations, and regression coefficients. The mathematical goal of factor analysis, like that of regression analysis, is to estimate the regression coefficients. (In the general case of oblique factors, the correlations between the factors must also be estimated.) However, unlike regression analysis, the coefficients in factor analysis relate the observed "dependent" variable to a set of explanatory factors, each of which is *unobserved.* Furthermore, there are multiple regression equations, one for each of the observed (dependent) variables.

As might be expected, the large number of unknowns creates a basic indeterminacy in factor analysis – the problem of rotation. There are infinitely many sets of factors that can equally well explain the correlations among the observed variables. Moreover, each set of factors defined by a particular rotation yields a distinctly different set of coefficients.

Thurstone proposed to deal with this indeterminacy by considering only those sets of rotations that result in a simple structure for the regression coefficients, thus facilitating the interpretation of the latent factors.[2] A variety of simple structures are possible, such as those resulting from varimax, equimax, and quartimax rotations, from flexible graphical rotation procedures, or more complex rotations strategies for oblique factors.[3]

In chapter 2 Jöreskog proposes and illustrates a different approach to the problem of rotation. He proposes that the researcher fix certain of the coefficients and estimate the rest using a constrained maximum likelihood approach. For example, in the case of nine observed test scores, consisting of three measures of verbal ability (factor 1), three measures of mathematical ability (factor 2), and three measures of spatial ability (factor 3), a researcher may formulate a three-factor model and hypothesize that the only nonzero coefficient associated with a given test measurement is for the factor it purports to measure. That is, the equations associated with each of the verbal tests, for example, would have zero coefficients for the mathematical and spatial factors; the equations associated with each math test would have zero coefficients for the verbal and spatial factors, and so on.

The additional information incorporated into the model by the zero restrictions on the coefficients is sufficient in this example to uniquely estimate the coefficients (factor loadings) and factor correlations. Chapter 2 presents a detailed discussion of the kinds of restrictions that insure unique solutions. Jöreskog gives two sufficient conditions for uniqueness, one for orthogonal factors, and one for oblique factors. These conditions are now known to be incorrect, and Jöreskog has therefore supplemented chapter 2 with a statement of what is currently known about conditions for obtaining unique solutions.

Chapter 3 illustrates this general maximum likelihood approach to a wide variety of applications in the social sciences. In addition to factor analysis models, the approach is extended to estimate variance components and correlated measurement errors.

---

[1] Charles Spearman, "General Intelligence Objectively Determined and Measured," *American Journal of Psychology* 15 (1904): 201-293.

[2] L. L. Thurstone, *Multiple Factor Analysis* (Chicago: University of Chicago Press, 1947).

[3] See H. H. Harmon, *Modern Factor Analysis,* 2nd ed. (Chicago: University of Chicago Press, 1967); and S. A. Mulaik, *The Foundations of Factor Analysis* (New York: McGraw-Hill, 1972).

Chapter 1

Basic Ideas of Factor and Component Analysis

Karl G. Jöreskog

1. Introduction

Factor analysis is the common term for a number of statistical techniques for the resolution of a set of variables in terms of a small number of hypothetical variables, called factors. Within the statistical framework factor analysis belongs to the field of multivariate analysis. In fact, as we shall see, factor analysis can be formulated in terms of partial correlations.

Though the models and methods of factor analysis are of a statistical nature, factor analysis has been developed mainly by psychologists, particularly for analyzing the observed scores of many individuals on a number of psychological tests such as aptitude and achievement tests. The phenomenon continually observed in this situation is that the tests correlate (positively) with each other. Factor analysis attempts to "explain" these correlations by an analysis that yields a smaller number of underlying factors that contain all the essential information about the linear interrelationships among the test scores.

Factor analysis is most often employed in the behavioral sciences for studying the measurement properties (reliabilities and validities) of the instruments used in these sciences. But the techniques of factor analysis are not limited to behavioral science applications. Factor analysis has been used in diverse fields such as meteorology, political science, medicine, geography, and business. For a comprehensive general description of the concepts, theories and techniques of factor analysis, see Harman (1967). For a statistical treatment of factor analysis, see Lawley and Maxwell (1971).

This chapter is an elementary introduction to the basic ideas and concepts of factor analysis, developed by means of a simple example. To begin with we shall deal with population properties and ignore any sampling aspects that may arise. Thus we are concerned here with explicating the meaning of the model rather than with fitting the model to empirical data.

Suppose six tests with test scores denoted by $x_1, x_2, \ldots, x_6$ have been administered to a large population of individuals. Let us assume that $x_1, x_2,$

and x_6 are three different measures of verbal ability and that x_3, x_4, and x_5 are three different measures of numerical ability. For simplicity we assume that the tests are measured in standard scores so that each test has zero mean and unit standard deviation in the population. Furthermore suppose the following intercorrelations are obtained.

	x_1	x_2	x_3	x_4	x_5	x_6
x_1	1.000	0.720	0.378	0.324	0.270	0.270
x_2	0.720	1.000	0.336	0.288	0.240	0.240
x_3	0.378	0.336	1.000	0.420	0.350	0.126
x_4	0.324	0.288	0.420	1.000	0.300	0.108
x_5	0.270	0.240	0.350	0.300	1.000	0.090
x_6	0.270	0.240	0.126	0.108	0.090	1.000

A typical correlation in this symmetric matrix is denoted ρ_{ij}. The symbol ρ indicates a population correlation as distinct from a sample correlation, denoted by r. For example $\rho_{32} = \rho_{23} = 0.336$ is the correlation between x_2 and x_3.

2. Factor Analysis

Factor analysis sets out to explain these correlations by introducing underlying factors $f_1, f_2, \ldots$ that account for the correlations. This is done in the following way. One first asks the question, Is there a factor f_1 such that if this is partialed out there remain no intercorrelations between the tests? If so the *partial correlation* between any pair of tests x_i and x_j after f_1 has been eliminated must vanish; that is,

$$\rho\,(x_i, x_j; f_1) = 0, \qquad i \neq j \tag{1.1}$$

This is the same as saying that there exist numbers $\lambda_1, \lambda_2, \ldots, \lambda_6$ and residual variables $e_1, e_2, \ldots, e_6$ such that

$$\begin{aligned} x_1 &= \lambda_1 f_1 + e_1 \\ x_2 &= \lambda_2 f_1 + e_2 \\ &\;\;\vdots \\ x_6 &= \lambda_6 f_1 + e_6 \end{aligned} \tag{1.2}$$

and such that the correlation between any pair of residuals e_i and e_j is zero; that is,

$$\rho\,(e_i, e_j) = 0 \tag{1.3}$$

Each equation in (1.2) represents *the linear least squares regression* of a test score on the factor f_1. Since all x-variables have been assumed to have zero means, we omit the constant terms in the regressions. A well-known property of least squares regression is that the residual e_i in the ith equation is uncorrelated with f_1 (for a proof, see Cramér, 1945, section 23.3). Without loss of generality we may assume that f_1 is scaled so that its mean is zero and its variance is one. Using these results, we can easily verify that

$$\rho_{ij} = \lambda_i \lambda_j, \quad i \neq j \tag{1.4}$$

Hence the correlations in any pair of rows or columns of the correlation matrix are proportional. This does not hold, however, for the given correlation matrix. For example, $\rho_{51}/\rho_{61} = 1$ but $\rho_{53}/\rho_{63} = 0.350/0.126 \neq 1$. We therefore conclude that no such factor f_1 exists.

It is then natural to try to explain the given correlations by means of two underlying factors instead of one. One then asks, Are there two factors f_1 and f_2 such that when these are partialed out, there remain no intercorrelations between the tests? If so, the partial correlation between x_i and x_j, given f_1 and f_2 is zero:

$$\rho\,(x_i, x_j; f_1, f_2) = 0, \quad i \neq j \tag{1.5}$$

and the linear least squares regressions of the tests on these factors may be written as

$$
\begin{aligned}
x_1 &= \lambda_{11} f_1 + \lambda_{12} f_2 + e_1 \\
x_2 &= \lambda_{21} f_1 + \lambda_{22} f_2 + e_2 \\
&\;\;\cdot \\
&\;\;\cdot \\
&\;\;\cdot \\
x_6 &= \lambda_{61} f_1 + \lambda_{62} f_2 + e_6
\end{aligned}
\tag{1.6}
$$

where the λ's are the regression coefficients and the e's are residuals in each equation that are uncorrelated with f_1 and f_2. As before, we may without loss of generality assume that $E(f_1) = E(f_2) = 0$ and $E(f_1^2) = E(f_2^2) = 1$, where $E(x)$ is the expected value, or mean, of a random variable x. The condition of *partial linear independence* expressed in (1.5) is now equivalent to

$$\rho\,(e_i, e_j) = 0, \quad i \neq j \tag{1.7}$$

If f_1 and f_2 are uncorrelated we have, using these results,

$$\begin{aligned}\rho(x_i, x_j) &= E(x_i x_j)\\ &= E\left[(\lambda_{i1}f_1 + \lambda_{i2}f_2 + e_i)(\lambda_{j1}f_1 + \lambda_{j2}f_2 + e_j)\right]\\ &= \lambda_{i1}\lambda_{j1}E(f_1^2) + \lambda_{i2}\lambda_{j2}E(f_2^2) + \lambda_{i1}\lambda_{j2}E(f_1f_2) + \lambda_{i2}\lambda_{j1}E(f_1f_2)\\ &= \lambda_{i1}\lambda_{j1} + \lambda_{i2}\lambda_{j2}\end{aligned} \tag{1.8}$$

In our example the following coefficient matrix $\Lambda = (\lambda_{im})$ satisfies this equation for all pairs of i and j:

$$\Lambda = \begin{bmatrix} 0.889 & -0.138\\ 0.791 & -0.122\\ 0.501 & 0.489\\ 0.429 & 0.419\\ 0.358 & 0.349\\ 0.296 & -0.046 \end{bmatrix}$$

This matrix of regression coefficients has actually been obtained by the method of maximum likelihood, but this should not be of any concern now. At this point it is sufficient to verify that this matrix does indeed reproduce the given correlations, as predicted in equation (1.8). For example, we may verify that

$$\begin{aligned}\lambda_{11}\lambda_{21} + \lambda_{12}\lambda_{22} &= 0.889 \times 0.791 + 0.138 \times 0.122\\ &= 0.7032 + 0.0168\\ &= 0.7200\\ &= \rho_{12}\end{aligned}$$

Thus we have established that for the particular correlation matrix at hand, there exist two uncorrelated factors f_1 and f_2 such that the representation

$$\begin{aligned}x_1 &= 0.889f_1 - 0.138f_2 + e_1\\ x_2 &= 0.791f_1 - 0.122f_2 + e_2\\ x_3 &= 0.501f_1 + 0.489f_2 + e_3\\ x_4 &= 0.501f_1 + 0.419f_2 + e_4\\ x_5 &= 0.358f_1 + 0.349f_2 + e_5\\ x_6 &= 0.296f_1 - 0.046f_2 + e_6\end{aligned} \tag{1.9}$$

holds with the condition 1.7 fulfilled for all pairs i and j. The factors f_1 and f_2 are not, however, the only factors that satisfy these requirements. For example, the following two factors also do it.

$$f_1^* = 0.988f_1 - 0.153f_2$$
$$f_2^* = 0.153f_1 + 0.988f_2 \tag{1.10}$$

It is easily verified that $E(f_1^*) = E(f_2^*) = 0$ and $\text{Var}(f_1^*) = \text{Var}(f_2^*) = 1$ and that f_1^* and f_2^* are uncorrelated if f_1 and f_2 have these properties. Solving equations 1.10 for f_1 and f_2 in terms of f_1^* and f_2^*, we get

$$f_1 = 0.988f_1^* + 0.153f_2^*$$
$$f_2 = -0.153f_1^* + 0.988f_2^* \tag{1.11}$$

Substituting (1.11) into (1.9), we get the regression of each test on the factors f_1^* and f_2^*, as

$$\begin{aligned}
x_1 &= 0.90f_1^* &&+ e_1\\
x_2 &= 0.80f_1^* &&+ e_2\\
x_3 &= 0.42f_1^* + 0.56f_2^* &&+ e_3\\
x_4 &= 0.36f_1^* + 0.48f_2^* &&+ e_4\\
x_5 &= 0.30f_1^* + 0.40f_2^* &&+ e_5\\
x_6 &= 0.30f_1^* &&+ e_6
\end{aligned} \tag{1.12}$$

The matrix of regression coefficients is now

$$\Lambda^* = \begin{bmatrix} 0.90 & 0 \\ 0.80 & 0 \\ 0.42 & 0.56 \\ 0.36 & 0.48 \\ 0.30 & 0.40 \\ 0.30 & 0 \end{bmatrix}$$

The difference between Λ^* and Λ is that Λ^* exhibits a kind of simplicity in the sense that some elements are zero. Factor analysts want to find factors such that the regression matrix Λ is as simple as possible in the sense that it contains as many zero elements as possible. In general, even greater simplicity may be obtained by allowing the factors to become correlated. For example, the following factors

$$f_1^{**} = f_1^*$$
$$f_2^{**} = 0.6f_1^* + 0.8f_2^* \tag{1.13}$$

with $\rho(f_1^{**}, f_2^{**}) = 0.6$, yield

$$\Lambda^{**} = \begin{bmatrix} 0.9 & 0 \\ 0.8 & 0 \\ 0 & 0.7 \\ 0 & 0.6 \\ 0 & 0.5 \\ 0.3 & 0 \end{bmatrix}$$

as is easily verified by solving (1.13) for f_1^* and f_2^* and substituting these into (1.12). Note that Λ^{**} has three zeroes in each column whereas Λ^* has only three zeroes in one column.

Geometrically the transformation from one set of factors to another, as in (1.10) or (1.11), corresponds to a rotation of coordinate axes in Euclidean space. The two elements in each row of the matrix Λ may be regarded as the coordinates of a point in two-dimensional Euclidean space, and these points may be plotted in this space as in figure 1.1. The points in the space are fixed, but the coordinate axes are arbitrary. The correlation between the two factors is equal to the cosine of the angle between the two coordinate axes. If the axes are orthogonal, the factors are uncorrelated. In figure 1.1, the axes labeled λ_{i1} and λ_{i2} correspond to the matrix Λ. If the axis λ_{i1} is rotated clockwise so that it passes through the points 1, 2, and 6 while maintaining the orthogonality of the axes, we get the coordinate system λ_{i1}^* and λ_{i2}^* corresponding to the matrix Λ^*. If the first axis is kept at λ_{i1}^* and the second axis rotated clockwise so that it passes through the points 3, 4, and 5, we obtain the coordinate system λ_{i1}^{**} and λ_{i2}^{**}, in which the cosine of the angle between the coordinate axes is 0.60. This corresponds to the matrix Λ^{**} and the correlation $\rho(f_1^{**}, f_2^{**}) = 0.6$. Note that zero elements in Λ^* or Λ^{**} correspond to points in the space lying on one coordinate axis.

We have demonstrated the two basic principles of factor analysis. The first one is *the principle of conditional linear independence*. This principle expresses the idea that the factors shall account for all linear relationships among the variables. Once the factors have been partialed out, there shall remain no correlation between the variables. In this sense factor analysis is a method for *classification of linear dependence*. In general, for p variables $x_1, x_2, \ldots, x_p$, once it has been established that correlation exists, we may ask whether we can find one factor f_1 such that

$$\rho\,(x_i, x_j; f_1) = 0, \quad i \neq j$$

for all pairs i and j. If the answer is yes, we say that $x_1, x_2, \ldots, x_p$ has *linear dependence of degree 1.* If, on the other hand, the answer is no, we ask for two factors f_1 and f_2 such that

$$\rho\,(x_i, x_j; f_1, f_2) = 0, \quad i \neq j$$

If such factors can be found, we say that $x_1, x_2, \ldots, x_p$ have *linear dependence of degree 2,* and so forth. It may be shown that this process always ends after having found some $k < p$ factors $f_1, f_2, \ldots, f_k$ such that

$$\rho\,(x_i, x_j; f_1, f_2, \ldots, f_k) = 0, \quad i \neq j \tag{1.14}$$

in which case we have *linear dependence of degree k* among the tests.

Equation (1.14) is equivalent to the *linear factor analytic model*

$$\begin{aligned} x_1 &= \lambda_{11}f_1 + \lambda_{12}f_2 + \ldots + \lambda_{1k}f_k + e_1 \\ x_2 &= \lambda_{21}f_1 + \lambda_{22}f_2 + \ldots + \lambda_{2k}f_k + e_2 \\ &\cdot \\ &\cdot \\ &\cdot \\ x_p &= \lambda_{p1}f_1 + \lambda_{p2}f_2 + \ldots + \lambda_{pk}f_k + e_p \end{aligned} \tag{1.15}$$

where

$$\rho\,(e_i, e_j) = 0, \quad i \neq j \tag{1.16}$$

for all pairs of i and j. Each equation in (1.15) represents *the linear least squares regression* of a variable on the factors $f_1, f_2, \ldots, f_k$. The coefficients λ_{im} are regression coefficients. In psychological terminology these are called *factor loadings.* The factors $f_1, f_2, \ldots, f_k$ are called *common factors* because they measure attributes that are common to two or more of the variables. The residuals $e_1, e_2, \ldots, e_p$ are called residuals or *unique factors.* They represent what is left over of each variable when the factors have been determined so as to account for all intercorrelations of the tests. A unique factor is usually regarded as a sum of an *error of measurement* and a *specific factor,* the specific factor having to do only with the particular variable. None of these parts contributes to any correlation between variables. Each equation in (1.15) decomposes the variable x_i into two uncorrelated parts,

$$x_i = c_i + e_i \tag{1.17}$$

where

$$c_i = \lambda_{i1}f_1 + \lambda_{i2}f_2 + \ldots + \lambda_{ik}f_k \tag{1.18}$$

is called the *common part* of x_i.

Once a set of k factors has been found that account for the intercorrelations of the tests, as in (1.15), these may be transformed to another set of k factors that account equally well for the correlations. In fact, any nonsingular linear transformation of the first set of factors yields a new set of factors with this property. For our particular data, (f_1, f_2), (f_1^*, f_2^*), and (f_1^{**}, f_2^{**}) are three sets of factors, each one accounting for the correlations in the sense that (1.15) and (1.16) are fulfilled. The regression of the tests on the factors, as represented by the matrix of factor loadings, for two different sets of factors, may be quite different. Since there are infinitely many sets of factors, this is a great indeterminacy in the model. However, Thurstone (1947) proposed giving attention only to the factors for which the variables have a simple representation. The matrix of factor loadings shall have as many zero elements as possible. If $\lambda_{im} = 0$, the mth factor does not enter into the ith test. A variable should not depend on all common factors but only on a small portion of them. Also the same factor should be involved only in a small portion of the variables. Such a matrix is regarded as giving the simplest structure and presumably the one with the most meaningful psychological interpretation. This is the second basic principle of factor analysis, *the principle of simple structure.* The factor analyst usually tries to meet these requirements when he chooses the tests to be included in the factorial study.

In our particular example the matrix Λ^* represents a simple structure with two uncorrelated factors, and Λ^{**} represents a simple structure with two correlated factors. Uncorrelated factors are also called *orthogonal,* and correlated factors are called *oblique.* Usually simple structure is better achieved with oblique factors than with orthogonal. In our example the interpretation of the matrix Λ^* might be that some general factor f_1^* is required to perform well in all six tests. In addition to this general factor f_1^* there is a second factor f_2^*, which has nothing to do with f_1^* but which is necessary to perform well on the numerical tests x_3, x_4, and x_5. This interpretation is not a very reasonable one. A better interpretation is based on Λ^{**}. Here the factors f_1^{**} and f_2^{**} are interpreted as a verbal and a numerical factor. These factors correlate 0.6 in the population of individuals examined. This means that individuals who perform well on verbal tests are likely to perform well on numerical tests too.

Consider the equation (1.17). Since the common part c_i and the unique part e_i are uncorrelated, this equation also partitions the variance of x_i as

$$\mathrm{Var}(x_i) = \mathrm{Var}(c_i) + \mathrm{Var}(e_i) \tag{1.19}$$

$\mathrm{Var}(c_i)$ is called the *common variance* or the *communality* of x_i and $\mathrm{Var}(e_i)$ is called the *unique variance* or the *uniqueness* of x_i. The communality of a variable is the portion of a variable's total variance that is accounted for by the common factors. The uniqueness is the portion left unexplained by the common factors.

To determine the communality of x_i, we use equation (1.18) and get

$$\begin{aligned} E(c_i^2) = {} & \lambda_{i1}^2 E(f_1^2) + \lambda_{i2}^2 E(f_2^2) + \ldots + \lambda_{ik}^2 E(f_k^2) \\ & + 2\lambda_{i1}\lambda_{i2} E(f_1 f_2) + 2\lambda_{i1}\lambda_{i3} E(f_1 f_3) + \ldots \\ & + 2\lambda_{i,\,k-1}\lambda_{ik} E(f_{k-1} f_k) \end{aligned}$$

If $f_1, f_2, \ldots, f_k$ all have unit variances, this equation becomes

$$\begin{aligned}\text{Var}(c_i) = {} & \lambda_{i1}^2 + \lambda_{i2}^2 + \ldots + \lambda_{ik}^2 \\ & + 2\lambda_{i1}\lambda_{i2}\rho(f_1, f_2) + 2\lambda_{i1}\lambda_{i3}\rho(f_1, f_3) + \ldots \\ & + 2\lambda_{i,\,k-1}\lambda_{jk}\rho(f_{k-1}, f_k)\end{aligned} \tag{1.20}$$

If the factors are uncorrelated, this reduces further to

$$\text{Var}(c_i) = \lambda_{i1}^2 + \lambda_{i2}^2 + \ldots + \lambda_{ik}^2 \tag{1.21}$$

which is simply the sum of squares of the elements in the ith row of Λ.

Using (1.21) and the matrix Λ^* in our example, we get

Test	*Communality*	*Uniqueness*
1	0.81	0.19
2	0.64	0.36
3	0.49	0.51
4	0.36	0.64
5	0.25	0.75
6	0.09	0.91
Total	2.64	3.36

(1.22)

The same result is obtained if we apply the general formula (1.20) to the matrix Λ^{**}. This shows that communalities and uniquenesses are unaffected by linear transformation of factors such as (1.13). That this property is general is evident from the fact that the unique factors themselves $e_1, e_2, \ldots, e_p$ are unaffected by linear transformations of the common factors. Hence their variances, the uniquenesses, and their counterparts, the communalities, are invariant under such transformations.

3. Factor Analysis versus Component Analysis

Factor analysis is often confused with principal component analysis. The two methods of analysis are similar to some extent but have entirely different aims. The distinction between factor and component analysis has been emphasized by Kendall and Lawley (1956), Lawley and Maxwell (1971), and others.

The first principal component of $x_1, x_2, \ldots, x_p$ is defined as the linear combination of $x_1, x_2, \ldots, x_p$,

$$u_1 = \beta_{11}x_1 + \beta_{21}x_2 + \ldots + \beta_{p1}x_p$$

that has maximal variance, subject to the restriction that

$$\beta_{11}^2 + \beta_{21}^2 + \ldots + \beta_{p1}^2 = 1$$

The coefficients $\beta_{11}, \beta_{21}, \ldots, \beta_{p1}$ are determined as the elements of the latent vector of the covariance matrix of $x_1, x_2, \ldots, x_p$ which corresponds to the largest latent root θ_1 (see Anderson, 1958, chapter 11). The variance of u_1 is θ_1.

The *second principal component* of $x_1, x_2, \ldots, x_p$ is that linear combination

$$u_2 = \beta_{12}x_1 + \beta_{22}x_2 + \ldots + \beta_{p2}x_p$$

which has maximal variance and is uncorrelated with u_1. The coefficients $\beta_{12}, \beta_{22}, \ldots, \beta_{p2}$ are the elements of the latent vector of the covariance matrix corresponding to the second largest latent root θ_2. The variance of u_2 is θ_2.

The third, fourth, . . . principal components of $x_1, x_2, \ldots, x_p$ are similarly defined. There will be as many principal components as there are positive latent roots $\theta_1, \theta_2, \ldots$ of the variance-covariance matrix. Usually the covariance matrix is positive definite and then all p roots are positive, thus giving rise to p principal components

$$u_i = \beta_{1i}x_1 + \beta_{2i}x_2 + \ldots + \beta_{pi}x_p, \qquad i = 1, 2, \ldots, p \tag{1.23}$$

The matrix of coefficients $B = (\beta_{ij})$ form an orthogonal matrix (see Anderson, 1958, chapter 11) and equation (1.23) can therefore be inverted to give

$$x_i = \beta_{i1}u_1 + \beta_{i2}u_2 + \ldots + \beta_{ip}u_p, \qquad i = 1, 2, \ldots, p \tag{1.24}$$

Equation (1.24) expresses each variable x_i as a linear combination of p uncorrelated variables with descending variances.

In some applications of principal component analysis a few principal components account for a large portion of the total variance of all the variables. Equation (1.24) may then be cut off after a certain number of terms, $k < p$ say, so that

$$x_i = \beta_{i1}u_1 + \beta_{i2}u_2 + \ldots + \beta_{ik}u_k + v_i \tag{1.25}$$

where

$$v_i = \beta_{i,\,k+1}u_{k+1} + \beta_{i,\,k+2}u_{k+2} + \ldots + \beta_{ip}u_p \tag{1.26}$$

is interpreted as a residual with small variance.

To bring out the formal similarity between factor and component analysis, we standardize the components $u_1, u_2, \ldots, u_k$ to unit variance by dividing them by $\sqrt{\theta_1}, \sqrt{\theta_2}, \ldots, \sqrt{\theta_k}$, respectively. If we write

$$u_i^* = u_i / \sqrt{\theta_i}$$

and

$$\beta_{ij}^* = \beta_{ij}\sqrt{\theta_j}$$

equation (1.25) becomes

$$x_i = \beta_{i1}^* u_1^* + \beta_{i2}^* u_2^* + \ldots + \beta_{ik}^* u_k^* + v_i \qquad (1.27)$$

an equation that formally resembles the ith equation in (1.15). However, the residuals $v_1, v_2, \ldots, v_p$ are not all uncorrelated as in (1.16). To show this, multiply (1.25) by β_{im}, sum over i from 1 to p and use (1.23) and the property that the matrix **B** is orthogonal. We then obtain, for $m = 1, 2, \ldots, k$,

$$\sum_{i=1}^{p} \beta_{im} v_i = \sum_{i=1}^{p} \beta_{im} x_i - \sum_{i=1}^{p} \sum_{n=1}^{k} \beta_{im}\beta_{in} u_n \qquad (1.28)$$

$$= u_m - u_m$$

$$= 0$$

The residuals $v_1, v_2, \ldots, v_p$ therefore satisfy the k linear equations (1.28) and hence cannot all be uncorrelated.

Summing up, we may say that factor analysis is correlation oriented and principal component analysis is variance oriented. Whereas factor analysis aims to reproduce the intercorrelations of the variables, principal component analysis aims to reproduce their total variance. Although in principal component analysis a few components may extract a large portion of the total variance, all components are required to reproduce the correlations exactly. In factor analysis, on the other hand, there are by definition a certain number of factors, fewer than the number of variables, that reproduce the intercorrelations exactly. These factors, however, do not account for as much variance as does the same number of principal components.

We may illustrate these remarks on the basis of the example considered in the previous section. The first two principal components (of unit variances) of $x_1, x_2, \ldots, x_6$ are

$$\begin{aligned} u_1^* &= 0.32x_1 + 0.30x_2 + 0.26x_3 + 0.24x_4 + 0.21x_5 + 0.15x_6 \\ u_2^* &= -0.29x_1 - 0.32x_2 + 0.34x_3 + 0.38x_4 + 0.42x_5 - 0.60x_6 \end{aligned} \qquad (1.29)$$

The equations corresponding to (1.27) are

$$\begin{aligned} x_1 &= 0.81u_1^* - 0.31u_2^* + v_1 \\ x_2 &= 0.78u_1^* - 0.34u_2^* + v_2 \\ x_3 &= 0.68u_1^* + 0.36u_2^* + v_3 \\ x_4 &= 0.62u_1^* + 0.40u_2^* + v_4 \\ x_5 &= 0.55u_1^* + 0.45u_2^* + v_5 \\ x_6 &= 0.39u_1^* - 0.60u_2^* + v_6 \end{aligned} \tag{1.30}$$

One can readily verify that these representations do not reproduce the intercorrelations as do (1.9). For example,

$$\begin{aligned} \beta_{i1}^*\beta_{j1}^* + \beta_{i2}^*\beta_{j2}^* &= 0.81 \times 0.78 + 0.31 \times 0.34 \\ &= 0.6318 + 0.1054 \\ &= 0.7372 \end{aligned}$$

which is not equal to ρ_{12}. The variance of each variable accounted for by the two principal components and the residual variances are

i	$\mathrm{Var}(\beta_{i1}^*u_1^* + \beta_{i2}^*u_2^*)$	$\mathrm{Var}(v_i)$	
1	0.75	0.25	
2	0.72	0.28	
3	0.59	0.41	
4	0.55	0.45	(1.31)
5	0.51	0.49	
6	0.51	0.49	
Total	3.63	2.37	

This is to be compared with the corresponding table (1.22). The residual variances in (1.31) are smaller than those of (1.22) except for x_1. The total variance accounted for by the two principal components is 3.63, compared with 2.64 for the two factors. This is a reflection of the specific factor included in each e_i but not in v_i.

In precisely the same way as in factor analysis, the two principal components may be transformed linearly to another set of components that together account for as much variance as the original principal components. In component analysis this may also be done to obtain some interesting interpretation. Principal components do not usually have a meaningful interpretation except in terms of their variance and correlation properties. Another difference between factor analysis and component analysis is that components are by definition linear combinations of the variables $x_1, x_2, \ldots, x_p$ whereas the common factors are not. Instead the factors are linear combinations of the common parts $c_1, c_2, \ldots, c_p$ of the variables. For example, the factors f_1^{**} and f_2^{**} in (1.13) have the following representations:

$$\begin{aligned} f_1^{**} &= (0.9^2 + 0.8^2 + 0.3^2)^{-1}(0.9c_1 + 0.8c_2 + 0.3c_6) \\ f_2^{**} &= (0.7^2 + 0.6^2 + 0.5^2)^{-1}(0.7c_3 + 0.6c_4 + 0.5c_5) \end{aligned} \qquad (1.32)$$

These are easily verified by substituting $c_1, c_2, \ldots, c_6$ from the equations corresponding to (1.18) into the right sides of (1.32).

Principal component analysis is not a model in the usual sense. It is merely a descriptive method of analysis that can be used to analyze all kinds of quantitative variables. Factor analysis, on the other hand, postulates a certain model (1.15), which is to be tested against empirical data. The equations (1.15) are not capable of direct verification, since the p variables x_i are expressed in terms of $p + k$ other variables which are not observable; but the equations imply a hypothesis that can be tested, namely that the correlations ρ_{ij} of the x's are of the form

$$\rho_{ij} = \lambda_{i1}\lambda_{j1} + \lambda_{i2}\lambda_{j2} + \ldots + \lambda_{ik}\lambda_{jk} \qquad (1.33)$$

If k is small, then (1.33) imposes restrictions on the ρ's. The smaller k is, the more restrictive the factor analysis model is.

4. Estimation

The basic model in factor analysis, as given by (1.15), can be written in matrix form as

$$\mathbf{x} = \Lambda\mathbf{f} + \mathbf{e} \qquad (1.34)$$

where $\mathbf{x}$ is a column vector of observations on p variables, $\mathbf{f}$ is a column vector of k common factors, $\mathbf{e}$ is a vector of p residuals, which represent the combined effects of specific factors and random measurement error, and $\Lambda = (\lambda_{im})$ is a $p \times k$ matrix of factor loadings.

The residuals in $\mathbf{e}$ are assumed to be uncorrelated with each other and with the common factors $\mathbf{f}$. All three vectors $\mathbf{x}$, $\mathbf{f}$, and $\mathbf{e}$ are assumed to have zero mean vectors and their covariance matrices are denoted respectively by Σ $(p \times p)$, Φ $(k \times k)$ and Ψ^2 $(p \times p)$. The matrix Ψ^2 is diagonal with elements ψ_{ii}^2, $i = 1, 2, \ldots, p$, which are the residual, or unique, variances of the variables. Since each column of Λ may be scaled arbitrarily, we may assume without loss of generality that the common factors have unit variances, so that the diagonal elements of Φ are unity. In addition, if, for $k > 1$ the common factors are orthogonal or uncorrelated, then the offdiagonal elements of Φ are zeroes and the matrix Φ is an identity matrix. Otherwise, if the factors are correlated, Φ is the correlation matrix of the factors.

In view of equation (1.34) and the assumptions just made, the covariance matrix Σ, of the observed variables, $\mathbf{x}$, is

$$\begin{aligned} \Sigma &= E(\mathbf{xx}') \\ &= E[(\Lambda\mathbf{f} + \mathbf{e})(\Lambda\mathbf{f} + \mathbf{e})'] \\ &= E(\Lambda\mathbf{ff}'\Lambda) + E(\Lambda\mathbf{fe}') + E(\mathbf{ef}'\Lambda') + E(\mathbf{ee}') \\ &= \Lambda\Phi\Lambda' + \Psi^2 \end{aligned} \tag{1.35}$$

since the second and third terms are zero because **f** and **e** were assumed to be uncorrelated.

Equations (1.34) and (1.35) represent a model for a population of individuals. This population is characterized by the parameters Λ, Φ, and Ψ^2. In practice these parameters are unknown and must be estimated from data on N individuals. Let x_{ai} be the observed value of variable i for individual a. Then the available data may be written as a data matrix $\mathbf{X}$ of order $N \times p$. From this we can compute the sample mean vector $\bar{\mathbf{x}}' = (\bar{x}_1, \bar{x}_2, \ldots, \bar{x}_p)$ and the sample covariance matrix $\mathbf{S} = (s_{ij})$, where

$$\bar{x}_i = (1/N) \sum_{a=1}^{N} x_{ai} \tag{1.36}$$

$$s_{ij} = (1/n) \sum_{a=1}^{N} (x_{ai} - \bar{x}_i)(x_{aj} - \bar{x}_j) \tag{1.37}$$

with $n = N - 1$.

The information provided by **S** may also be represented by a correlation matrix $\mathbf{R} = (r_{ij})$ and set of standard deviations $s_1, s_2, \ldots, s_p$ where $s_i = \sqrt{s_{ii}}$ and $r_{ij} = s_{ij}/s_i s_j$.

In most applications both the origin and the unit of measurement in the observed variables $\mathbf{x}$ are arbitrary or irrelevant, and then only the correlation matrix **R** is of any interest. In such cases one takes **S** to be a correlation matrix **R** in what follows. This is what we did in the previous sections.

In practical work with empirical data the factor analysis model does not fit the data perfectly, as it did in the artificial example used in the previous sections. The statistical problem then is how to fit the model matrix Σ of the form (1.35) to a sample covariance matrix **S**. This assumes that the number of factors k is known or specified a priori. However, in most exploratory factor studies this is not the case; instead the investigator wants to determine the smallest k for which the model fits the data. This is usually done by a sequential procedure testing increasing values of k until a sufficiently good fit has been obtained (see Lawley and Maxwell, 1971).

Several methods have been developed for estimating the parameters of the factor analysis model. Three different methods of fitting Σ to $\mathbf{S}$ will be considered here, namely the method of *unweighted least squares* (ULS), which minimizes the sum of squares of all elements of $\mathbf{S} - \Sigma$:

$$U = \text{tr}(\mathbf{S} - \Sigma)^2 \tag{1.38}$$

the method of *generalized least squares* (GLS), which minimizes the sum of squares of all the elements of $\mathbf{I} - \mathbf{S}^{-1}\Sigma$:

$$G = \text{tr}(\mathbf{I} - \mathbf{S}^{-1}\Sigma)^2 \tag{1.39}$$

and the method of *maximum likelihood* (ML), which minimizes

$$M = \text{tr}(\Sigma^{-1}\mathbf{S}) - \log|\Sigma^{-1}\mathbf{S}| - p \tag{1.40}$$

where tr($\mathbf{A}$) and $|\mathbf{A}|$ denote the trace and the determinant of $\mathbf{A}$, respectively. The last method is equivalent to the maximization of the likelihood of the observations under multinormality of $\mathbf{x}$, hence the name maximum likelihood method.

Each fitting function U, G, and M is to be minimized with respect to Λ, Φ, and Ψ. Derivations and justifications of these methods are found in Anderson (1959), Jöreskog (1967), Lawley and Maxwell (1971), and Jöreskog and Goldberger (1972). All three functions may be minimized numerically by basically the same algorithm. For details of the minimization procedure see Jöreskog (1977).

The GLS and ML methods are scale-free in the sense that analyses of the same variables in two different sets of scales are related by proper scale factors in the rows of Λ and Ψ. This property does not hold for ULS. When $\mathbf{x}$ has a multivariate normal distribution, both GLS and ML yield estimates that have good properties in large samples. Both GLS and ML require a positive definite matrix $\mathbf{S}$, while ULS will work even on a matrix which is non-Gramian.

When $k > 1$, so that there is more than one common factor, it is necessary to remove an element of indeterminacy in the basic model before the procedure for minimizing the fitting function can be applied. As demonstrated in the previous section, this indeterminacy arises because there exists a nonsingular linear transformation of the common factors which changes Λ and in general also Φ but leaves Σ and therefore the function unaltered. Hence to obtain a unique set of parameters and a corresponding unique set of estimates, we must impose some additional conditions. These have the effect of selecting a particular set of factors and thus of defining the parameters uniquely.

The usual way to choose the conditions in exploratory factor analysis is to choose $\Phi = I$, $\Lambda'\Lambda$ to be diagonal in ULS and $\Lambda'\Psi^{-2}\Lambda$ to be diagonal in GLS and ML, and to estimate Λ and Ψ subject to these conditions. This leads to an arbitrary set of factors that may then be subjected to a rotation to another set of factors to facilitate a more meaningful interpretation. The rotation is guided by the principle of simple structure. Techniques for rotation of factors are given by Harman (1967).

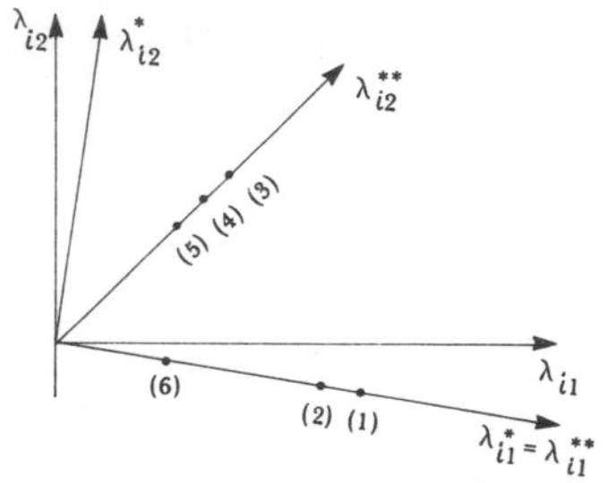

Figure 1. Plots of factor 2 versus factor 1 in three coordinate systems

References

Anderson, T. W. 1958. *An introduction to multivariate statistical analysis.* New York: Wiley.

Anderson T. W. 1959. Some scaling models and estimation procedures in the latent class model. In *Probability and Statistics,* The Harald Cramér Volume, ed. U. Grenander. New York: Wiley, pp. 9-38.

Cramér, H. 1957. *Mathematical methods of statistics.* Princeton, N.J.: Princeton University Press.

Harman, H. H. 1967. *Modern factor analysis,* 2nd ed. Chicago: University of Chicago Press.

Jöreskog, K. G. 1967. Some contributions to maximum likelihood factor analysis. *Psychometrika* 32: 443-482.

Jöreskog, K. G. 1977. Factor analysis by least squares and maximum likelihood methods. In *Statistical methods for digital computers*, ed. K. Enslein, A. Ralston, and H. S. Wilf. New York: Wiley, pp. 125-153.

Jöreskog, K. G., and Goldberger, A. S. 1972. Factor analysis by generalized least squares. *Psychometrika* 37: 243-259.

Kendall, M. G., and Lawley, D. N. 1956. The principles of factor analysis. *Journal of the Royal Statistical Society, Series A,* 119: 83-84.

Lawley, D. N., and Maxwell, A. E. 1971. *Factor analysis as a statistical method,* 2nd ed. London: Butterworth & Co.

Thurstone, L. L. 1947. *Multiple factor analysis.* Chicago: University of Chicago Press.

Chapter 2

A General Approach to Confirmatory Maximum Likelihood Factor Analysis

with Addendum

Karl G. Jöreskog

We describe a general procedure by which any number of parameters of the factor analytic model can be held fixed at any values and the remaining free parameters estimated by the maximum likelihood method. The generality of the approach makes it possible to deal with all kinds of solutions: orthogonal, oblique and various mixtures of these. By choosing the fixed parameters appropriately, factors can be defined to have desired properties and make subsequent rotation unnecessary. The goodness of fit of the maximum likelihood solution under the hypothesis represented by the fixed parameters is tested by a large sample χ^2 test based on the likelihood ratio technique. A by-product of the procedure is an estimate of the variance-covariance matrix of the estimated parameters. From this, approximate confidence intervals for the parameters can be obtained. Several examples illustrating the usefulness of the procedure are given.

1. *Introduction and Summary*

We shall describe a general procedure for performing factor analysis in the following way. Any values may be specified in advance for any number of factor loadings, factor correlations and unique variances. The remaining free parameters, if any, are estimated by the maximum likelihood method. A typical application of the procedure is in confirmatory factor studies, where the experimenter has already obtained a certain amount of knowledge about the variables measured and is therefore in a position to formulate a hypothesis that specifies some of the factors involved. For exploratory maximum likelihood factor analysis a computer program has been made available earlier [Jöreskog, 1967(a–b)]. This can be used to determine an appropriate number of factors to use and a preliminary interpretation of the data. The present procedure can then be used for a more precise analysis. We shall give examples of how a preliminary interpretation of the factors can be successively modified to determine a final solution that is acceptable from the point of view of both goodness of fit and psychological interpretation. It is highly desirable that a hypothesis that has been generated in this way should subsequently be confirmed or disproved by obtaining new data and subjecting these to a confirmatory analysis. Jöreskog and Lawley [1967], giving an expository

This work was supported by a grant (NSF-GB 1985) from the National Science Foundation to Educational Testing Service.

account of both exploratory and confirmatory methods, present an example where the original sample was randomly divided into two halves, where one half was used to generate a hypothesis and the other half was used to test this hypothesis.

The approach of this paper is similar to those of Howe [1955], Anderson and Rubin [1956], Lawley [1958] and Jöreskog [1966], but is more general in that it is possible to deal with all kinds of solutions: orthogonal, oblique and various mixtures of these. The fixed elements need not necessarily be zeros and the restrictions need not even be sufficient to make the solution unique. Factors can be defined to have desired properties and if a preliminary interpretation is available, the restrictions can be chosen to make any subsequent rotation unnecessary. Several examples are given in Sections 4 and 5 to illustrate the usefulness of the procedure.

The computational procedure is based on the minimization method of Fletcher and Powell [1963] and yields as a by-product an estimate of the variance-covariance matrix of the estimated parameters. From this, approximate confidence intervals for the parameters can easily be obtained.

2. Preliminary Considerations

In factor analysis the basic model is

$$y = \Lambda x + z, \tag{1}$$

where y is a vector of order p of observed test scores, x is a vector of order $k < p$ of latent common factor scores, z is a vector of order p of unique scores, and $\Lambda = (\lambda_{im})$ is a $p \times k$ matrix of factor loadings. It is assumed that $E(x) = E(z) = 0$, $E(xx') = \Phi$ and $E(zz') = \Psi$, a diagonal matrix. From these assumptions one deduces that the dispersion matrix of y, $\Sigma = E(yy')$, is

$$\Sigma = \Lambda\Phi\Lambda' + \Psi. \tag{2}$$

The basic idea of the model is that the common factors x shall account for all correlations between the y's. Once factors x have been partialed out, there shall remain no correlation between the tests.

The elements of Λ, Φ and Ψ are parameters of the model to be estimated from the data. Suppose that from a random sample of $n + 1$ observations of y we find the matrix S whose elements are the usual estimates of variances and covariances of the components of y. If y has a multivariate normal distribution, the elements of S follow a Wishart distribution with n degrees of freedom. The log-likelihood function, neglecting a function of the observations, is given by

$$\log L = -\tfrac{1}{2}n[\log |\Sigma| + \mathrm{tr}\,(S\Sigma^{-1})]. \tag{3}$$

Let

$$F(\Lambda, \Phi, \Psi) = \log |\Sigma| + \mathrm{tr}\,(S\Sigma^{-1}) - \log |S| - p. \tag{4}$$

Then maximizing log L is equivalent to minimizing F, and n times the minimum value of F is equal to the likelihood ratio test statistic of goodness of fit [see e.g., Jöreskog, 1967b]. It should be noted, however, that if T is any nonsingular $k \times k$ matrix, then

$$F(\Lambda T^{-1}, T\Phi T', \Psi) = F(\Lambda, \Phi, \Psi). \tag{5}$$

This means that the parameters in Λ and Φ are not independent of one another, and in order to make the maximum likelihood estimates of Λ and Φ unique, we must impose k^2 independent restrictions on Λ and Φ. In an exploratory factor analysis, where no hypotheses concerning the factors are involved, it is convenient to choose these restrictions so that $\Phi = I$ and $\Lambda'\Psi^{-1}\Lambda$ is diagonal [see e.g., Lawley & Maxwell, 1963; Jöreskog, 1967b]. In a confirmatory factor analysis, on the other hand, the investigator has certain hypotheses as to which factors are to be involved in certain tests, and it is therefore convenient to choose the restrictions by requiring that certain elements of Λ and Φ have values specified in advance. For example, $\lambda_{im} = 0$ means that the m-th factor does not enter into the i-th test and $\phi_{rs} = 0$ means that factors r and s are uncorrelated. Values other than zero could be used also. Depending on the number, values and positions of the fixed elements in Λ, Φ and Ψ, we may distinguish between two kinds of solutions: unrestricted and restricted. An *unrestricted solution* is one that does not restrict the common factor space, i.e., one that leaves $\Lambda\Phi\Lambda'$ unrestricted. All such solutions can be obtained by a rotation of an arbitrary unrestricted orthogonal maximum likelihood solution, for example, one obtained by the computational procedure of Jöreskog [1967a–b]. An unrestricted solution will usually result if the number of fixed elements in Λ and Φ is at most k^2 and if these elements are properly distributed over all factors. All unrestricted solutions for the same data will have the same communalities and the same uniquenesses, and they will all yield the same fit to the observed variances and covariances in S. In an unrestricted solution no element of Ψ can be held fixed, since clearly a restriction on Ψ is a restriction of the common factor space. A *restricted solution*, on the other hand, imposes restrictions on the whole factor space, and such a solution, therefore, cannot be obtained by a rotation of an unrestricted solution. Communalities and uniquenesses will not be the same for an unrestricted and a restricted solution for the same data. The fit to the observed variances and covariances in S, as measured by the function F, will, in general, be better for the unrestricted than for the restricted solution. However, if differences in number of estimated parameters are taken into account this may not be so.

Both unrestricted and restricted solutions may or may not be unique. This depends on the positions and values of the fixed parameters. A solution is unique if all linear transformations of the factors that leave the fixed parameters unchanged also leave the free parameters unchanged. Various sufficient conditions for a unique solution have been given by Reiersøl [1950], Howe [1955] and Anderson and Rubin [1956].

Two simple sufficient conditions, as given by Howe [1955], are as follows. In the orthogonal case, let $\Phi = I$ and let the columns of Λ be arranged so that, for $s = 1, 2, \cdots, k$, column s contains at least $s - 1$ fixed elements. In the oblique case, let diag $\Phi = I$ and let each column of Λ have at least $k - 1$ fixed elements. It should be noted that in the orthogonal case there are $\frac{1}{2}k(k + 1)$ conditions on Φ and a minimum of $\frac{1}{2}k(k - 1)$ conditions on Λ. In the oblique case there are k normalizations in Φ and a minimum of $k(k - 1)$ conditions on Λ. Thus, in both cases, there is a minimum of k^2 specified elements in Λ and Φ. Let Λ be a solution under any such conditions and let $\Lambda^{(s)}$ be the submatrix of Λ, consisting of those rows of Λ, that has fixed elements in the s-th column. Then Λ is unique if for all $s = 1, 2, \cdots, k$, $\Lambda^{(s)}$ has rank equal to the smallest of the numbers m_s and k, where m_s is the number of fixed elements in the s-th column of Λ. This condition is usually fulfilled in practice. It should be noted that even a restricted solution need not be unique. For example, an orthogonal solution with no restrictions on the first two columns of Λ and with more than $\frac{1}{2}k(k - 1)$ fixed elements in the other columns is restricted but not unique. Any orthogonal rotation in the plane of the first two factors will change these and leave all the fixed elements unchanged. In general, if a solution is not unique, transformations may exist that change the free parameters while leaving the fixed parameters unchanged. To make the solution unique, additional restrictions must be imposed.

In addition to the two main kinds of restrictions on Φ, *orthogonal* and *oblique*, various mixtures of these can be used. For example, one factor may be postulated to be uncorrelated with all the others—these factors being correlated among themselves. It is convenient to refer to all such solutions as *mixed* solutions.

Examples of many different kinds of solutions are given in Sections 4 and 5.

3. Minimization Procedure

The minimization problem is that of minimizing the function $F(\Lambda, \Phi, \Psi)$ with respect to the free parameters, keeping the others fixed at the given values. A numerical procedure for obtaining the maximum likelihood estimates, under certain special cases, was first given by Howe [1955]. A very similar procedure was later proposed by Lawley [1958] and referred to subsequently by Lawley and Maxwell [1963]. In both cases the derivatives of F are equated to zero and, after some simplification, a numerical solution of the resulting equations is sought. Recent work suggests, however, that these procedures do not always converge [Jöreskog, 1966]. Even when convergence does occur, it is usually very slow. A better method, for which ultimate convergence is assured, was given by Jöreskog [1966]. Experience with this method has revealed that it is sometimes still difficult to obtain a very accurate solution unless many iterations are performed. Efficient minimization of $F(\Lambda, \Phi, \Psi)$ seems impossible without the use of second-order derivatives.

The present procedure is based on the method of Fletcher and Powell [1963], which was used successfully in the unrestricted maximum likelihood problem [Jöreskog, 1967b]. It is a rapidly converging iterative procedure for minimizing a function of several variables when analytical expressions for the first-order derivatives are available. The efficiency of the method is obtained by the use of a matrix E, which is evaluated in each iteration. Initially, E is any positive definite matrix approximating the inverse of the matrix of second-order derivatives. In subsequent iterations E is improved, using the information built up about the function, so that ultimately E converges to the inverse of the second-order derivative matrix, evaluated at the minimum. If the number of parameters is large, the number of iterations required may still be large, but this can be decreased considerably by the provision of a good starting point and good initial estimates of second-order derivatives. Since the iteration equations have been given in detail by Jöreskog [1967b], they are not repeated here. The function F is considered only in the region where $\psi_{ii} \geq \epsilon$, $i = 1, 2, \cdots, p$, for some arbitrary small positive number ϵ. The treatment of these boundary conditions is the same as is given in the paper by Jöreskog.

The real advantage with the Fletcher and Powell method, as compared to the Newton–Raphson method, is that once an initial estimate of E has been obtained, the successive modifications of E throughout the iterations can be done very rapidly. One variant of the Newton–Raphson method computes the matrix of second-order derivatives and its inverse in each iteration, but this can be very time-consuming, especially when the number of parameters is large. Another variant of the Newton–Raphson method computes the matrix of second-order derivatives and its inverse only once and uses this inverse in all iterations. Such a procedure may require a large number of iterations to converge, especially if the starting point is not close to the solution point. The Fletcher and Powell method is a compromise between these two extremes. It improves the inverse of the matrix of second order derivatives in each iteration at very little cost.

In the unrestricted maximum likelihood problem it is possible to eliminate the parameters in Λ analytically so that the method of Fletcher and Powell is applied to a function of p variables only. Unfortunately, such a two-stage minimization procedure is not possible in this case, except under certain special conditions. The function F has, therefore, to be minimized simultaneously with respect to all free parameters. In a factor analysis of 40 tests and 10 factors, say, the number of free parameters may be almost 400 and, consequently, the matrix E, which must be evaluated in each iteration, is of order 400 × 400. The handling of such matrices in a computer presents many difficulties, even with present-day computers.

The function F is considered as a function of the free parameters in Λ, the free parameters in the *lower half of* Φ, *including the diagonal*, and the free parameters in the diagonal of Ψ. Expressions for the first-order derivatives of F have been given elsewhere [see e.g., Lawley & Maxwell, 1963; Jöreskog, 1966]. These expressions are

(6) $\partial F/\partial\Lambda = 2\Sigma^{-1}(\Sigma - S)\Sigma^{-1}\Lambda\Phi$

(7) $\partial F/\partial\Phi = c\Lambda'\Sigma^{-1}(\Sigma - S)\Sigma^{-1}\Lambda \begin{Bmatrix} c = 1 & \text{for diagonal elements} \\ c = 2 & \text{for non-diagonal elements} \end{Bmatrix}$

(8) $\partial F/\partial\Psi = \text{diag}\,[\Sigma^{-1}(\Sigma - S)\Sigma^{-1}]$,

with the understanding that elements in the matrices on the left that correspond to fixed values of Λ, Φ and Ψ are zero. The above matrices may be simplified for computational purposes by use of the identities

$$\Sigma^{-1} = \Psi^{-1} - \Psi^{-1}\Lambda(I + \Phi\Lambda'\Psi^{-1}\Lambda)^{-1}\Phi\Lambda'\Psi^{-1} \tag{9}$$

$$\Sigma^{-1}\Lambda = \Psi^{-1}\Lambda(I + \Phi\Lambda'\Psi^{-1}\Lambda)^{-1}. \tag{10}$$

The free parameters in Λ, Φ, Ψ are arranged in a vector θ as follows. Let θ_i, $i = 1, 2, \cdots, k$, be a vector containing the free parameters in column i of Λ and let θ_{k+1} and θ_{k+2} be vectors containing the free parameters in Φ and Ψ respectively. Then $\theta' = (\theta_1', \theta_2', \cdots, \theta_{k+2}')$. Formally we may now regard F as a function of θ and write $F(\theta)$. Similarly we may arrange all nonvanishing derivatives as a gradient vector $\partial F/\partial\theta$. If there are q free parameters all together, θ and $\partial F/\partial\theta$ are vectors of order q.

The Fletcher and Powell method requires repeated computation of function values $F(\theta)$ and gradient vectors $\partial F/\partial\theta$. For the iterative procedure to work satisfactorily it is necessary that this be done rapidly and accurately. The following method is used.

1. Compute $\Gamma = \Lambda'\Psi^{-1}\Lambda$ and $I + \Phi\Gamma$. The last matrix is unsymmetric, nonsingular and of order $k \times k$. Invert this to give $A = (I + \Phi\Gamma)^{-1}$. This also gives $|I + \Phi\Gamma|$.
2. Compute

$$|\Sigma| = \left(\prod_{i=1}^{p} \psi_{ii}\right) |I + \Phi\Gamma|$$

3. Compute $B = A\Phi$
4. Compute $C = \Psi^{-1} - \Psi^{-1}\Lambda B\Lambda'\Psi^{-1} = \Sigma^{-1}$
5. Compute $D = SC = S\Sigma^{-1}$ and $\text{tr}\, D = \text{tr}\,(S\Sigma^{-1})$

6. Compute F from (4) using $|\Sigma|$ from step 2 and tr $(S\Sigma^{-1})$ from step 5. (The quantity $\log |S| + p$ is a constant computed before the minimization begins.)
7. Compute $E = C - CD = \Sigma^{-1}(\Sigma - S)\Sigma^{-1}$
8. Compute $G = E\Lambda$
9. Compute $2G\Phi = \partial F/\partial \Lambda$
10. Compute $c\Lambda' G = \partial F/\partial \Phi$
11. Compute diag $E = \partial F/\partial \Psi$
12. Form the vector $\partial F/\partial \theta$ from the quantities of steps 9–11.

Formulas for large-sample approximations of second-order derivatives of F were derived by Lawley [1967] and independently by Lockhart [1967]. Lawley derived them by differentiation of the elements of the first order derivative matrices ignoring contributions arising from the differentiation of elements of $\Sigma - S$. Another way to derive these formulas is to make use of the covariance structure of a Wishart matrix S [see e.g., Anderson, 1958, Theorem 4.2.4],

$$nE[(s_{gh} - \sigma_{gh})(s_{ij} - \sigma_{ij})] = \sigma_{gi}\sigma_{hj} + \sigma_{gj}\sigma_{hi} \,. \tag{11}$$

As shown by Lawley [1967], the formulas for second-order derivatives can best be expressed in terms of the elements of the following matrices

$$\xi = \Sigma^{-1}\Lambda \tag{12}$$

$$\eta = \Sigma^{-1}\Lambda\Phi = \xi\Phi \tag{13}$$

$$\alpha = \Lambda'\Sigma^{-1}\Lambda = \Lambda'\xi \tag{14}$$

$$\beta = \Phi\Lambda'\Sigma^{-1}\Lambda = \Phi\alpha \tag{15}$$

$$\gamma = \Phi\Lambda'\Sigma^{-1}\Lambda\Phi = \beta\Phi. \tag{16}$$

The formulas then become

$$E(\partial^2 F/\partial\lambda_{ir}\partial\lambda_{js}) = 2(\sigma^{ij}\gamma_{rs} + \eta_{is}\eta_{jr}) \tag{17}$$

$$E(\partial^2 F/\partial\lambda_{ir}\partial\phi_{st}) = (2 - \delta_{st})(\xi_{is}\beta_{rt} + \xi_{it}\beta_{rs}) \tag{18}$$

$$E(\partial^2/\partial\lambda_{ir}\partial\psi_{jj}) = 2\sigma^{ij}\eta_{ir} \tag{19}$$

$$E(\partial^2 F/\partial\phi_{rs}\partial\phi_{tu}) = \tfrac{1}{2}(2 - \delta_{rs})(2 - \delta_{tu})(\alpha_{rt}\alpha_{su} + \alpha_{ru}\alpha_{st}) \tag{20}$$

$$E(\partial^2 F/\partial\phi_{rs}\partial\psi_{ii}) = (2 - \delta_{rs})\xi_{ir}\xi_{is} \tag{21}$$

$$E(\partial^2 F/\partial\psi_{ii}\partial\psi_{jj}) = (\sigma^{ij})^2. \tag{22}$$

Here δ_{rs} is Kronecker's delta, which is 1 if $r = s$ and zero otherwise. Except for (18), (20) and (21), these formulas agree with the results obtained by Lockhart who gives these results without the factors involving the δ's.

However, these factors are necessary if some factor is not scaled to unit variance. This will usually be the case whenever there are *fixed non-zero* values in Λ.

As an example of how these formulas are derived from (11), we prove the last formula. The i-th and j-th diagonal elements of $\partial F/\partial \Psi$ are

$$\partial F/\partial \psi_{ii} = \sum_{\alpha=1}^{p} \sum_{\beta=1}^{p} \sigma^{i\alpha}(s_{\alpha\beta} - \sigma_{\alpha\beta})\sigma^{\beta i} \tag{23}$$

$$\partial F/\partial \psi_{jj} = \sum_{u=1}^{p} \sum_{v=1}^{p} \sigma^{ju}(s_{uv} - \sigma_{uv})\sigma^{vj}. \tag{24}$$

Multiplying these and using (11) then gives

$$\begin{aligned} E(\partial^2 F/\partial \psi_{ii} \partial \psi_{jj}) &= (n/2)E(\partial F/\partial \psi_{ii}\, \partial F/\partial \psi_{jj}) \\ &= (\tfrac{1}{2}) \sum_{\alpha} \sum_{\beta} \sum_{u} \sum_{v} \sigma^{i\alpha}\sigma^{\beta i}\sigma^{ju}\sigma^{vj} nE[(s_{\alpha\beta} - \sigma_{\alpha\beta})(s_{uv} - \sigma_{uv})] \\ &= \sum_{\alpha} \sum_{\beta} \sum_{u} \sum_{v} \sigma^{i\alpha}\sigma^{\beta i}\sigma^{ju}\sigma^{vj}\sigma_{\alpha u}\sigma_{\beta v} \\ &= (\sigma^{ij})^2, \end{aligned}$$

where the first step follows from a general formula for likelihood functions [see e.g., Kendall & Stuart, 1961, eq. 18.60].

The computational method starts with arbitrary initial estimates of Λ, Φ, Ψ. The better these are, the fewer iterations will be required. If most of the restricted parameters are in Λ and these are sufficient to define Λ uniquely, such initial estimates can be obtained by rotating an unrotated orthogonal factor matrix using some Procrustes method, e.g. that of Lawley and Maxwell [1964]. In our computer program [Jöreskog & Gruvaeus, 1967], we generate initial estimates by a modified centroid method, if they are not provided by the user. From the initial starting point five steepest descent iterations are performed. Steepest descent iterations have been found to be very effective in the beginning but very ineffective in the neighborhood of the minimum. After these five iterations one has usually come so close to the minimum that it is worthwhile to compute the above approximations for second-order derivatives.

When suitably arranged these form a symmetric positive definite matrix G. The inverse matrix G^{-1} then serves as an initial approximation to E, and in subsequent iterations the Fletcher and Powell method is employed and the matrix E is modified accordingly.

When the number of free parameters to be estimated is large so that the order of the matrix G is large, the inversion of G is very difficult and time-consuming. In this case an approximation for G^{-1} which is sufficient to determine an initial estimate of E can be obtained as follows. Let

$$B = \begin{bmatrix} G_{11} & \cdots & 0 & 0 & 0 \\ \vdots & \ddots & \vdots & \vdots & \vdots \\ 0 & \cdots & G_{kk} & & \\ 0 & \cdots & \cdot & G_{k+1,k+1} & \cdot \\ 0 & \cdots & \cdot & \cdot & G_{k+2,k+2} \end{bmatrix} \tag{25}$$

where

$$G_{ii} = \partial^2 F/\partial\theta_i \partial\theta_i', \qquad i = 1, 2, \cdots, k + 2. \tag{26}$$

Then B^{-1} can be taken as an initial estimate of E. This is reasonable because earlier studies on sampling variability have indicated that a factor loading in one column of Λ correlates little with a factor loading in another column, with a factor correlation and with a unique variance. This reduces the problem to that of inverting $k + 2$ small matrices instead of one large matrix. In the Fletcher and Powell iterations the full E matrix is used.

The above method has been programmed in FORTRAN IV and tested out on the IBM 7044 [Jöreskog & Gruvaeus, 1967]. The program performs all computations in memory and is limited to at most 30 variables, 10 factors and 120 free parameters. The program is quite feasible for all the sets of data that it can handle. In computers with larger storage capacities than that of the IBM 7044, the above limits can be increased so that larger data can be handled.

When the minimum of F has been found, the minimizing values of Λ, Φ and Ψ are the maximum likelihood estimates $\hat{\Lambda}$, $\hat{\Phi}$ and $\hat{\Psi}$, and the hypothesis implied by the fixed parameters can be statistically tested. The maximum likelihood estimate of Σ under the hypothesis is

$$\hat{\Sigma} = \hat{\Lambda}\hat{\Phi}\hat{\Lambda}' + \hat{\Psi}, \tag{27}$$

and the likelihood ratio test statistic for testing the hypothesis is

$$n[\log|\hat{\Sigma}| - \log|S| + \operatorname{tr}(S\hat{\Sigma}^{-1}) - p]. \tag{28}$$

This is simply n times the minimum value $F(\hat{\Lambda}, \hat{\Phi}, \hat{\Psi})$ of $F(\Lambda, \Phi, \Psi)$. If n is large, this is distributed as χ^2 with

$$d = \tfrac{1}{2}p(p + 1) - pk - \tfrac{1}{2}k(k + 1) - p + m_\Psi + \sum_{i=1}^{k} \max\{m_i, k\} \tag{29}$$

degrees of freedom, where m_i is the number of independent restrictions on factor i, including restrictions on the ϕ_{ii} and the ϕ_{ij}, and m_Ψ is the number of fixed parameters in Ψ. In some cases the evaluation of the last term of (29) may be ambiguous. For example a restriction on ϕ_{ij} may be counted as a restriction either on factor i or on factor j. The rule to follow is to distribute

TABLE 1

Examples of Various Maximum Likelihood Solutions for Nine Psychological Variables

(Data from Holzinger & Swineford [1939]. An asterisk indicates that the parameter was fixed at this value.)

(a) Unrestricted Orthogonal Solution (m = 6)

1. **Visual Perception**	0.51	0.49	-0.02				0.50
2. Cubes	0.35	0.36	-0.10				0.74
3. Lozenges	0.52	0.44	-0.09				0.54
4. Paragraph Comprehension	0.81	-0.15	0.28	1.00*			0.24
5. Sentence Completion	0.74	-0.12	0.38	0.00*	1.00*		0.30
6. Word Meaning	0.77	-0.13	0.25	0.00*	0.00*	1.00*	0.32
7. Addition	0.04	0.26	0.74				0.39
8. Counting Dots	0.05	0.58	0.58				0.32
9. Straight-Curved Capitals	0.36	0.54	0.35				0.46

χ^2 = 9.77 with 12 degrees of freedom

P = 0.64

(b) Unrestricted Orthogonal Solution (m = 9)
Three-factor General Triangular Solution

1. Visual Perception	0.39	0.59	0.00*				0.50
2. Cubes	0.24	0.44	-0.07				0.74
3. Lozenges	0.38	0.56	-0.08				0.54
4. Paragraph Comprehension	0.87	0.00*	0.00*	1.00*			0.24
5. Sentence Completion	0.83	-0.01	0.12	0.00*	1.00*		0.30
6. Word Meaning	0.82	0.01	-0.01	0.00*	0.00*	1.00*	0.32
7. Addition	0.22	0.08	0.75				0.39
8. Counting Dots	0.14	0.42	0.70				0.32
9. Straight-Curved Capitals	0.36	0.51	0.39				0.46

χ^2 = 9.77 with 12 degrees of freedom

P = 0.64

(c) Unrestricted Oblique Solution (m = 9)
Reference Variables Solution

1. Visual Perception	0.71	0.00*	0.00*				0.50
2. Cubes	0.54	-0.03	-0.08				0.74
3. Lozenges	0.67	0.04	-0.09				0.54
4. Paragraph Comprehension	0.00*	0.87	0.00*	1.00*			0.24
5. Sentence Completion	-0.03	0.81	0.13	0.54	1.00*		0.30
6. Word Meaning	0.01	0.82	-0.01	0.24	0.28	1.00*	0.32
7. Addition	0.00*	0.00*	0.78				0.39
8. Counting Dots	0.42	-0.30	0.73				0.32
9. Straight-Curved Capitals	0.56	-0.06	0.41				0.46

χ^2 = 9.77 with 12 degrees of freedom

P = 0.64

(d) Restricted Oblique Solution (m = 21)
Independent Cluster Solution

1. Visual Perception	0.68	0.00*	0.00*				0.54
2. Cubes	0.52	0.00*	0.00*				0.73
3. Lozenges	0.69	0.00*	0.00*				0.52
4. Paragraph Comprehension	0.00*	0.87	0.00*	1.00*			0.25
5. Sentence Completion	0.00*	0.83	0.00*	0.54	1.00*		0.31
6. Word Meaning	0.00*	0.83	0.00*	0.52	0.34	1.00*	0.31
7. Addition	0.00*	0.00*	0.66				0.57
8. Counting Dots	0.00*	0.00*	0.80				0.37
9. Straight-Curved Capitals	0.00*	0.00*	0.70				0.51

χ^2 = 51.19 with 24 degrees of freedom

P = 0.00

such ambiguous restrictions so as to make the last term in (29) minimum. Examples of this rule are given in the next section.

The final matrix E that has been built up during the iterations is an approximation to the inverse of the matrix of second-order derivatives at the minimum. When multiplied by $(2/n)$ this gives an estimate of the variance-covariance matrix of the maximum likelihood estimates of the free parameters. Substantial experience with the method, however, has revealed that this estimate is not sufficiently accurate for most purposes. This is due to the fact that the matrix E is built up assuming the function to be exactly quadratic [see Fletcher & Powell, 1963]. Our function F is not quadratic, however, but is approximately so in a small region around the minimum. It should be noted that the matrix E is used essentially for the purpose of obtaining fast convergence and not of providing an estimate of the information matrix. Convergence is usually obtained in less than q iterations, often much less. However, at least q iterations in a quadratic region are necessary to build up numerically an accurate estimate of the inverse of the second-order derivative matrix. To obtain an accurate estimate of the variance-covariance matrix of the estimated parameters it is best, therefore, to recompute the second-order derivative matrix at the minimum and invert this. Denoting the matrix so obtained by E^*, only the diagonal elements of this, corresponding to variances of estimates, would normally be of interest. For any parameter θ_i with corresponding maximum likelihood estimate θ_i and variance of estimate e_{ii}^* , an approximate 95% confidence interval is

$$\hat{\theta}_i - 2\sqrt{(2/n)e_{ii}^*} < \theta_i < \hat{\theta}_i + 2\sqrt{(2/n)e_{ii}^*} \, . \tag{30}$$

This formula should be used only when the restrictions are such that the solution is unique.

4. An Analysis of Nine Mental Ability Tests

We shall illustrate the preceding ideas and methods and indicate some possible uses of the approach on the basis of two sets of empirical data. The analysis of the first data is reported in this section and that of the second in the next section.

The first set of data consists of nine mental ability tests selected from a battery of 26 tests previously analysed by Holzinger and Swineford [1939]. The nine tests are listed in Table 1. They can be thought of as measuring essentially visualization (tests 1, 2, 3), verbal intelligence (tests 4, 5, 6) and speed (tests 7, 8, 9). A full description of the tests is given in the above reference, where miscellaneous descriptive statistics are given also. Data were obtained on seventh- and eighth-grade children from two different schools. Only the data from the Grant–White school sample of 145 children are used here. The correlations were computed directly from the raw scores given by Holzinger and Swineford.

TABLE 1 (Contd)

(e) Restricted Mixed Solution (m = 20)

1. Visual Perception	0.73	0.00*	0.00*				0.47
2. Cubes	0.50	0.00*	0.00*				0.76
3. Lozenges	0.67	0.00*	0.00*				0.56
4. Paragraph Comprehension	0.00*	0.86	0.00*	1.00*			0.25
5. Sentence Completion	0.00*	0.82	0.00*	0.49	1.00*		0.31
6. Word Meaning	0.00*	0.82	0.00*	0.00*	0.24	1.00*	0.32
7. Addition	0.00*	0.00*	0.84				0.30
8. Counting Dots	0.29	0.00	0.67				0.44
9. Straight-Curved Capitals	0.56	0.00*	0.45				0.44

χ^2 = 26.47 with 23 degrees of freedom

P = 0.28

(f) Unrestricted Orthogonal Solution (m = 16)
Four-factor General Triangular Solution

1. Visual Perception	0.38	0.58	0.00*	0.00*					0.52
2. Cubes	0.24	0.41	0.35	0.00*					0.65
3. Lozenges	0.38	0.53	0.30	-0.03					0.48
4. Paragraph Comprehension	0.87	0.00*	0.03	0.00*	1.00*				0.25
5. Sentence Completion	0.83	0.01	-0.13	0.06	0.00*	1.00*			0.28
6. Word Meaning	0.83	0.01	0.04	-0.02	0.00*	0.00*	1.00*		0.32
7. Addition	0.24	0.02	0.00*	0.95	0.00*	0.00*	0.00*	1.00*	0.05
8. Counting Dots	0.15	0.43	-0.13	0.57					0.44
9. Straight-Curved Capitals	0.36	0.59	-0.22	0.34					0.36

χ^2 = 2.75 with 6 degrees of freedom

P = 0.84

(g) Restricted Orthogonal Solution (m = 16)

1. Visual Perception	0.38	0.62	0.00*				0.47
2. Cubes	0.24	0.42	0.00*				0.77
3. Lozenges	0.38	0.53	0.00*				0.57
4. Paragraph Comprehension	0.87	0.00*	0.00*	1.00*			0.24
5. Sentence Completion	0.83	0.00*	0.00*	0.00*	1.00*		0.32
6. Word Meaning	0.82	0.00*	0.00*	0.00*	0.00*	1.00*	0.32
7. Addition	0.25	0.00*	0.78				0.32
8. Counting Dots	0.16	0.37	0.70				0.36
9. Straight-Curved Capitals	0.37	0.49	0.42				0.45

χ^2 = 13.81 with 19 degrees of freedom

P = 0.80

Tables 1a–g show seven different maximum likelihood solutions for these data obtained under different identification conditions imposed on the factors. The fixed values of the parameters are marked with asterisks, and the number of fixed parameters is denoted m and is listed above each table. The value of χ^2 and the degrees of freedom as computed by (28) and (29) respectively are given below each solution. The probability level P of the χ^2 value is also given. This is defined as the probability of getting a χ^2 value larger than the value actually obtained, given that the hypothesized pattern is true. Thus, small values of P correspond to poor fit and large values to good fit.

In Table 1a we have postulated that the factors be uncorrelated and have unit variances. The latter, of course, is just an arbitrary scaling of the factors and will be used throughout. This gives $m = 6$ fixed parameters. Since $k = 3$, we should normally require $m = k^2 = 9$ independent restrictions to obtain a unique solution. The solution of Table 1a is therefore not unique but is just an arbitrary orthogonal solution. It happens to be the first one that the computer program found. It can be rotated orthogonally or obliquely to any other unrestricted maximum likelihood solution for the same data. To obtain an orthogonal solution, for example, we postmultiply the factor matrix by an orthogonal matrix of order 3×3. Since this matrix has three independent elements, it can be chosen to satisfy three independent restrictions on the factor loadings.

A particular set of such restrictions is used in Table 1b. It is seen that three factor loadings have been postulated to be zero. Test 1 (Visual Perception) is postulated not to load on factor 3, and test 4 (Paragraph Comprehension) is postulated to load only on factor 1, the idea being that factor 3 should be a speed factor, factor 2 a visualization factor and factor 1 a general factor. The solution of Table 1b represents an unrestricted orthogonal solution with nine fixed parameters. Given these fixed values, the solution is unique. It can be verified readily that the conditions for uniqueness, given in Section 2, are satisfied. The rows of the factor matrix can be permuted so that the zeros appear in the upper right triangle, hence the term general triangular solution. Table 1b serves to illustrate one important point, namely, that one has nothing to lose but may have much to gain by fixing the three zero loadings. By choosing these zero loadings appropriately, the solution will be directly interpretable, thus making subsequent rotations unnecessary. However, if further rotation is still found to be preferable, the solution can be rotated orthogonally or obliquely by any of the available methods. If the zero loadings are not too unreasonable, the rotation can probably be done by hand. We shall return to the interpretation of the solution of Table 1b after examining some other solutions.

Another unrestricted solution is shown in Table 1c. Here the factors are permitted to be correlated. To make the solution uniquely determined, we have chosen three reference variables, tests 1, 4 and 7, to represent the factors. These reference tests are postulated to be pure in their respective factors. Thus Table 1c represents an unrestricted oblique unique solution with $m = 9$ fixed elements. That the solution is unrestricted and unique is readily verified by showing that there is a unique transformation from the solution of Table 1a to that of Table 1c. Reference variables solutions are particularly useful since they impose just enough restrictions to make the solution unique, and if the reference tests are carefully chosen to tap the factors of interest, the solution will be directly interpretable in most cases. The solution in Table 1c exhibits a fairly clear simple structure. It is seen that the first seven tests are all loaded in one factor only, whereas the last two tests are more complex, being loaded not only in the speed factor but also in the visualization factor. One interpretation of this might be that the material in tests 8 and 9 consists of configurations in the plane in a way similar to tests 1 and 3. That test 9, in particular, should involve some visualization factor is evident from the fact that it is necessary here to break up a letter into straight and curved lines. Test 8, on the other hand, has its highest loading on the speed factor.

The solutions of Tables 1a, 1b and 1c are three different unrestricted solutions for the same data. This is reflected in two ways in the tables: the unique variances are the same, and the χ^2 values and the corresponding degrees of freedom are also the same for the three solutions. The probability level of 0.64, in this case, merely indicates that three factors are sufficient to account for the intercorrelations between the tests.

Let us now assume that it has been hypothesized in advance that the tests form three independent clusters so that tests 1, 2 and 3 are loaded in the visualization factor only, tests 4, 5 and 6 in the verbal factor only and tests 7, 8 and 9 in the speed factor only. This leads to the solution of Table 1d, where the free parameters have been estimated by the maximum likelihood method. This is a restricted oblique solution with $m = 21$ fixed parameters. Only nine of these restrictions are necessary to make the solution unique; the additional 12 restrictions affect the whole factor space. This is demonstrated by the differences in unique variances between Tables 1c and 1d and also by the large increase in χ^2. The χ^2 value is 51.19 which, with 24 degrees of freedom, is highly significant. The hypothesized factor structure must therefore be rejected as being untenable.

The question now arises as to what is causing the poor fit of the solution in Table 1d. In general, poor fit may be due to the fact that either the number of factors is untenable or the hypothesized structure is untenable, or both. In this case, we know from any one of the solutions of Tables 1a, 1b and 1c that three factors are tenable, so the poor fit must be due to a too restrictive structure. Which of the hypothesized zero loadings are then untenable? There are several ways to find this out. One way is to examine the residual

TABLE 2

Approximate 95% Confidence Intervals for the Parameters in the Solution of Table 1b

0.71 ± 0.17	0.00*	0.00*				0.50 ± 0.18
0.54 ± 0.25	-0.03 ± 0.14	-0.08 ± 0.19				0.74 ± 0.20
0.67 ± 0.24	0.04 ± 0.19	-0.09 ± 0.23				0.54 ± 0.19
0.00*	0.87 ± 0.18	0.00*	1.00*			0.24 ± 0.10
-0.03 ± 0.21	0.81 ± 0.15	0.13 ± 0.22	0.54 ± 0.22	1.00*		0.30 ± 0.11
0.01 ± 0.25	0.82 ± 0.18	-0.01 ± 0.22	0.24 ± 0.30	0.28 ± 0.23	1.00*	0.32 ± 0.11
0.00*	0.00*	0.78 ± 0.23				0.39 ± 0.23
0.42 ± 0.24	-0.30 ± 0.18	0.73 ± 0.22				0.32 ± 0.19
0.56 ± 0.22	-0.06 ± 0.15	0.41 ± 0.19				0.46 ± 0.14

correlations obtained after the three factors have been removed. A better, more direct, way is to compute approximate confidence intervals for all the free parameters of the solution of Table 1c. These confidence intervals are given in Table 2. It is seen that several of the loadings that were set equal to zero in Table 1d are indeed not significantly different from zero but a few of them are. Relaxing two of the zeros in the first factor of Table 1d and adding the restriction $\phi_{13} = 0$ yields the solution of Table 1e. Since one of the factor correlations is postulated to be zero, whereas the other factor correlations are free, this represents a restricted mixed solution with $m = 20$ fixed parameters. It should be noted that we have relaxed two of the zero factor loadings but have added a zero factor correlation. Thus the number of fixed parameters is only one less than before. The value of χ^2 is now 26.47 with 23 degrees of freedom. This has a probability level of 0.28. The solution is therefore acceptable, and it has also been verified that the values of all the free parameters are statistically significantly different from zero. The interpretation of the solution is very similar to that of Table 1c. In terms of the three factors, visualization, verbal and speed, the first seven tests are pure tests, whereas the last two tests are more complex. The loadings of these on the visualization factor have already been commented upon. It should be noted that factors 1 and 3 are uncorrelated but both of them correlated with factor 2. This suggests that the verbal factor is more general than the other two factors and that one might try to split this factor up into a general factor which influences all the tests and a more specific factor which is associated only with tests 4, 5 and 6.

The solution of Table 1f represents the results of such an attempt. It was postulated that there should be an orthogonal solution with a general factor and three group factors, the first one not loaded in test 4, the second one not loaded in tests 1 and 7 and the third one not loaded in tests 1, 2 and 4. It is seen that the visualization and speed factors appear as before, but the

TABLE 3

An Analysis of Two Batteries of Tests into Interbattery and Battery Specific Factors

(Original data from Thurstone & Thurstone [1941]. An asterisk indicates that the parameter was fixed at this value.)

(a) Restricted Orthogonal Solution (m = 19)

1. Prefixes	0.70	-0.12	0.15	0.00*					0.48
2. Suffixes	0.74	-0.08	0.22	0.00*					0.41
3. Vocabulary	0.39	0.81	0.33	0.00*					0.09
4. Sentences	0.37	0.75	0.08	0.00*	1.00*				0.31
5. First and Last Letters	0.65	-0.03	0.00*	0.37	0.00*	1.00*			0.44
6. First Letters	0.72	-0.05	0.00*	0.15	0.00*	0.00*	1.00*		0.46
7. Four-letter Words	0.60	0.09	0.00*	0.35	0.00*	0.00*	0.00*	1.00*	0.52
8. Completion	0.51	0.65	0.00*	0.02					0.32
9. Same or Opposite	0.48	0.67	0.00*	-0.12					0.32

χ^2 = 6.73 with 10 degrees of freedom

P = 0.75

(b) Restricted Mixed Solution (m = 20)

1. Prefixes	0.73	-0.08	0.15	0.00*					0.48
2. Suffixes	0.75	-0.04	0.22	0.00*					0.41
3. Vocabulary	0.00*	0.90	0.33	0.00*					0.09
4. Sentences	0.00	0.83	0.08	0.00*	1.00*				0.31
5. First and Last Letters	0.64	0.01	0.00*	0.37	0.38	1.00*			0.44
6. First Letters	0.72	0.00*	0.00*	0.15	0.00*	0.00*	1.00*		0.46
7. Four-letter Words	0.54	0.14	0.00*	0.35	0.00*	0.00*	0.00*	1.00*	0.52
8. Completion	0.19	0.74	0.00*	0.02					0.32
9. Same or Opposite	0.15	0.75	0.00*	-0.12					0.32

χ^2 = 6.73 with 10 degrees of freedom

P = 0.75

(c) Restricted Mixed Solution (m = 24)

1. Prefixes	0.67	0.00*	0.21	0.00*					0.51
2. Suffixes	0.71	0.00*	0.39	0.00*					0.35
3. Vocabulary	0.00*	0.91	0.17	0.00*					0.15
4. Sentences	0.00*	0.84	0.03	0.00*	1.00*				0.29
5. First and Last Letters	0.68	0.00*	0.00*	0.33	0.35	1.00*			0.43
6. First Letters	0.75	0.00*	0.00*	0.05	0.00*	0.00*	1.00*		0.44
7. Four-letter words	0.56	0.14	0.00*	0.29	0.00*	0.00*	0.00*	1.00*	0.52
8. Completion	0.20	0.73	0.00*	-0.01					0.32
9. Same or opposite	0.14	0.75	0.00*	-0.13					0.33

χ^2 = 9.43 with 14 degrees of freedom

P = 0.80

(d) Restricted Mixed Solution (m = 27)

1. Prefixes	0.70	0.00*	-0.15	0.00*					0.49
2. Suffixes	0.75	0.00*	-0.20	0.00*					0.39
3. Vocabulary	0.00*	0.89	0.12	0.00*					0.20
4. Sentences	0.00*	0.82	0.36	0.00*	1.00*				0.20
5. First and Last Letters	0.63	0.00*	0.00*	0.34	0.47	1.00*			0.48
6. First Letters	0.68	0.00*	0.00*	0.26	0.00*	0.00*	1.00*		0.46
7. Four-letter words	0.58	0.00*	0.00*	0.35	0.00*	0.00*	0.00*	1.00*	0.53
8. Completion	0.00*	0.81	0.00*	0.25					0.27
9. Same or opposite	0.00*	0.81	0.00*	0.06					0.33

χ^2 = 34.26 with 17 degrees of freedom

P = 0.01

attempt to isolate one general and one verbal factor was not successful. Tests 4, 5 and 6 load only in the general factor, and the second group factor is a very weak one having no psychological interpretation. This solution, therefore, is an example of overfactoring. If the second group factor is omitted and zero loadings postulated to identify the other two group factors as before, we arrive at the unrestricted orthogonal three-factor general triangular solution as in Table 1b. In this solution, all the small loadings are not significant. If these are set to zero, we get the restricted orthogonal solution of Table 1g.

The solutions of Tables 1e and 1g are two alternative final solutions for these data. Both solutions are final in the sense that all nonzero parameters are significant. The solution of Table 1e is slightly more restrictive than that of Table 1g, but the latter has a much better fit. The choice between the solutions, of course, is a matter of psychological theory [e.g., cf. Thurstone, 1947; Vernon, 1951].

The above examples show how the procedure of Section 3 can be used in an exploratory way to determine a solution that is reasonable from the point of view of both goodness of fit and psychological interpretation.

5. *Interbattery Analysis*

The second example serves to illustrate how our method can be used to analyse two batteries of tests into interbattery factors and battery specific factors. Tucker [1958] developed a method for determining factors that is common to the two batteries. These factors, called interbattery factors, account for the correlations between batteries but may not account for correlations within batteries. This example shows how factors that are specific to each battery can also be determined.

We shall use the same data as were used by Tucker. The data consist of nine tests from Thurstone and Thurstone [1941] listed in Tables 3a–d. The first four tests constitute battery 1 and the last five battery 2. Within and between correlations are given in Tucker's Table 2. The correlations are based on a sample of 710. Tucker found two interbattery factors that account for the correlations between batteries. This was also supported by a statistical test. However, two factors do not adequately account for all the correlations in the whole 9×9 correlation matrix. We checked this by performing an unrestricted maximum likelihood analysis with two factors. This gave $\chi^2 = 50.10$ with 19 degrees of freedom which is highly significant. The hypothesis of only two common factors in the two batteries must therefore be rejected. This suggests that there are factors specific to each battery but common to two or more tests within the battery. It is therefore postulated that there are four factors in the two batteries, that the first two are interbattery factors, that the third is specific to battery 1 and that the fourth is specific to battery 2. The two battery specific factors are postulated to be uncorrelated and uncorrelated with the two interbattery factors. Otherwise they would contribute to correlations between tests and hence not be battery specific. In addition we arbitrarily set the correlation between factors 1 and 2 equal to zero. The maximum likelihood solution under the above restrictions is given in Table 3a. This solution has acceptable fit, thus confirming the hypothesis that correlations between and within the two batteries can be accounted for by four factors with the above structure. It should be noted that this solution, regarded as a solution for all the nine tests, is a restricted orthogonal nonunique solution. It is restricted because too many restrictions have been imposed on factors 3 and 4, and it is not unique because too few restrictions have been imposed on factors 1 and 2. The latter can be rotated, orthogonally or obliquely, without changing the former. For example, such a rotation can be done to give two correlated interbattery factors with one zero loading for each factor. An example of this kind of solution is given in Table 3b, where test 3 (vocabulary) and test 6 (first letters) have been used as reference variables. A refinement of this solution, obtained by setting to zero the small loadings for the interbattery factors, is given in Table 3c. This imposes additional restrictions, and, as a consequence, factor loadings for the battery specific factors and the unique variances are changed. The solution of Table 3c fits the data well, and the two interbattery factors can be interpreted as a word-fluency and a verbal factor. Considering these two factors only, the first six tests are loaded on only one of them, whereas the last three are loaded on both. The three loadings $\lambda_{72} = 0.14$, $\lambda_{81} = 0.20$ and $\lambda_{91} = 0.14$, though small, appear to be significant. If all three are postulated to be zero, the solution of Table 3d is obtained.

The interbattery factors of Table 3d agree fairly well with the rotated solution obtained by Tucker [1958], but there are some differences for some of the loadings. Although battery specific factors may not be of direct interest, this example shows that they may be important in determining the interbattery factors.

6. The Question of Goodness of Fit

It was stated in Section 3 that n times the minimum value of F can be used as a large sample χ^2 statistic to test the hypothesis that the population variance-covariance matrix Σ is of the form (2) with specified values for certain parameters in Λ, Φ, Ψ. Such a hypothesis may be quite unrealistic in most empirical work with test data. If a sufficiently large sample were obtained this χ^2 statistic would, no doubt, indicate that any such non-trivial hypothesis is statistically untenable. The hypothesis of the experimenter rather is that (2) represents the variance-covariance matrix of the major factors that the experimenter is interested in, but that there are also a lot of minor factors which influence the test scores and which the experimenter has little or no control over [Tucker *et al.*, 1968]. These minor factors cause the lack of agreement between the formal mathematical model (2) and the variance-covariance matrix of the entire population. From this point of view the statistical problem is not one of testing a given hypothesis but rather one of fitting models with different numbers of parameters and of deciding when to stop fitting. The meaning and use of χ^2 in such problems are as follows. If a value of χ^2 is obtained, which is large compared to the number of degrees of freedom, this is an indication that more information can be extracted from the data. One may then try to relax the model somewhat by introducing more parameters. This can be done by relaxing some restrictions on the common factor space or by introducing additional factors or both. If, on the other hand, a value of χ^2 is obtained which is close to the number of degrees of freedom, this is an indication that the model "fits too well". Such a model is not likely to remain stable in future samples and all parameters may not have real meaning. When to stop fitting additional parameters cannot be decided on a purely statistical basis. This is largely a matter of the experimenter's interpretations of the data based on substantive theoretical and conceptual considerations. Ultimately the criteria for goodness of the model depends on the usefulness of it and the results it produces.

Examining the solutions of Table 1 from this point of view, we may say that for these data the solutions of Tables 1a, b, c, f, g are examples of overfitting whereas the solution of Table 1d is too restrictive. The solution of Table 1e represents a reasonable compromise. For the other data, the solution of Table 3d is the most reasonable, the other three solutions being fitted too well.

REFERENCES

Anderson, T. W. *An introduction to multivariate statistical analysis*. New York: Wiley, 1958.

Anderson, T. W., & Rubin, H. Statistical inference in factor analysis. In J. Neyman (Ed.), *Proceedings of the Third Berkeley Symposium on Mathematical Statistics and Probability*, Vol. V. Berkeley: University of California Press, 1956. pp. 111–150.

Fletcher, R., & Powell, M. J. D. A rapidly convergent descent method for minimization. *Computer Journal*, 1963, **2**, 163–168.

Holzinger, K. J., & Swineford, F. *A study in factor analysis: The stability of a bi-factor solution*. University of Chicago: Supplementary Educational Monographs, No. 48, 1939.

Howe, W. G. Some contributions to factor analysis. Report No. ORNL-1919, Oak Ridge National Laboratory, Oak Ridge, Tennessee, 1955.

Jöreskog, K. G. Testing a simple structure hypothesis in factor analysis. *Psychometrika*, 1966, **31**, 165–178.

Jöreskog, K. G. UMLFA - A computer program for unrestricted maximum likelihood factor analysis. Research Memorandum 66–20. Princeton, N. J.: Educational Testing Service. Revised Edition, 1967. (a)

Jöreskog, K. G. Some contributions to maximum likelihood factor analysis. *Psychometrika*, 1967, **32**, 443–482. (b)

Jöreskog, K. G., & Gruvaeus, G. RMLFA - A computer program for restricted maximum likelihood factor analysis. Research Memorandum 67–21. Princeton, N. J.: Educational Testing Service, 1967.

Jöreskog, K. G., & Lawley, D. N. New methods in maximum likelihood factor analysis. *British Journal of Mathematical and Statistical Psychology*, 1968, **21**, in press.

Kendall, M. G., & Stuart, A. *The advanced theory of statistics*, vol. 2. London: Charles Griffin and Co., 1961.
Lawley, D. N. Estimation in factor analysis under various initial assumptions. *British Journal of Statistical Psychology*, 1958, **11**, 1–12.
Lawley, D. N. Some new results in maximum likelihood factor analysis. *Proceedings of the Royal Society of Edinburgh*, Section A, 1967, **67**, 256–264.
Lawley, D. N., & Maxwell, A. E. *Factor analysis as a statistical method*. London: Butterworth, 1963.
Lawley, D. N., & Maxwell, A. E. Factor transformation methods. *British Journal of Statistical Psychology*, 1964, **17**, 97–103.
Lockhart, R. S. Asymptotic sampling variances for factor analytic models identified by specified zero parameters. *Psychometrika*, 1967, **32**, 265–277.
Reiersøl, O. On the identifiability of parameters in Thurstone's multiple factor analysis. *Psychometrika*, 1950, **15**, 121–149.
Thurstone, L. L. *Multiple-factor analysis*. Chicago: University of Chicago Press, 1947.
Thurstone, L. L., & Thurstone, T. G. Factorial studies of intelligence. *Psychometric Monographs No. 2*, 1941. Chicago: University of Chicago Press.
Tucker, L. R. An inter-battery method of factor analysis. *Psychometrika*, 1958, **23**, 111–136.
Tucker, L. R, Koopman, R. F., & Linn, R. L. Evaluation of factor analytic research procedures by means of simulated correlation matrices. *Psychometrika*, 1968, in press.
Vernon, P. E. *The structure of human abilities*. London: Methuen, 1951.

Reprinted from: *Psychometrika*, June 1969, Vol. 34, No. 2.

Author's Addendum
February, 1979

In this article, written in 1967, I quoted Howe (1955) and gave two conditions for the uniqueness, under factor rotation, of a factor matrix Λ containing specified fixed elements. Howe (1955) gave these conditions for the case of fixed *zero* elements, but I was assuming that they would be valid also for the case of nonzero fixed elements. However, this has been shown to be incorrect by Jennrich (1978). Also, in trying to formulate the conditions for orthogonal and oblique solutions at the same time, I made a mistake, so that Howe's formulation in the oblique case with fixed zero elements is incorrectly stated in my article. To clarify the issues it seems best to consider the four cases separately: (i) orthogonal solution with fixed zero elements, (ii) orthogonal solution with arbitrary fixed elements, (iii) oblique solution with fixed zero elements, (iv) oblique solution with arbitrary fixed elements.

Case (i) was considered by Dunn (1973) who gave a counterexample to show that the original condition, given by Howe (1955) and correctly quoted by me, is not sufficient. He also stated and proved a substitute condition for sufficiency. In Dunn's counterexample there are two columns of Λ with fixed zeroes in the same rows. Such a Λ cannot be unique, since it can be rotated orthogonally in the plane of these two columns without affecting the fixed zero elements.

Case (ii) was considered by Jennrich (1978) who gave an example of two orthogonally equivalent Λ-matrices with $k\ (k - 1)/2$ fixed nonzero elements. These two Λ-matrices have different elements in the nonfixed positions and hence are transparently different. In case (i) Dunn's conditions are sufficient for uniqueness up to column sign changes. Thus the specification of fixed zero elements does not lead to uniqueness but to a transparent form of uniqueness (column sign changes). However, when the specified values are not necessarily zero, one may be led to much less transparent forms of nonuniqueness. In both cases the specification of $k\ (k - 1)/2$ loadings reduces the indeterminacy of Λ considerably, from an infinite number of solutions to 2^k solutions. This suggests that while one may not obtain unique solutions using $k\ (k - 1)/2$ specified values, one will probably obtain solutions that are at least locally unique; indeed, one usually sees computer confirmation of this in the form of a positive definite information matrix.

Case (iii) is by far the most interesting case in practice, and I shall therefore restate and prove the original sufficiency conditions given by Howe (1955).

(α) Let Φ be a symmetric positive definite matrix with diag $\Phi = \mathbf{I}$.

(β) Let Λ have at least $k - 1$ fixed zeroes in each column.

(γ) Let Λ_s have rank $k - 1$, where Λ_s, $s = 1, 2, \ldots, k$, is the submatrix of Λ, consisting of the rows of Λ which have fixed zero elements in the sth column.

Then conditions α, β, and γ are sufficient for uniqueness of Λ.

The fixed unities in the diagonal of Φ merely fix the unit of measurement of the factors. An alternative way of doing this is to fix one nonzero value in each column of Λ instead. Conditions α and β are therefore equivalent to

(δ) Λ has at least $k - 1$ fixed zeroes in each column and one fixed nonzero value in each column, the latter values being in different rows.

I shall prove that conditions δ and γ are sufficient to define Λ uniquely. Let $\mathbf{B} = \Lambda\mathbf{T}$, where $\mathbf{T}$ is an arbitrary nonsingular matrix of order $k \times k$. I shall prove that if $\mathbf{B}$ has the same fixed elements as Λ and if conditions δ and γ hold, then $\mathbf{T}$ must be an identity matrix. Let Λ_s^* be the submatrix of Λ consisting of the rows of Λ that have fixed elements (including the fixed nonzero value) in the sth column. This is of the order $m_s + 1$ by k, where m_s is the number of fixed zeroes in column s. With a suitable permutation of the rows of Λ, Λ_s^* will be of the form

$$\Lambda_s^* = \begin{bmatrix} a & \gamma' \\ 0 & \\ 0 & \\ \cdot & \Lambda_{(s)} \\ \cdot & \\ \cdot & \\ 0 & \end{bmatrix}$$

where a is the fixed nonzero value in column s, λ' is a row vector of the remaining $k - 1$ values in the same row as a and $\Lambda_{(s)}$ of order m_s by $k - 1$ is the matrix Λ_s with the zero column omitted. Since **B** has the same fixed values as Λ, we must have

$$\Lambda_s^* \mathbf{t}_s = \begin{bmatrix} a \\ 0 \\ \cdot \\ \cdot \\ \cdot \\ 0 \end{bmatrix} \tag{1.41}$$

where $\mathbf{t}_s = (\mathrm{t}_{1s}' \ \mathrm{t}_{2s}')'$ is the sth column of **T**. Equation (1) is equivalent to

$$a t_{1s} + \lambda' \mathrm{t}_{2s} = \mathrm{a} \tag{1.42}$$

$$\Lambda_{(s)} \mathrm{t}_{2s} = \mathbf{0} \tag{1.43}$$

Since $m_s \geqslant k - 1$ and Λ_s has rank $k - 1$, the omission of the zero column of Λ_s will not change the rank. So the rank of $\Lambda_{(s)}$ is also $k - 1$. Therefore the only solution for $\mathbf{t}_{2s}$ satisfying (2a) is $\mathbf{t}_{2s} = \mathbf{0}$. With $\mathbf{t}_{2s} = \mathbf{0}$ and $a \neq 0$, (2a) implies that $\mathbf{t}_{1s} = 1$. Hence $\mathbf{t}_s$ is equal to a column of the identity matrix. The condition that the fixed nonzero values in Λ are in different rows will guarantee that a different column $\mathbf{t}_s$ will be obtained for $s = 1, 2, \ldots, k$. Hence $\mathbf{T} = \mathbf{I}$.

It is obvious that the same conclusion does not follow if the fixed zeroes in (1) are replaced by fixed nonzero values. In view of the results in case (ii) it is clear that much more research need to be done in case (iv) in order to clarify the issues and resolve the identification problem.

References

Dunn, J. E. 1973. A note on a sufficiency condition for uniqueness of a restricted factor matrix. *Psychometrika* 38: 141-143.

Howe, W. G. 1955. Some contributions to factor analysis. Report No. ORNL-1919 (Oak Ridge, Tenn.: Oak Ridge National Laboratory).

Jennrich, R, I. 1978. Rotational equivalence of factor loading matrices with specified values. *Psychometrika* 43: 421-426.

Chapter 3

Analyzing Psychological Data by Structural Analysis of Covariance Matrices

Karl G. Jöreskog

1. INTRODUCTION

Quantitative studies in psychology, and in other behavioral sciences as well, usually involve several measurements that correlate with each other in various ways. Structural analysis of covariance matrices is a general method for analyzing such measurements in order to detect and assess latent sources of variation and covariation in the observed measurements. When successfully applied this technique yields latent variables that account for the intercorrelations between the observed variables and that contain all the essential information about the linear interrelationships among these variables.

It is convenient to distinguish between exploratory and confirmatory studies. In confirmatory studies the experimenter has obtained such knowledge about the variables measured that he is in a position to formulate a hypothesis that specifies the latent variables on which the observed variables depend. Such a hypothesis may arise because of a specified theory, a given experimental design, known experimental conditions, or as a result of previous studies based on extensive data. In exploratory analysis, on the other hand, no such knowledge is available, and the main object is to find a simple, but meaningful, interpretation of the experimental results.

In practice, the above distinction is not always clear cut. Many investigations are to some extent both exploratory and confirmatory, since they involve some variables of known and other variables of unknown composition. The former should be chosen with great care so that as much information as possible about the latter may be extracted. It is highly desirable that a hypothesis suggested by mainly exploratory procedures should subsequently be confirmed, or disproved, by obtaining new data and subjecting them to more rigorous statistical techniques.

In this chapter I describe a general method for analysis of covariance structures and give many examples of areas and problems in the behavioral sciences where this method is useful. Although the method may be used for exploratory analysis, it is most useful for confirmatory analysis and I shall mainly deal with this situation. In Section 2, the model is treated as a purely formal model without substantive interpretations, and the problems of estimation and testing of this model are briefly reviewed. Examples of particular models, as special cases of the general model, are presented in Sections 3 through 6 of the chapter.

2. THE GENERAL MODEL

Definition of the General Model

I shall describe a general method for analyzing data according to a general model for covariance structures. The model assumes that the population variance-covariance matrix $\boldsymbol{\Sigma}(p \times p) = (\sigma_{ij})$ of a set of variables has the form

$$\boldsymbol{\Sigma} = \mathbf{B}(\boldsymbol{\Lambda}\boldsymbol{\Phi}\boldsymbol{\Lambda}' + \boldsymbol{\Psi}^2)\mathbf{B}' + \boldsymbol{\Theta}^2, \tag{1}$$

where $\mathbf{B}(p \times q) = (\beta_{ik})$, $\boldsymbol{\Lambda}(q \times r) = (\lambda_{km})$, the symmetric matrix $\boldsymbol{\Phi}(r \times r) = (\varphi_{mn})$, and the diagonal matrices $\boldsymbol{\Psi}(q \times q) = (\delta_{kl}\psi_k)$ and $\boldsymbol{\Theta}(p \times p) = (\delta_{ij}\theta_i)$ are parameter matrices (δ_{ij} denotes Kronecker's delta, which is one if $i = j$ and zero otherwise). It is assumed that the mean vector of the variables is unconstrained so that the information about the covariance structure is provided by the usual sample variance-covariance matrix $\mathbf{S}(p \times p) = (s_{ij})$, which may be taken to be a correlation matrix if the model is scale free and if the units of measurements in the variables are arbitrary or irrelevant.

The covariance structure of Equation 1 arises when the observed variables $\mathbf{x}(p \times 1)$ are of the form

$$\mathbf{x} = \boldsymbol{\mu} + \mathbf{B}\boldsymbol{\Lambda}\boldsymbol{\xi} + \mathbf{B}\boldsymbol{\zeta} + \mathbf{e}, \tag{2}$$

where $\boldsymbol{\xi}(r \times 1)$, $\boldsymbol{\zeta}(q \times 1)$, and $\mathbf{e}(p \times 1)$ are uncorrelated random latent vectors, in general unobserved, with zero mean vectors and dispersion matrices $\boldsymbol{\Phi}$, $\boldsymbol{\Psi}^2$, and $\boldsymbol{\Theta}^2$, respectively, and where $\boldsymbol{\mu}$ is the mean vector of $\mathbf{x}$.

In any application of the model of Equation 1, the number of variables p is given by the data, and q and r are given by the particular application that the investigator has in mind. In any such application any parameter in $\mathbf{B}$, $\boldsymbol{\Lambda}$, $\boldsymbol{\Phi}$, $\boldsymbol{\Psi}$, or $\boldsymbol{\Theta}$ may be known a priori and one or more subsets of the remaining parameters may have identical but unknown values. Thus, parameters are of three kinds: (a) *fixed parameters* that have been assigned given values, (b) *constrained parameters* that are unknown, but equal to one or more other parameters, and (c) *free parameters* that are unknown and not constrained to be equal to any other parameter. The advantage of such an approach is the great generality and flexibility obtained by the various specifications that may be imposed. Thus the general model contains a wide range of specific models.

In the general model (Eq. 1), it should be noted that if $\mathbf{B}$ is replaced by $\mathbf{BT}_1^{-1}$, $\boldsymbol{\Lambda}$ by $\mathbf{T}_1\boldsymbol{\Lambda}\mathbf{T}_2^{-1}$, $\boldsymbol{\Phi}$ by $\mathbf{T}_2\boldsymbol{\Phi}\mathbf{T}_2'$, and $\boldsymbol{\Psi}^2$ by $\mathbf{T}_1\boldsymbol{\Psi}^2\mathbf{T}_1'$ while $\boldsymbol{\Theta}$ is left unchanged, then $\boldsymbol{\Sigma}$ is unaffected. This holds for all nonsingular matrices $\mathbf{T}_1(q \times q)$ and $\mathbf{T}_2(r \times r)$ such that $\mathbf{T}_1\boldsymbol{\Psi}^2\mathbf{T}_1'$ is diagonal. Hence in order to obtain a unique set of parameters and a corresponding unique set of estimates, some restrictions must be imposed. This may be done by defining certain fixed and constrained parameters. In some cases these restrictions are given in a natural way by the particular application intended. In other cases they can be chosen in an almost arbitrary way. To make sure that all indeterminacies have been eliminated, one should verify that the only transformations $\mathbf{T}_1$ and $\mathbf{T}_2$ that preserve the specifications about fixed and constrained parameters are identity matrices.

Estimation and Testing of the Model

To determine the estimates of the unknown parameters, two different methods of fitting the model to the observed data may be used. One is the *generalized least squares method* (GLS) that minimizes

$$G = \mathrm{tr}(\mathbf{I} - \mathbf{S}^{-1}\boldsymbol{\Sigma})^2; \tag{3}$$

the other is the maximum-likelihood method (ML) that minimizes

$$M = \log|\boldsymbol{\Sigma}| + \mathrm{tr}(\mathbf{S}\boldsymbol{\Sigma}^{-1}) - \log|\mathbf{S}| - p. \tag{4}$$

Both G and M are regarded as functions of the independent elements of $\mathbf{B}$, $\boldsymbol{\Lambda}$, $\boldsymbol{\Phi}$, $\boldsymbol{\Psi}$, $\boldsymbol{\Theta}$. For derivation and justification of the GLS method, see Jöreskog and Goldberger (1972). The function M is a transform of the likelihood function obtained under the assumption that the observed variables have a multivariate normal distribution (see, e.g., Jöreskog, 1969). In a large sample of size N, $(N - 1)$ times the minimum value of G or M may be used as a χ^2 to test the goodness of fit of the model and for both methods approximate standard errors may be obtained for each estimated parameter by computing the inverse of the information matrix. The χ^2 test is a test of the specified model against the most general alternative that $\boldsymbol{\Sigma}$ is any positive definitive matrix.

Suppose H_0 represents one model under given specifications of fixed, free, and constrained parameters. Then it is possible, in large samples, to test the model H_0 against any more general model H_1, by estimating each of them separately and comparing their χ^2 goodness-of-fit values. The difference in χ^2 is asymptotically a χ^2 with degrees of freedom equal to the corresponding difference in degrees of freedom. In many situations, it is possible to set up a sequence of hypotheses such that each one is a special case of the preceding and to test these hypotheses sequentially.

The values of χ^2 should be interpreted very cautiously. In most empirical work many of the hypotheses may not be realistic. If a sufficiently large sample were obtained, the test statistic would, no doubt, indicate that any such hypothesis is statistically untenable. The hypothesis should rather be that Equation 1 represents a reasonable approximation to the population variance-covariance matrix. From this point of view the statistical problem is not one of testing a given hypothesis (which a priori may be considered false), but rather one of fitting various models with different numbers of parameters and of deciding when to stop fitting. In other words, the problem is to extract as much information as possible out of a sample of given size without going so far that the result is affected to a large extent by 'noise.' It is reasonable and likely that more information can be extracted from a large sample than from a small sample. In such a problem the differences between χ^2 values matter rather than the χ^2 values themselves. In an exploratory study, if a value of χ^2 is obtained, which is large compared to the number of degrees of freedom, the fit may be examined by an inspection of the residuals, i.e., the discrepancies between observed and reproduced values. Often the results of an analysis, an inspection of residuals or other considerations will suggest ways to relax the model somewhat by introducing more parameters. The new model usually yields a smaller χ^2. A large drop in χ^2, compared to the difference in degrees of freedom, indicates that the changes made in the model represent a real improvement. On the other hand, a drop in χ^2 close to the difference in number of degrees of freedom indicates that the improvement in fit is obtained by 'capitalizing on chance,' and the added parameters may not have real significance and meaning.

The function G or M is minimized numerically with respect to the independent parameters using a modification of the iterative method of Fletcher and Powell (1963) (see Gruvaeus & Jöreskog, 1970). The minimization method makes use of the first-order derivatives and large sample approximations to the elements of the matrix of second-order derivatives. The essential formulas and the basic algorithm are given in Jöreskog (1970a), and a computer program ACOVS is described in Jöreskog, Gruvaeus and van Thillo (1970).

In the following sections, several examples are given of models that are useful in the behavioral sciences, and some of these models are illustrated by means of real data. All analyses presented are based on the ML method and all χ^2 values have been obtained as $(N - 1)$ times the minimum value of M. Sometimes a probability level P for a χ^2 value is given that refers to the probability of obtaining a χ^2 larger than that actually obtained, given that the hypothesized model holds.

In presenting some models, it is sometimes convenient to use a path diagram in which the observed variables are enclosed in squares and latent variables in circles. Residuals (errors in equations) and errors of measurement are not enclosed. A one-way arrow between two variables indicates a direct causal influence of one variable on another, whereas a two-way arrow indicates correlation or covariation between two variables not dependent on other variables in the model.

3. MODELS FOR SETS OF CONGENERIC TESTS

Test Theory Models

Most measurements employed in the behavioral sciences contain sizable errors of measurements and any adequate theory or model must take this fact into account. Of particular importance is the study of congeneric measurements, i.e., those measurements that are assumed to measure the same thing. Jöreskog (1971) considered several models for sets of congeneric tests and some material in this section is drawn from that paper.

Classical test theory (Lord & Novick, 1968) assumes that a test score x is the sum of a true score τ and an error score e, where e and τ are uncorrelated. A set of test scores $x_1, \ldots, x_p$ with true scores $\tau_1, \ldots, \tau_p$ is said to be congeneric if every pair of true scores τ_i and τ_j have unit correlation. Such a set of test scores can be represented as

$$\mathbf{x} = \boldsymbol{\mu} + \boldsymbol{\beta}\tau + \mathbf{e},$$

where $\mathbf{x}' = (x_1, \ldots, x_p)$, $\boldsymbol{\beta}' = (\beta_1, \ldots, \beta_p)$ is a vector of regression coefficients, $\mathbf{e}' = (e_1, \ldots, e_p)$ is the vector of error scores, $\boldsymbol{\mu}$ is the mean vector of $\mathbf{x}$, and τ is a true score, scaled to zero mean and unit variance for convenience. The elements of $\mathbf{x}$, $\mathbf{e}$, and τ are regarded as random variables for a population of examinees. Let $\theta_1^2, \ldots, \theta_p^2$ be the variances of $e_1, \ldots, e_p$, respectively, i.e., the error variances. The corresponding true score variances are $\beta_1^2, \ldots, \beta_p^2$. One important problem is that of estimating these quantities. The variance-covariance matrix of $\mathbf{x}$ is

$$\boldsymbol{\Sigma} = \boldsymbol{\beta}\boldsymbol{\beta}' + \boldsymbol{\Theta}^2, \tag{5}$$

where $\boldsymbol{\Theta} = \text{diag}(\theta_1, \ldots, \theta_p)$. This is a special case of Equation 1 obtained by specifying $q = r = 1$, $\mathbf{B} = \boldsymbol{\beta}$, $\boldsymbol{\Lambda} = \boldsymbol{\Phi} = 1$, and $\boldsymbol{\Psi} = 0$. The congeneric test model with four tests is illustrated in Figure 1.

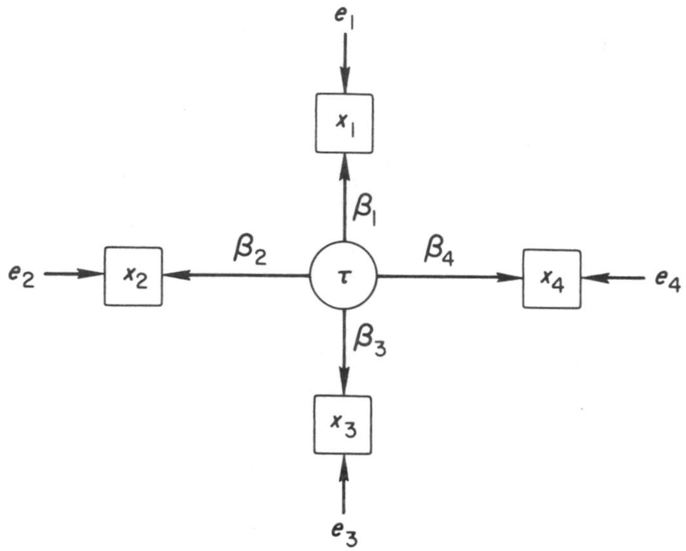

FIGURE 1.
The congeneric test model with four tests.

Parallel tests and tau-equivalent tests, in the sense of Lord and Novick (1968), are special cases of congeneric tests. Parallel tests have equal true score variances and equal error variances, i.e.,

$$\beta_1^2 = \cdots = \beta_p^2, \theta_1^2 = \cdots = \theta_p^2.$$

Tau-equivalent tests have equal true score variances, but possibly different error variances. These two models are obtained from Equation 1 by specification of equality of the corresponding set of parameters.

Parallel and tau-equivalent tests are homogenous in the sense that all covariances between pairs of test scores are equal. Scores on such tests are directly comparable, i.e., they represent measurements on the same scale. For tests composed of binary items this can hold only if the tests have the same number of items and are administrated under the same time limits. Congeneric tests, on the other hand, need not satisfy such strong restrictions. They need not even be tests consisting of items but can consist of ratings, for example, or even measurements produced by different measuring instruments.

Recently Kristof (1971) developed a model for tests that differ only in length. This model assumes that there is a 'length' parameter β_i associated with each test score x_i in such a way that the true score variance is proportional to β_i^4 and that the error variance is proportional to β_i^2. It can be shown that the covariance structure for this model is of the form

$$\boldsymbol{\Sigma} = \mathbf{D}_\beta(\boldsymbol{\beta}\boldsymbol{\beta}' + \psi^2\mathbf{I})\mathbf{D}_\beta,$$

where $\mathbf{D}_\beta = \text{diag}(\beta_1, \beta_2, \ldots, \beta_p)$, and $\boldsymbol{\beta}' = (\beta_1, \beta_2, \ldots, \beta_p)$. This is a special case of Equation 1, obtained by specifying $q = p$, $r = 1$, $\mathbf{B} = \mathbf{D}_\beta$, $\boldsymbol{\Lambda} = \boldsymbol{\beta}$, $\boldsymbol{\Phi} = 1$, $\boldsymbol{\Psi}^2 = \psi^2\mathbf{I}$, and $\boldsymbol{\Theta} = \mathbf{0}$. It should be noted that this model specifies equality constraints between the diagonal elements of $\mathbf{B}$ and the elements of the column vector $\boldsymbol{\Lambda}$, and also the equality of all the diagonal elements of $\boldsymbol{\Psi}$. The model has $p + 1$ independent parameters and is less restrictive than the parallel model, but is more restrictive than the congeneric model. A summary of the various test theory models and their number of parameters is given in Table 1. In the table, $\mathbf{j}$ denotes a column vector with all elements equal to one.

TABLE 1
Various test theory models

Model	Covariance Structure	No. of Parameters
Parallel	$\boldsymbol{\Sigma} = \beta^2\mathbf{jj}' + \theta^2\mathbf{I}$	2
Tau-equivalent	$\boldsymbol{\Sigma} = \beta^2\mathbf{jj}' + \boldsymbol{\theta}^2$	$p + 1$
Variable-length	$\boldsymbol{\Sigma} = \mathbf{D}_\beta(\boldsymbol{\beta}\boldsymbol{\beta}' + \psi^2\mathbf{I})\mathbf{D}_\beta$	$p + 1$
Congeneric	$\boldsymbol{\Sigma} = \boldsymbol{\beta}\boldsymbol{\beta}' + \boldsymbol{\theta}^2$	$2p$

As an illustration, consider the following variance-covariance matrix $\mathbf{S}$ taken from Kristof (1971):

$$\mathbf{S} = \begin{bmatrix} 54.85 & & \\ 60.21 & 99.24 & \\ 48.42 & 67.00 & 63.81 \end{bmatrix}.$$

This is based on candidates ($N = 900$) who took the January, 1969, administration of the Scholastic Aptitude Test (SAT). The first test, Verbal Omnibus, was administered in 30 minutes, and the second test, Reading Comprehension, in 45 minutes. These two tests contained 40 and 50 items, respectively. The third test is an additional section of the SAT not normally administered.

The following maximum-likelihood estimates were obtained with the program ACOVS: $\hat{\beta}_1 = 2.58$, $\hat{\beta}_2 = 3.03$, $\hat{\beta}_3 = 2.69$, and $\hat{\psi} = 1.60$. The goodness-of-fit test yielded $\chi^2 = 4.93$ with 2 degrees of freedom and a probability level of $P = 0.09$.

A Statistical Model for Several Sets of Congeneric Test Scores

The previous model generalizes immediately to several sets of congeneric test scores. If there are q sets of such tests, with $m_1, m_2, \ldots, m_q$ tests, respectively, we write $\mathbf{x}' = (\mathbf{x}_1', \mathbf{x}_2', \ldots, \mathbf{x}_q')$, where $\mathbf{x}_g'$, $g = 1, 2, \ldots, q$ is the vector of observed scores for the gth set. Associated with the vector $\mathbf{x}_g$ is a true score τ_g and vectors $\boldsymbol{\mu}_g$ and $\boldsymbol{\beta}_g$ defined as in the previous section, so that

$$\mathbf{x}_g = \boldsymbol{\mu}_g + \boldsymbol{\beta}_g \tau_g + \mathbf{e}_g.$$

As before we may, without loss of generality, assume that τ_g is scaled to zero mean and unit variance. If the different true scores $\tau_1, \tau_2, \ldots, \tau_q$ are all mutually uncorrelated, then each set of tests can be analyzed separately as in the previous section. However, in most cases these true scores correlate with each other, and an overall analysis of the entire set of tests must be made. Let $p = m_1 + m_2 + \cdots + m_q$ be the total number of tests. Then $\mathbf{x}$ is of order p. Let $\boldsymbol{\mu}$ be the mean vector of $\mathbf{x}$, and let $\mathbf{e}$ be the vector of error scores. Furthermore, let

$$\boldsymbol{\tau}' = (\tau_1, \tau_2, \ldots, \tau_q)$$

and let $\mathbf{B}$ be the matrix of order $p \times q$, partitioned as

$$\mathbf{B} = \begin{bmatrix} \boldsymbol{\beta}_1 & 0 & \cdots & 0 \\ 0 & \boldsymbol{\beta}_2 & \cdots & 0 \\ \cdot & \cdot & \cdots & \cdot \\ \cdot & \cdot & \cdots & \cdot \\ 0 & 0 & \cdots & \boldsymbol{\beta}_q \end{bmatrix}. \qquad (6)$$

Then $\mathbf{x}$ is represented as

$$\mathbf{x} = \boldsymbol{\mu} + \mathbf{B}\boldsymbol{\tau} + \mathbf{e}.$$

Let $\boldsymbol{\Gamma}$ be the correlation matrix of $\boldsymbol{\tau}$. Then the variance-covariance matrix $\boldsymbol{\Sigma}$ of $\mathbf{x}$ is

$$\boldsymbol{\Sigma} = \mathbf{B}\boldsymbol{\Gamma}\mathbf{B}' + \boldsymbol{\Theta}^2, \qquad (7)$$

where $\boldsymbol{\Theta}^2$ is a diagonal matrix of order p containing the error variances. One such model is illustrated in Figure 2.

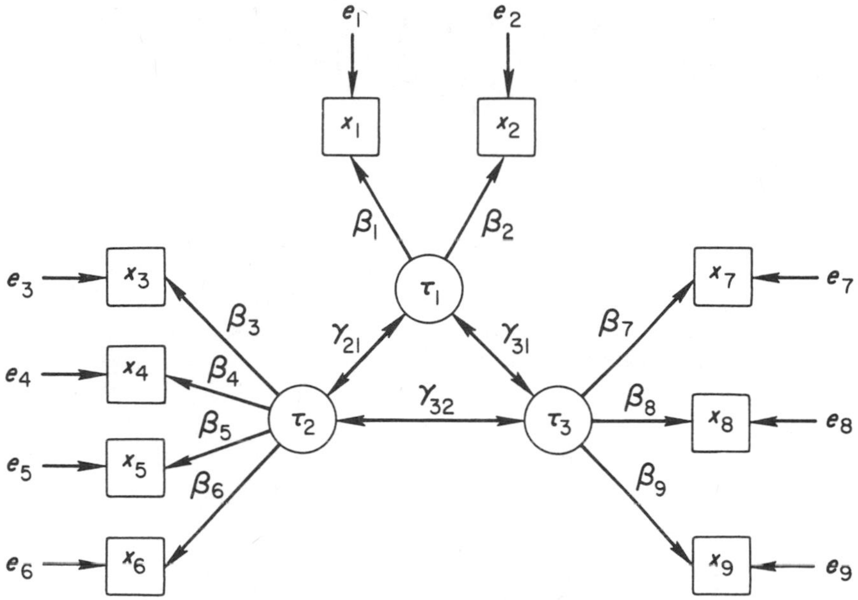

FIGURE 2.
A model with three sets of congeneric tests.

Testing the Hypothesis that the Correlation Coefficient Corrected for Attenuation Is Unity

The correlation coefficient corrected for attenuation between two tests x and y is the correlation between their true scores. If, on the basis of a sample of examinees, the corrected coefficient is near unity, the experimenter concludes that the two tests measure the same trait. Lord (1957) and McNemar (1958) have developed two different tests of the hypothesis that the population correlation coefficient corrected for attenuation $\rho_{T_x T_y}$ is equal to one. Lord's test is based on the large sample χ^2 approximation to the likelihood ratio statistic. McNemar's test is based on analysis of variance assuming homogeneity of error variances. Both tests employ two parallel forms x_1 and x_2 of x and y_1 and y_2 of y. McNemar's test, in addition, assumes that x and y are equally reliable. I formulate these two tests in terms of the model of Equation 7 and show how the hypothesis can also be tested under less restrictive assumptions.

Let x_1, x_2, y_1, and y_2 be four tests with zero means and let them satisfy the following model:

$$\begin{pmatrix} x_1 \\ x_2 \\ y_1 \\ y_2 \end{pmatrix} = \begin{pmatrix} \beta_1 & 0 \\ \beta_2 & 0 \\ 0 & \beta_3 \\ 0 & \beta_4 \end{pmatrix} \begin{pmatrix} \tau_x \\ \tau_y \end{pmatrix} + \begin{pmatrix} e_1 \\ e_2 \\ e_3 \\ e_4 \end{pmatrix},$$

with variance-covariance matrix

$$\boldsymbol{\Sigma} = \begin{pmatrix} \beta_1 & 0 \\ \beta_2 & 0 \\ 0 & \beta_3 \\ 0 & \beta_4 \end{pmatrix} \begin{pmatrix} 1 & \rho \\ \rho & 1 \end{pmatrix} \begin{pmatrix} \beta_1 & \beta_2 & 0 & 0 \\ 0 & 0 & \beta_3 & \beta_4 \end{pmatrix} + \begin{bmatrix} \theta_1^2 & 0 & 0 & 0 \\ 0 & \theta_2^2 & 0 & 0 \\ 0 & 0 & \theta_3^2 & 0 \\ 0 & 0 & 0 & \theta_4^2 \end{bmatrix}. \quad (8)$$

This model is shown in Figure 3.

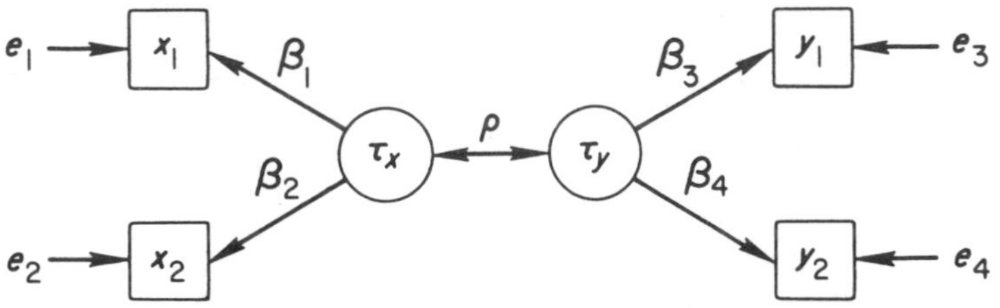

FIGURE 3.
A model with two sets of two congeneric tests.

McNemar's test is: Given $\beta_1 = \beta_2$, $\beta_3 = \beta_4$, $\theta_1^2 = \theta_2^2$, $\theta_3^2 = \theta_4^2$, $\beta_1^2/(\beta_1^2 + \theta_1^2) = \beta_3^2/(\beta_3^2 + \theta_3^2)$, test the hypothesis that $\rho = 1$. Lord's test is: Given that $\beta_1 = \beta_2$, $\beta_3 = \beta_4$, $\theta_1^2 = \theta_2^2$, $\theta_3^2 = \theta_4^2$, test the hypothesis that $\rho = 1$. McNemar's procedure requires a common metric for all four tests, whereas Lord's requires a common metric for x_1 and x_2 and also for y_1 and y_2. The conditions for these two tests are unnecessarily restrictive. It is possible to test the hypothesis assuming only that x_1 and x_2, as well as y_1 and y_2, are congeneric. This amounts to testing $\rho = 1$ under the model of Equation 8.

To illustrate the above ideas, I use some data from Lord (1957). The first two tests x_1 and x_2 are 15-item vocabulary tests administered under liberal time limits. The last two tests y_1 and y_2 are highly speeded 75-item vocabulary tests. The variance-covariance matrix is given in Table 2. I analyze these data under four different hypotheses:

H_1: $\beta_1 = \beta_2$, $\beta_3 = \beta_4$, $\theta_1^2 = \theta_2^2$, $\theta_3^2 = \theta_4^2$, $\rho = 1$;

H_2: $\beta_1 = \beta_2$, $\beta_3 = \beta_4$, $\theta_1^2 = \theta_2^2$, $\theta_3^2 = \theta_4^2$;

H_3: $\rho = 1$;

H_4: $\boldsymbol{\Sigma}$ is of the form of Equation 8 with β_1, β_2, β_3, β_4, θ_1, θ_2, θ_3, θ_4, and ρ unconstrained.

TABLE 2
Lord's vocabulary test data

Variance-Covariance Matrix, $N = 649$

	x_1	x_2	y_1	y_2
x_1	86.3979			
x_2	57.7751	86.2632		
y_1	56.8651	59.3177	97.2850	
y_2	58.8986	59.6683	73.8201	97.8192

Summary of Analyses

Hypothesis	No. Par.	χ^2	d.f.	P
H_1	4	37.33	6	0.00
H_2	5	1.93	5	0.86
H_3	8	36.21	2	0.00
H_4	9	0.70	1	0.70

Tests of Hypotheses

	Parallel	Congeneric	
$\rho = 1$	$\chi^2_6 = 37.33$	$\chi^2_2 = 36.21$	$\chi^2_4 = 1.12$
$\rho \neq 1$	$\chi^2_5 = 1.93$	$\chi^2_1 = 0.70$	$\chi^2_4 = 1.23$
	$\chi^2_1 = 35.40$	$\chi^2_1 = 35.51$	

The results are shown under Summary of Analyses. Each hypothesis is tested against the general alternative that Σ is unconstrained. To consider various hypotheses that can be tested, I recorded the four χ^2 values in a 2-×-2 table under Tests of Hypotheses. Lord's test is equivalent to testing H_1 against H_2, which gives $\chi^2 = 35.40$ with 1 degree of freedom. An alternative test is H_3 against H_4, which gives $\chi^2 = 35.51$ with 1 degree of freedom. Thus, regardless of whether we treat the two pairs of tests as parallel or congeneric, the hypothesis $\rho = 1$ is rejected. There is strong evidence that the unspeeded and speeded tests do not measure the same trait. I shall return to this question in the next section. The hypothesis of parallelism of the two pairs of tests can also be tested by means of the Tests of Hypotheses, which give $\chi^2 = 1.12$ or $\chi^2 = 1.23$ with 4 degrees of freedom, depending on whether we assume $\rho = 1$ or $\rho \neq 1$. Thus we cannot reject the hypothesis that the two pairs of tests are parallel. It appears that H_2 is the most reasonable of the four hypotheses. The maximum-likelihood estimate of ρ under H_2 is $\hat{\rho} = 0.899$ with a standard error of 0.019. An approximate 95-percent confidence interval for ρ is $0.86 < \rho < 0.94$.

Analysis of Speeded and Unspeeded Tests

Table 3 shows correlations and standard deviations for 15 tests. The three types of tests, vocabulary, intersections, and arithmetic reasoning, each has two levels of unspeeded or very little speeded tests (L) and three levels of highly speeded tests (S). These tests, together with several others, were analyzed by Lord (1956) with the objective being to isolate and to identify speed factors and their relationships to academic grades. Lord lists among others the following questions that his study was designed to answer: "Is speed on cognitive tests a unitary trait? Or are there different kinds of speed for different kinds of tasks? If so, how highly correlated are these kinds of speed? How highly correlated are speed and level on the same task? [p. 31]." Lord used maximum-likelihood factor analysis and oblique rotations, and verbal-speed and spatial-speed factors were clearly identified, but no arithmetic-reasoning speed factor was found. The smaller battery in Table 3 will be used here to answer the same kinds of questions by means of a series of confirmatory analyses.

I begin with the hypothesis that the five tests of each type are all congeneric. In principle, one could start by examining the more restrictive hypotheses of parallelism and tau-equivalence, but in view of the highly different time limits used for the unspeeded and the speeded tests, these hypotheses seem a priori implausible. If all tests of each kind are congeneric, it would mean that if errors of measurement were not present, speeded and unspeeded tests would measure the same thing. This, of course, is not very likely, a fact

TABLE 3
Lord's speed test data

Type of Test		Standard Deviations	Correlations ($N = 649$)											
Vocabulary	L	0.930	1.00											
	L	0.929	0.67	1.00										
	S	0.986	0.62	0.65	1.00									
	S	0.961	0.69	0.70	0.77	1.00								
	S	0.989	0.64	0.65	0.76	0.85	1.00							
Intersections	L	0.955	0.15	0.18	0.16	0.12	0.09	1.00						
	L	0.967	0.06	0.14	0.14	0.06	0.07	0.72	1.00					
	S	0.980	0.12	0.17	0.19	0.13	0.14	0.70	0.71	1.00				
	S	0.983	0.09	0.13	0.17	0.11	0.10	0.70	0.74	0.80	1.00			
	S	0.988	0.08	0.12	0.16	0.10	0.11	0.70	0.75	0.79	0.83	1.00		
Arith. Reasoning	L	0.905	0.30	0.32	0.28	0.29	0.27	0.34	0.30	0.32	0.31	0.33	1.00	
	L	0.967	0.25	0.31	0.29	0.27	0.29	0.36	0.30	0.30	0.29	0.31	0.54	1.00
	S	0.991	0.28	0.26	0.36	0.35	0.38	0.29	0.28	0.31	0.29	0.36	0.55	0.54	1.00		
	S	0.985	0.27	0.29	0.33	0.35	0.35	0.34	0.28	0.35	0.34	0.37	0.53	0.55	0.63	1.00	
	S	0.983	0.28	0.30	0.38	0.36	0.39	0.26	0.23	0.27	0.25	0.31	0.51	0.53	0.64	0.61	1.00

TABLE 4
Lord's speed test data. Summary of analyses

	Hypothesis	No. Par.	d.f.	χ^2	*P*
1	Three sets of congeneric tests	33	87	264.35	0.000
2	Four factors	42	78	140.50	0.000
3	Six sets of parallel tests	27	93	210.17	0.000
4	Six sets of tau-equivalent tests	36	84	138.72	0.000
5	Six sets of congeneric tests	45	75	120.57	0.001
6	Six factors	45	75	108.37	0.007

that is confirmed by the goodness-of-fit statistic in Table 4, the value of χ^2 being 264.35 with 87 degrees of freedom. This is an indication that the speeded tests measure something that the unspeeded tests do not. We therefore consider the hypothesis that there is a unitary factor associated with speed on which only the speeded tests are loaded. This leads to a matrix **B** with four columns, where the first three columns have nonzero loadings in the same position as the previous analysis and where the fourth column has nonzero loadings for the nine speeded tests. An analysis under this model yields a χ^2 of 140.50 with 78 degrees of freedom. Since this χ^2 is still large, we are led to consider the hypothesis that different kinds of speed are associated with different kinds of tests. We therefore consider the hypothesis that the two unspeeded and the three speeded tests of each kind are congeneric, i.e., we regard the tests as being six sets of congeneric tests. However, since tests in each set are equally speeded or unspeeded, we first examine the hypotheses of parallelism and tau-equivalence. It is seen in Table 4 that the hypothesis of parallelism has a very poor fit and that, when the restriction of equality of error variances is relaxed, the value of χ^2 drops significantly. However, the value 138.72 of χ^2 is still large compared to the degrees of freedom 84, so the restriction of equality of true score variances is also relaxed and the previously mentioned hypothesis of six sets of congeneric tests is examined. This gives $\chi^2 = 120.57$, a drop of 18.15 from the previous χ^2 with a corresponding drop of 9 in the degrees of freedom. Since this decrease in χ^2 is not very large compared to the drop in the degrees of freedom, there is an indication that the hypothesis of tau-equivalence does not fit much worse than the hypothesis of congenerism and that the true score variances are fairly equal. That this is indeed so can be seen in Table 5, which gives the solution under the hypothesis of congenerism. In fact, the variation between factor loadings in each column of $\hat{\mathbf{B}}$ is, in most cases, of the same size as the standard errors of estimate. An inspection of the correlations in the matrix $\hat{\Gamma}$ reveals that true scores on speeded and unspeeded tests of the same kind are very highly correlated, whereas true scores on speeded tests of different kinds correlate

TABLE 5
Lord's speed test data. Results of analysis 5

i	Type of Test		$\hat{\beta}_{i1}$	$\hat{\beta}_{i2}$	$\hat{\beta}_{i3}$	$\hat{\beta}_{i4}$	$\hat{\beta}_{i5}$	$\hat{\beta}_{i6}$	$\hat{\theta}_i$
1	Vocabulary	L	0.75	0.*	0.*	0.*	0.*	0.*	0.55
2		L	0.77	0.*	0.*	0.*	0.*	0.*	0.52
3		S	0.*	0.*	0.*	0.82	0.*	0.*	0.54
4		S	0.*	0.*	0.*	0.90	0.*	0.*	0.33
5		S	0.*	0.*	0.*	0.90	0.*	0.*	0.42
6	Intersections	L	0.*	0.80	0.*	0.*	0.*	0.*	0.53
7		L	0.*	0.84	0.*	0.*	0.*	0.*	0.48
8		S	0.*	0.*	0.*	0.*	0.86	0.*	0.48
9		S	0.*	0.*	0.*	0.*	0.89	0.*	0.41
10		S	0.*	0.*	0.*	0.*	0.90	0.*	0.40
11	Reasoning	L	0.*	0.*	0.66	0.*	0.*	0.*	0.62
12		L	0.*	0.*	0.71	0.*	0.*	0.*	0.65
13		S	0.*	0.*	0.*	0.*	0.*	0.80	0.58
14		S	0.*	0.*	0.*	0.*	0.*	0.77	0.61
15		S	0.*	0.*	0.*	0.*	0.*	0.77	0.62

$\hat{\gamma}_{kl}$

k	l = 1	2	3	4	5	6
1	1.*					
2	0.19	1.*				
3	0.49	0.51	1.*			
4	0.90	0.12	0.42	1.*		
5	0.16	0.93	0.47	0.15	1.*	
6	0.43	0.41	0.93	0.50	0.45	1.*

Note: Asterisks denote parameter values specified by hypothesis.

only moderately; the same holds for true scores on unspeeded tests of different kinds. This suggests that speeded and unspeeded tests of the same kind have a large common element that we may interpret as 'power' and that we should try to separate this power factor from the speed factor. If the hypothesis of congenerism had yielded a reasonably good fit, it would be natural to do this by a second-order analysis, i.e., by a factor analysis of $\hat{\Gamma}$ into one power factor and one speed factor. However, since $\chi^2 = 120.57$ with 75 degrees of freedom does not represent a good fit, it is more appropriate to separate speed from power at the first-order level by relaxing the hypothesis of congenerism, which is done by adding loadings for the three speeded tests on the first three factors. The resulting solution, given in Table 6, represents three power factors and three speed factors. The value of χ^2 for this solution

TABLE 6
Lord's speed test data. Results of analysis 6

i	Type of Test		$\hat{\beta}_{i1}$	$\hat{\beta}_{i2}$	$\hat{\beta}_{i3}$	$\hat{\beta}_{i4}$	$\hat{\beta}_{i5}$	$\hat{\beta}_{i6}$	$\hat{\theta}_i$
1	Vocabulary	L	0.75	0.*	0.*	0.*	0.*	0.*	0.55
2		L	0.77	0.*	0.*	0.*	0.*	0.*	0.51
3		S	0.76	0.*	0.*	0.31	0.*	0.*	0.54
4		S	0.82	0.*	0.*	0.37	0.*	0.*	0.35
5		S	0.78	0.*	0.*	0.48	0.*	0.*	0.38
6	Intersections	L	0.*	0.80	0.*	0.*	0.*	0.*	0.53
7		L	0.*	0.84	0.*	0.*	0.*	0.*	0.48
8		S	0.*	0.82	0.*	0.*	0.75	0.*	0.49
9		S	0.*	0.84	0.*	0.*	0.32	0.*	0.41
10		S	0.*	0.84	0.*	0.*	0.34	0.*	0.40
11	Reasoning	L	0.*	0.*	0.67	0.*	0.*	0.*	0.61
12		L	0.*	0.*	0.72	0.*	0.*	0.*	0.65
13		S	0.*	0.*	0.73	0.*	0.*	0.37	0.57
14		S	0.*	0.*	0.72	0.*	0.*	0.28	0.61
15		S	0.*	0.*	0.69	0.*	0.*	0.34	0.61

$\hat{\gamma}_{kl}$

k	l = 1	2	3	4	5	6
1	1.*					
2	0.17	1.*				
3	0.47	0.48	1.*			
4	0.*	0.*	0.*	1.*		
5	0.*	0.*	0.*	0.26	1.*	
6	0.*	0.*	0.*	0.70	0.39	1.*

Note: Asterisks denote parameter values specified by hypothesis.

is 108.37, which is 12.20 below that of Table 5. It should be noted, however, that the degrees of freedom for the two solutions are the same, due to the fact that the same number of parameters has been estimated in both solutions. Thus it appears that the solution of Table 6 is a better one than that of Table 5, both from the point of view of goodness of fit and of interpretation of the data. Table 6 clearly shows that three power factors and three speed factors can be isolated from these data. Speed factors are defined to be uncorrelated with the power factors. For each test, speed is defined to be the part that remains after the power factor has been eliminated. The three power factors correlate moderately and so do the three speed factors. If factor loadings and standard deviations in Table 6 are squared, one gets a decomposition of the unit variance of each test into components directly associated with power, speed, and error, respectively.

Analysis of Multitrait-Multimethod Data

A particular instance when sets of congeneric tests are employed is in multitrait-multimethod studies, where each of a number of traits is measured with a number of different methods or measuring instruments (see, e.g., Campbell & Fiske, 1959). One objective may be to find the best method of measuring each trait. In particular, one would like to get estimates of the trait, method, and error variance involved in each measure. A second objective is to study the internal relationships between the measures employed, in particular between the traits and between the methods.

Data from multitrait-multimethod studies are usually summarized in a correlation matrix giving correlations for all pairs of trait-method combinations. If there are m methods and n traits, this correlation matrix is of order $mn \times mn$. In analyzing such a correlation matrix, it seems natural to begin with the hypothesis that all methods are equivalent in measuring each trait, in the sense that scores obtained for a given trait with the different methods are congeneric. This hypothesis implies that all variation and covariation in the multitrait-multimethod matrix is due to trait factors only and may be tested by using a factor matrix $\mathbf{B}$ of order $mn \times n$ with one column for each trait. If the measurements are arranged with methods within traits, $\mathbf{B}$ is of the form of Equation 6. If, on the other hand, measurements are arranged with traits within methods, $\mathbf{B}$ has the form

$$\mathbf{B} = \begin{bmatrix} \mathbf{\Delta}_1 \\ \mathbf{\Delta}_2 \\ \vdots \\ \mathbf{\Delta}_m \end{bmatrix}, \qquad (9)$$

where each $\boldsymbol{\Delta}_i$ is a diagonal matrix of order $n \times n$. In both cases, the model is given by Equation 7, where $\boldsymbol{\Gamma}$ is the correlation matrix for the trait factors and $\boldsymbol{\Theta}^2$ is the diagonal matrix of error variances. If this model fits the data, the interrelationships between the trait factors may be analyzed further by a factoring of $\boldsymbol{\Gamma}$ as described in Section 4. However, if the hypothesis of equivalent methods does not fit the data, this is an indication that method factors are present. It then seems best to postulate the existence of one method factor for each method. This leads to a factor matrix $\mathbf{B}$ of order $mn \times (m+n)$ of the form (with traits within methods)

$$\mathbf{B} = \begin{bmatrix} \boldsymbol{\Delta}_1 & \boldsymbol{\beta}_1 & \mathbf{0} & \cdots & \mathbf{0} \\ \boldsymbol{\Delta}_2 & \mathbf{0} & \boldsymbol{\beta}_2 & \cdots & \mathbf{0} \\ \vdots & \vdots & \vdots & \cdots & \vdots \\ \boldsymbol{\Delta}_m & \mathbf{0} & \mathbf{0} & \cdots & \boldsymbol{\beta}_m \end{bmatrix}, \tag{10}$$

where the $\boldsymbol{\Delta}$s are as before and each $\boldsymbol{\beta}_i$ is a column vector of order n. The correlation matrix $\boldsymbol{\Gamma}$ of the factors may be specified as

$$\boldsymbol{\Gamma} = \begin{pmatrix} \boldsymbol{\Gamma}_1 & \mathbf{0} \\ \mathbf{0} & \boldsymbol{\Gamma}_2 \end{pmatrix}, \tag{11}$$

where $\boldsymbol{\Gamma}_1$ is the correlation matrix for the trait factors and $\boldsymbol{\Gamma}_2$ is the correlation matrix for the method factors. One such model is illustrated in Figure 4. In this model it is assumed that trait factors and method factors are uncorrelated. This defines each method factor to be independent of the particular traits that the method is used to measure. However, the trait factors may also be allowed to correlate with the method factors, in which case $\boldsymbol{\Gamma}$ is defined as a full correlation matrix. Substituting Equations 10 and 11 into Equation 7 gives the variance-covariance matrix $\boldsymbol{\Sigma}$ under this model. An analysis of data under this model yields estimates of $\mathbf{B}$, $\boldsymbol{\Gamma}$, and $\boldsymbol{\Theta}$. If the two factor loadings in each row of $\mathbf{B}$ and the corresponding element of $\boldsymbol{\Theta}$ are squared, one obtains a partition of the total variance of each measurement into components due to traits, methods, and error, respectively. If the fit of the model is good, and there are many traits and/or methods, one may analyze the interrelationships in $\boldsymbol{\Gamma}_1$ and $\boldsymbol{\Gamma}_2$ further in a way similar to that of Section 4.

In analyzing data in accordance with the above model it sometimes happens that one or more correlations in $\hat{\boldsymbol{\Gamma}}_2$ are close to unity or else that $\hat{\boldsymbol{\Gamma}}_2$ is not Gramian. This means that two or more method factors are collinear and have to be combined into one factor.

Analysis of Assessment Ratings of Clinical Psychologists

Clinical psychology students rated themselves and three teammates on each of several traits or characteristics. The median of the three teammate ratings

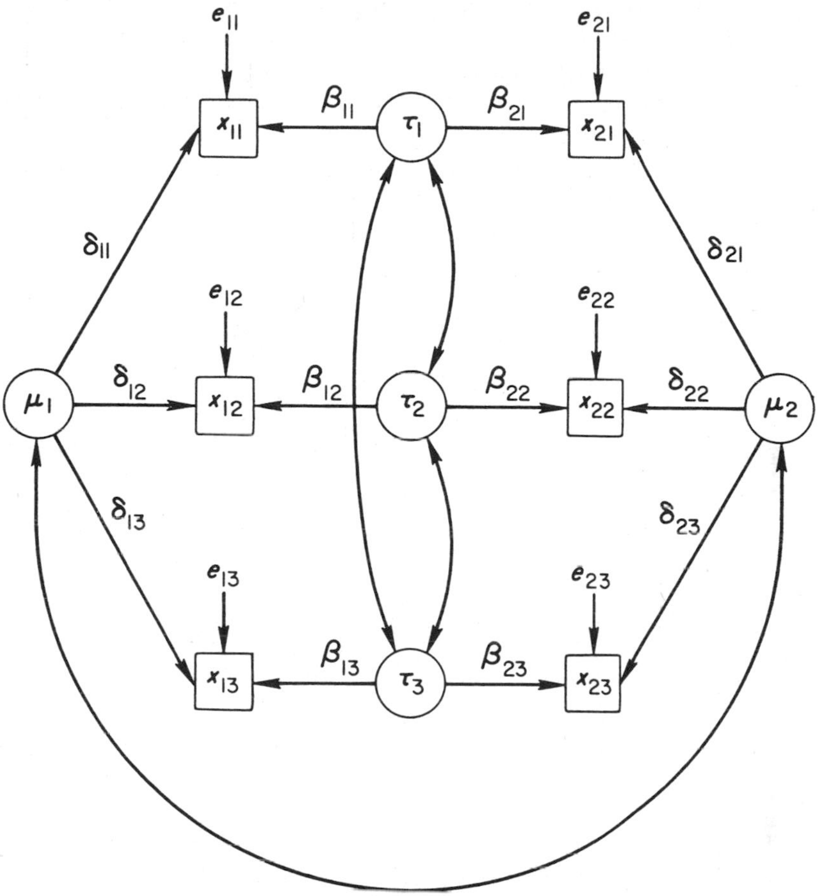

FIGURE 4.
A model for three traits measured by two methods.

was used as the score for this method of measurement. An assessment staff also rated the students, the ratings of the three members of the staff being pooled for a score on a given subject. The three methods of measurement thus are staff ratings, teammate ratings, and self-ratings. The study was conducted by Kelly and Fiske (1951) and full descriptions of the traits, methods, and data are contained in their article. On the basis of validity, Campbell and Fiske (1959) selected five traits—Assertive, Cheerful, Serious, Unshakable Poise, and Broad Interests—and published the 15 × 15 correlation matrix obtained from measurements on each of the five traits by each of the three methods. This correlation matrix is reproduced in Table 7 and is analyzed as outlined in the previous section.

TABLE 7
Ratings from assessment study of clinical psychologists

$N = 124$

		Staff Ratings					Teammate Ratings					Self-Ratings				
		A_1	B_1	C_1	D_1	E_1	A_2	B_2	C_2	D_2	E_2	A_3	B_3	C_3	D_3	E_3
Staff Ratings																
Assertive	A_1															
Cheerful	B_1	0.37														
Serious	C_1	−0.24	−0.14													
Unshakable Poise	D_1	0.25	0.46	0.08												
Broad Interests	E_1	0.35	0.19	0.09	0.31											
Teammate Ratings																
Assertive	A_2	0.71	0.35	−0.18	0.26	0.41										
Cheerful	B_2	0.39	0.53	−0.15	0.38	0.29	0.37									
Serious	C_2	−0.27	−0.31	0.43	−0.06	0.03	−0.15	−0.19								
Unshakable Poise	D_2	0.03	−0.05	0.03	0.20	0.07	0.11	0.23	0.19							
Broad Interests	E_2	0.19	0.05	0.04	0.29	0.47	0.33	0.22	0.19	0.29						
Self-Ratings																
Assertive	A_3	0.48	0.31	−0.22	0.19	0.12	0.46	0.36	−0.15	0.12	0.23					
Cheerful	B_3	0.17	0.42	−0.10	0.10	−0.03	0.09	0.24	−0.25	−0.11	−0.03	0.23				
Serious	C_3	−0.04	−0.13	0.22	−0.13	−0.05	−0.04	−0.11	0.31	0.06	0.06	−0.05	−0.12			
Unshakable Poise	D_3	0.13	0.27	−0.03	0.22	−0.04	0.10	0.15	0.00	0.14	−0.03	0.16	0.26	0.11		
Broad Interests	E_3	0.37	0.15	−0.22	0.09	0.26	0.27	0.12	−0.07	0.05	0.35	0.21	0.15	0.17	0.31	

The hypothesis that the three measurements of each trait are congeneric yields a χ^2 of 140.46 with 80 degrees of freedom. Since this does not represent a good fit, three method factors are added to the five trait factors, which yields a χ^2 of 57.91 with 62 degrees of freedom and represents an acceptable fit. However, the maximum-likelihood solution reveals that the two method factors for staff ratings and self-ratings correlate unity. These two factors are therefore combined into one. The final solution is shown in Table 8, which has a χ^2 of 61.51 with 64 degrees of freedom. Variance components of each measurement are given in Table 9.

It is interesting to see how Tables 8 and 9 can be interpreted directly in terms of what method best measures a given trait. The trait Assertiveness is best measured by staff ratings, but is also measured well by teammate ratings. When self-ratings are used, a less reliable measurement is obtained. Whatever method is used to measure Assertiveness, a very small method variance is involved. Thus it appears that all three methods produce valid measurements of Assertiveness. Also the trait Cheerfulness is validly measured by all three methods but the staff ratings are the most reliable measurements, though less reliable than staff ratings of Assertiveness. It should also be noted that Cheerfulness has fairly high correlations with Assertiveness and Unshakable Poise and is therefore likely to be confused with these traits. The trait Seriousness is best measured by teammate ratings, but cannot be measured as reliably as the other traits. The trait Unshakable Poise can be measured by staff ratings, but not with the other two methods. Broad Interests is reliably measured by self-ratings, but only with a large method variance.

4. FACTOR ANALYSIS MODELS

Factor analysis is a widely used technique, especially among psychologists and other behavioral scientists. The basic idea is that for a given set of response variates $x_1, \ldots, x_p$ one wants to find a set of underlying or latent factors $f_1, \ldots, f_k$, fewer in number than the observed variates, that will account for the intercorrelations of the response variates, in the sense that when the factors are partialled out from the observed variates no correlation remains between them. This leads to the model

$$\mathbf{x} = \boldsymbol{\mu} + \boldsymbol{\Lambda}\mathbf{f} + \mathbf{z}, \tag{12}$$

where $E(\mathbf{x}) = \boldsymbol{\mu}$, $E(\mathbf{f}) = \mathbf{0}$ and $E(\mathbf{z}) = \mathbf{0}$, $\mathbf{z}$ being uncorrelated with $\mathbf{f}$. Let $\boldsymbol{\Phi} = E(\mathbf{ff}')$, which may be taken as a correlation matrix and $\boldsymbol{\Psi}^2 = E(\mathbf{zz}')$, which is diagonal. Then the variance-covariance matrix $\boldsymbol{\Sigma}$ of $\mathbf{x}$ becomes

$$\boldsymbol{\Sigma} = \boldsymbol{\Lambda\Phi\Lambda}' + \boldsymbol{\Psi}^2. \tag{13}$$

TABLE 8
Ratings from assessment study of clinical psychologists.
Maximum-likelihood solution

		Factor Matrix							
		A	B	C	D	E	1–3	2	θ
Staff Ratings									
Assertive	A_1	0.87	0.*	0.*	0.*	0.*	0.11	0.*	0.49
Cheerful	B_1	0.*	0.84	0.*	0.*	0.*	0.02	0.*	0.55
Serious	C_1	0.*	0.*	0.57	0.*	0.*	−0.30	0.*	0.76
Unshakable Poise	D_1	0.*	0.*	0.*	0.78	0.*	−0.25	0.*	0.56
Broad Interests	E_1	0.*	0.*	0.*	0.*	0.69	−0.34	0.*	0.63
Teammate Ratings									
Assertive	A_2	0.83	0.*	0.*	0.*	0.*	0.*	0.16	0.54
Cheerful	B_2	0.*	0.70	0.*	0.*	0.*	0.*	0.29	0.68
Serious	C_2	0.*	0.*	0.72	0.*	0.*	0.*	0.32	0.59
Unshakable Poise	D_2	0.*	0.*	0.*	0.21	0.*	0.*	0.53	0.82
Broad Interests	E_2	0.*	0.*	0.*	0.*	0.60	0.*	0.44	0.65
Self-Ratings									
Assertive	A_3	0.55	0.*	0.*	0.*	0.*	0.11	0.*	0.83
Cheerful	B_3	0.*	0.45	0.*	0.*	0.*	0.22	0.*	0.87
Serious	C_3	0.*	0.*	0.43	0.*	0.*	0.23	0.*	0.88
Unshakable Poise	D_3	0.*	0.*	0.*	0.43	0.*	0.38	0.*	0.82
Broad Interests	E_3	0.*	0.*	0.*	0.*	0.70	0.62	0.*	0.41

	Factor Intercorrelations						
	A	B	C	D	E	1–3	2
A	1.*						
B	0.56	1.*					
C	−0.37	−0.44	1.*				
D	0.38	0.66	−0.08	1.*			
E	0.55	0.29	−0.02	0.43	1.*		
1–3	0.*	0.*	0.*	0.*	0.*	1.*	
2	0.*	0.*	0.*	0.*	0.*	−0.21	1.*

Note: Asterisks denote parameter values specified by hypothesis.

TABLE 9
Ratings from assessment study of clinical psychologists

		Variance Components		
		Trait	Method	Error
Staff Ratings				
Assertive	A_1	0.76	0.01	0.24
Cheerful	B_1	0.70	0.00	0.30
Serious	C_1	0.33	0.09	0.58
Unshakable Poise	D_1	0.61	0.06	0.32
Broad Interests	E_1	0.48	0.11	0.40
Teammate Ratings				
Assertive	A_2	0.69	0.03	0.29
Cheerful	B_2	0.48	0.09	0.46
Serious	C_2	0.52	0.10	0.35
Unshakable Poise	D_2	0.05	0.28	0.68
Broad Interests	E_2	0.36	0.19	0.43
Self-Ratings				
Assertive	A_3	0.30	0.01	0.69
Cheerful	B_3	0.21	0.05	0.75
Serious	C_3	0.18	0.05	0.77
Unshakable Poise	D_3	0.18	0.14	0.68
Broad Interests	E_3	0.49	0.39	0.17

If $(p - k)^2 < p + k$, this relationship can be tested statistically, unlike Equation 12, which involves hypothetical variates and cannot be verified directly. Equation 13 may be obtained from the general model (Eq. 1) by specifying $\mathbf{B} = \mathbf{I}$ and $\boldsymbol{\Theta} = \mathbf{0}$.

When $k > 1$ there is an indeterminacy in Equation 14 arising from the fact that a nonsingular linear transformation of $\mathbf{f}$ changes $\boldsymbol{\Lambda}$ and in general also $\boldsymbol{\Phi}$ but leaves $\boldsymbol{\Sigma}$ unchanged. The usual way to eliminate this indeterminacy in exploratory factor analysis (see, for example, Lawley & Maxwell, 1963; Jöreskog, 1967; Jöreskog & Lawley, 1968) is to choose $\boldsymbol{\Phi} = \mathbf{I}$ and $\boldsymbol{\Lambda}'\boldsymbol{\Psi}^{-1}\boldsymbol{\Lambda}$ to be diagonal and to estimate the parameters in $\boldsymbol{\Lambda}$ and $\boldsymbol{\Psi}$ subject to these conditions. This leads to an arbitrary set of factors that may then be subjected to a rotation or a linear transformation to another set of factors that can be given a more meaningful interpretation.

In terms of the general model (Eq. 1), the indeterminacy in Equation 13 may be eliminated by assigning zero values, or any other values, to k^2 elements in $\boldsymbol{\Lambda}$ and/or $\boldsymbol{\Phi}$ in such a way that the assigned values will be destroyed by all nonsingular transformations of the factors except the identity transformation. There may be an advantage in eliminating the indeterminacy this way, in that, if the fixed parameters are chosen in a reasonable way, the resulting solution will be directly interpretable and the subsequent rotation of factors may be avoided.

Specification of parameters a priori may also be used in a confirmatory factor analysis, where the experimenter has already obtained a certain amount of knowledge about the variates measured and is in a position to formulate a hypothesis that specifies the factors on which the variates depend. Such a hypothesis may be specified by assigning values to some parameters in $\boldsymbol{\Lambda}$, $\boldsymbol{\Phi}$, and $\boldsymbol{\Psi}$; (see, e.g., Jöreskog & Lawley, 1968; Jöreskog, 1969). If the number of fixed parameters in $\boldsymbol{\Lambda}$ and $\boldsymbol{\Phi}$ exceeds k^2, the hypothesis represents a restriction of the common factor space, and a solution obtained under such a hypothesis cannot be obtained by a rotation of an arbitrary solution such as that obtained in an exploratory analysis.

The model of Equation 5 is formally equivalent to a factor analytic model with one common factor, and the model of Equation 7 is equivalent to a factor analytic model with q correlated nonoverlapping factors. In the latter case the factors are the true scores $\boldsymbol{\tau}' = (\tau_1, \ldots, \tau_q)$ of the tests. These true scores may themselves satisfy a factor analytic model, i.e.,

$$\boldsymbol{\tau} = \boldsymbol{\Lambda}\mathbf{f} + \mathbf{s},$$

where $\mathbf{f}$ is a vector of order k of common true score factors, $\mathbf{s}$ is a vector of order q of specific true score factors, and $\boldsymbol{\Lambda}$ is a matrix of order $q \times k$ of factor loadings. Let $\boldsymbol{\Phi}$ be the variance-covariance matrix of $\mathbf{f}$, and let $\boldsymbol{\Psi}^2$ be a diagonal matrix whose diagonal elements are the variances of the specific true score factors $\mathbf{s}$. Then $\boldsymbol{\Gamma}$, the variance-covariance matrix of $\boldsymbol{\tau}$, becomes

$$\boldsymbol{\Gamma} = \boldsymbol{\Lambda}\boldsymbol{\Phi}\boldsymbol{\Lambda}' + \boldsymbol{\Psi}^2. \tag{14}$$

Substituting Equation 14 into Equation 7 gives $\boldsymbol{\Sigma}$ as

$$\boldsymbol{\Sigma} = \mathbf{B}(\boldsymbol{\Lambda}\boldsymbol{\Phi}\boldsymbol{\Lambda}' + \boldsymbol{\Psi}^2)\mathbf{B}' + \boldsymbol{\Theta}^2. \tag{15}$$

This model is a special case of Equation 1 by specifying zero values in $\mathbf{B}$ as before. To define $\boldsymbol{\Lambda}$ and $\boldsymbol{\Phi}$ uniquely, we must impose k^2 independent conditions on these to eliminate the indeterminacy due to rotation. The model of Equation 15 is a special case of the second-order factor analytic model. An example of a second-order factor analysis model is given in Figure 5.

5. VARIANCE AND COVARIANCE COMPONENTS

Estimation of Variance Components

Several authors (Bock, 1960; Bock & Bargmann, 1966; Wiley, Schmidt, & Bramble, 1973) have considered covariance structure analysis as an approach

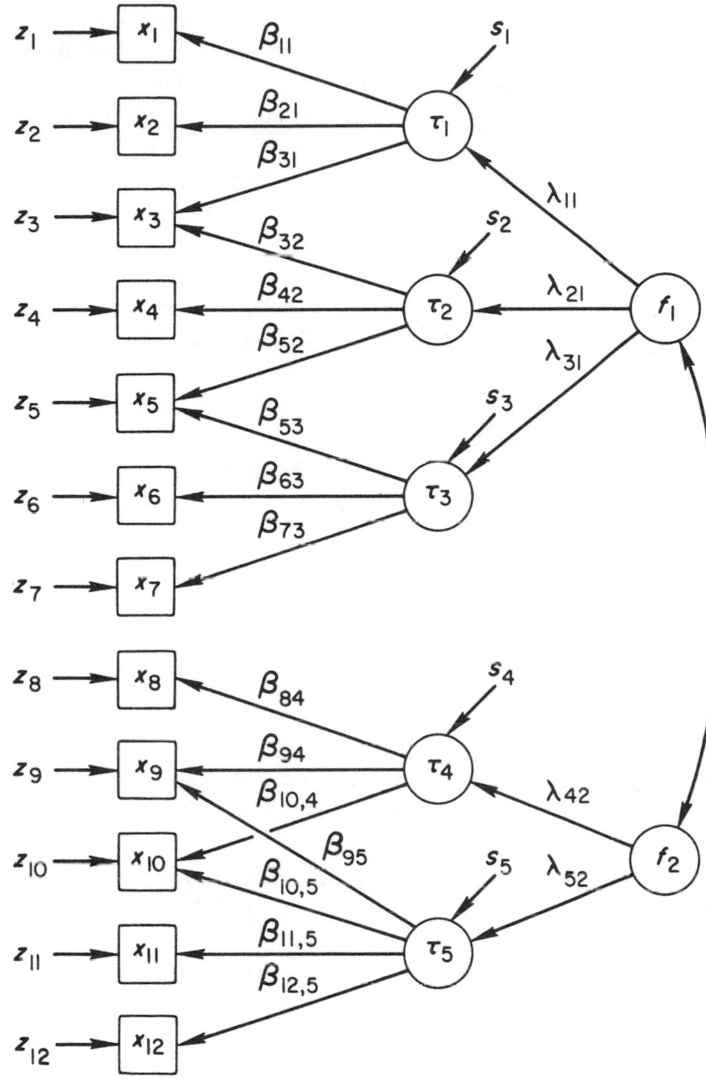

FIGURE 5.
A second-order factor analysis model.

to study differences in test performances when the tests have been constructed by assigning items or subtests according to objective features of content or format to subclasses of a factorial or hierarchical classification.

The idea of analyzing a design of test items or subtests is due to Burt (1947), who applied analysis of variance techniques to determine the effects on the test scores of the different classifications. Bock (1960) suggested that the scores of N subjects on a set of tests classified in 2^n factorial design may be viewed as data from an $N \times 2^n$ experimental design, where the subjects represent a random mode of classification and the tests represent n fixed modes of classification. Bock pointed out that conventional mixed-model analysis of variance gives useful information about the psychometric properties of the tests. In particular, the presence of nonzero variance components for the random mode of classification and for the interaction of the random and fixed modes of classification provides information about the number of dimensions in which the tests are able to discriminate among subjects. The relative size of these components measures the power of the tests to discriminate among subjects along the respective dimensions.

The multitrait-multimethod matrix of Campbell and Fiske (1959) is an example of a factorial design of tests. A more complex design is Guilford's (1956) structure of intellect. This design is based on a cross-classification of test items, not all of which may exist or be employed in any one study. Thus the classification scheme may be incomplete.

Consider an experimental design that has one random way of classification $\nu = 1, 2, \ldots, N$, one fixed way of classification $i = 1, 2, 3$ and another fixed way of classification $j = 1, 2, 3$ for $i = 1, 2$ and $j = 1, 2$ for $i = 3$. One model that may be considered is

$$x_{\nu ij} = \mu_{ij} + a_\nu + b_{\nu i} + c_{\nu j} + e_{\nu ij}, \tag{16}$$

where μ_{ij} is the mean of $x_{\nu ij}$ and where a_ν, $b_{\nu i}$, $c_{\nu j}$, and $e_{\nu ij}$ are uncorrelated random variables with zero means and variances σ_a^2, $\sigma_{b_i}^2$, $\sigma_{c_j}^2$, and $\sigma_{e_{ij}}^2$, respectively. Writing $\mathbf{x}_\nu' = (x_{\nu 11}, x_{\nu 12}, x_{\nu 13}, x_{\nu 21}, x_{\nu 22}, x_{\nu 23}, x_{\nu 31}, x_{\nu 32})$, $\mathbf{u}_\nu' = (a_\nu, b_{\nu 1}, b_{\nu 2}, b_{\nu 3}, c_{\nu 1}, c_{\nu 2}, c_{\nu 3})$ and

$$\mathbf{A} = \begin{bmatrix} 1 & 1 & 0 & 0 & 1 & 0 & 0 \\ 1 & 1 & 0 & 0 & 0 & 1 & 0 \\ 1 & 1 & 0 & 0 & 0 & 0 & 1 \\ 1 & 0 & 1 & 0 & 1 & 0 & 0 \\ 1 & 0 & 1 & 0 & 0 & 1 & 0 \\ 1 & 0 & 1 & 0 & 0 & 0 & 1 \\ 1 & 0 & 0 & 1 & 1 & 0 & 0 \\ 1 & 0 & 0 & 1 & 0 & 1 & 0 \end{bmatrix},$$

we may write Equation 16 as

$$\mathbf{x}_\nu = \boldsymbol{\mu} + \mathbf{A}\mathbf{u}_\nu + \mathbf{e}_\nu,$$

where $\boldsymbol{\mu}$ is the mean vector and $\mathbf{e}_\nu$ is a random error vector both of the same form as $\mathbf{x}_\nu$. The variance-covariance matrix of $\mathbf{x}_\nu$ is

$$\boldsymbol{\Sigma} = \mathbf{A}\boldsymbol{\Phi}\mathbf{A}' + \boldsymbol{\Psi}^2, \tag{17}$$

where $\boldsymbol{\Phi}$ is a diagonal matrix whose diagonal elements are σ_a^2, $\sigma_{b_1}^2$, $\sigma_{b_2}^2$, $\sigma_{b_3}^2$, $\sigma_{c_1}^2$, $\sigma_{c_2}^2$, and $\sigma_{c_3}^2$, and $\boldsymbol{\Psi}^2$ is a diagonal matrix whose elements are the $\sigma_{e_{ij}}^2$. In terms of the general model (Eq. 1), this model may be represented by choosing $\mathbf{B} = \mathbf{I}$, $\boldsymbol{\Lambda} = \mathbf{A}$, and $\boldsymbol{\Theta} = \mathbf{0}$. Matrices $\boldsymbol{\Phi}$ and $\boldsymbol{\Psi}^2$ are as defined in Equation 17. However, in this case the matrix $\boldsymbol{A}$ has rank 5, and only 5 linearly independent combinations of the components of $\mathbf{u}$, are estimable (see, e.g., Graybill, 1961, pp. 228–229). In conventional mixed-model analysis of variance one usually makes the assumptions that $\sigma_{b_i}^2 = \sigma_b^2$ for all $i = 1, 2, 3$, $\sigma_{c_j}^2 = \sigma_c^2$ for all $j = 1, 2, 3$, and $\sigma_{e_{ij}}^2 = \sigma_e^2$ for all i and j, but all these assumptions are not necessary.

In general, if $\mathbf{A}$ is of order $p \times r$ and of rank k, one may choose k independent linear functions, each one linearly dependent on the rows of $\mathbf{A}$ and estimate the mean vector and variance-covariance matrix of these functions. It is customary to choose linear combinations that are mutually uncorrelated, but this is not necessary in the analysis by our method. Let $\mathbf{L}$ be the matrix of coefficients of the chosen linear functions, and let $\mathbf{K}$ be any matrix such that $\mathbf{A} = \mathbf{KL}$. For example, $\mathbf{K}$ may be obtained from

$$\mathbf{K} = \mathbf{AL}'(\mathbf{LL}')^{-1}. \tag{18}$$

The model may then be reparameterized to full rank by defining $\mathbf{u}^* = \mathbf{Lu}$. We then have $\mathbf{x} = \mathbf{Au} + \mathbf{e} = \mathbf{KLu} + \mathbf{e} = \mathbf{Ku}^* + \mathbf{e}$. The variance-covariance matrix of $\mathbf{x}$ is represented as

$$\boldsymbol{\Sigma} = \mathbf{K}\boldsymbol{\Phi}^*\mathbf{K}' + \boldsymbol{\Psi}^2, \tag{19}$$

where $\boldsymbol{\Phi}^*$ is the variance-covariance matrix of $\mathbf{u}^*$ and $\boldsymbol{\Psi}^2$ is as before. The general method of analysis yields estimates of $\boldsymbol{\Psi}^2$ and $\boldsymbol{\Phi}^*$. The last matrix may be taken to be diagonal if desired.

A 2 × 2 Factorial Design with Replications

As an illustration, consider an example from Bock (1960). This example employs four subtests of an experimental form of the Language Modalities Survey (1959) to be used with aphasic subjects. The tests are classified according as the stimulus is a word or a picture and the response is oral or graphic:

		Response	
		Oral ($j = 1$)	Graphic ($j = 2$)
Stimulus	Picture ($i = 1$)	11	12
	Word ($i = 2$)	21	22

Let us assume that two tau-equivalent forms (replications) of each subtest are available. A suitable model is then

$$x_{\nu ijk} = \tau_{\nu ij} + e_{\nu ijk}, \tag{20}$$

where $x_{\nu ijk}$ is the score of person ν on the kth replication of the test classified as ij, and $\tau_{\nu ij}$ and $e_{\nu ijk}$ are the true and error scores, respectively. The true score $\tau_{\nu ij}$ is represented as

$$\tau_{\nu ij} = \mu_{ij} + a_\nu + b_{\nu i} + c_{\nu j} + d_{\nu ij}, \tag{21}$$

where μ_{ij} is the mean of $\tau_{\nu ij}$ in the population of individuals, a_ν is a component of score specific to individual ν and general to all tests, $b_{\nu i}$ and $c_{\nu j}$ are components of score specific to the performance of individual ν on tests of the 'Stimulus' and 'Response' classification, respectively, and $d_{\nu ij}$ is a component of score for individual ν specific to each test. The components a_ν, $b_{\nu i}$, $c_{\nu j}$, and $d_{\nu ij}$ are assumed to be uncorrelated random variables with zero mean and variances σ_a^2, σ_b^2, σ_c^2, and σ_d^2, respectively.

Equations 20 and 21 may be written in matrix form as

$$\begin{pmatrix} x_{\nu 111} \\ x_{\nu 112} \\ x_{\nu 121} \\ x_{\nu 122} \\ x_{\nu 211} \\ x_{\nu 212} \\ x_{\nu 221} \\ x_{\nu 222} \end{pmatrix} = \begin{bmatrix} 1 & 0 & 0 & 0 \\ 1 & 0 & 0 & 0 \\ 0 & 1 & 0 & 0 \\ 0 & 1 & 0 & 0 \\ 0 & 0 & 1 & 0 \\ 0 & 0 & 1 & 0 \\ 0 & 0 & 0 & 1 \\ 0 & 0 & 0 & 1 \end{bmatrix} \begin{pmatrix} \tau_{\nu 11} \\ \tau_{\nu 12} \\ \tau_{\nu 21} \\ \tau_{\nu 22} \end{pmatrix} + \begin{pmatrix} e_{\nu 111} \\ e_{\nu 112} \\ e_{\nu 121} \\ e_{\nu 122} \\ e_{\nu 211} \\ e_{\nu 212} \\ e_{\nu 221} \\ e_{\nu 222} \end{pmatrix},$$

$$\begin{pmatrix} \tau_{\nu 11} \\ \tau_{\nu 12} \\ \tau_{\nu 21} \\ \tau_{\nu 22} \end{pmatrix} = \begin{pmatrix} \mu_{11} \\ \mu_{12} \\ \mu_{21} \\ \mu_{22} \end{pmatrix} + \begin{bmatrix} 1 & 1 & 0 & 1 & 0 & 1 & 0 & 0 & 0 \\ 1 & 1 & 0 & 0 & 1 & 0 & 1 & 0 & 0 \\ 1 & 0 & 1 & 1 & 0 & 0 & 0 & 1 & 0 \\ 1 & 0 & 1 & 0 & 1 & 0 & 0 & 0 & 1 \end{bmatrix} \begin{pmatrix} a_\nu \\ b_{\nu 1} \\ b_{\nu 2} \\ c_{\nu 1} \\ c_{\nu 2} \\ d_{\nu 11} \\ d_{\nu 12} \\ d_{\nu 21} \\ d_{\nu 22} \end{pmatrix},$$

or

$$\mathbf{x}_\nu = \mathbf{B}\boldsymbol{\tau}_\nu + \mathbf{e}_\nu, \tag{22}$$

$$\boldsymbol{\tau}_\nu = \boldsymbol{\mu} + \mathbf{A}\mathbf{u}_\nu. \tag{23}$$

The variance-covariance matrix of $\mathbf{x}_\nu$ is

$$\boldsymbol{\Sigma} = \mathbf{B}\mathbf{A}\boldsymbol{\Phi}\mathbf{A}'\mathbf{B}' + \boldsymbol{\Theta}^2, \tag{24}$$

where

$$\boldsymbol{\Phi} = \operatorname{diag}(\sigma_a^2, \sigma_b^2, \sigma_b^2, \sigma_c^2, \sigma_c^2, \sigma_d^2, \sigma_d^2, \sigma_d^2, \sigma_d^2)$$

and $\boldsymbol{\Theta}^2$ is the diagonal matrix of error variances. The variance components in $\boldsymbol{\Phi}$ and the error variances in $\boldsymbol{\Theta}^2$ may be estimated by specifying the model as in Equation 24, by imposing the constraints on $\boldsymbol{\Phi}$ just described and by using the sample variance-covariance matrix $\mathbf{S}$ of all the eight test forms. Bock (1960), assuming the replicate forms to be parallel and the error variances for different tests to be homogenous, estimated the error variances separately and the variance components from the dispersion matrix of the means of the replicate scores. Our approach uses less restrictive assumptions and gives more efficient estimates of the variance components.

In Equation 24, the matrix $\mathbf{A}$ is of order 4×9 and of rank 4. However, reparameterization is not necessary if one is willing to assume that $\sigma^2_{b_{\nu i}} = \sigma^2_b$, $\sigma^2_{c_{\nu i}} = \sigma^2_c$, and $\sigma^2_{d_{\nu ij}} = \sigma^2_d$ for all i and j, since these constraints can be introduced directly in $\boldsymbol{\Phi}$. An alternative approach is to choose four independent linear functions of the components in $\boldsymbol{\tau}$ and to estimate the variance-covariance matrix of these. For example, one may choose the following functions:

$$u^*_{\nu 1} = 2a_\nu + b_{\nu 1} + b_{\nu 2} + c_{\nu 1} + c_{\nu 2} + \tfrac{1}{2}(d_{\nu 11} + d_{\nu 12} + d_{\nu 21} + d_{\nu 22}),$$

$$u^*_{\nu 2} = b_{\nu 1} - b_{\nu 2} + \tfrac{1}{2}(d_{\nu 11} + d_{\nu 12} - d_{\nu 21} - d_{\nu 22}),$$

$$u^*_{\nu 3} = c_{\nu 1} - c_{\nu 2} + \tfrac{1}{2}(d_{\nu 11} - d_{\nu 12} + d_{\nu 21} - d_{\nu 22}),$$

$$u^*_{\nu 4} = \tfrac{1}{2}(d_{\nu 11} - d_{\nu 12} - d_{\nu 21} + d_{\nu 22});$$

i.e., we choose $\mathbf{L}$ as

$$\mathbf{L} = \begin{bmatrix} 2 & 1 & 1 & 1 & 1 & \tfrac{1}{2} & \tfrac{1}{2} & \tfrac{1}{2} & \tfrac{1}{2} \\ 0 & 1 & -1 & 0 & 0 & \tfrac{1}{2} & \tfrac{1}{2} & -\tfrac{1}{2} & -\tfrac{1}{2} \\ 0 & 0 & 0 & 1 & -1 & \tfrac{1}{2} & -\tfrac{1}{2} & \tfrac{1}{2} & -\tfrac{1}{2} \\ 0 & 0 & 0 & 0 & 0 & \tfrac{1}{2} & -\tfrac{1}{2} & -\tfrac{1}{2} & \tfrac{1}{2} \end{bmatrix}.$$

Then from Equation 18, $\mathbf{K}$ becomes

$$\mathbf{K} = \tfrac{1}{2}\begin{bmatrix} 1 & 1 & 1 & 1 \\ 1 & 1 & -1 & -1 \\ 1 & -1 & 1 & -1 \\ 1 & -1 & -1 & 1 \end{bmatrix},$$

Equation 23 becomes

$$\boldsymbol{\tau}_\nu = \boldsymbol{\mu} + \mathbf{K}\mathbf{u}^*_\nu$$

and the variance-covariance matrix of $\mathbf{x}$ is

$$\boldsymbol{\Sigma} = \mathbf{B}\mathbf{K}\boldsymbol{\Phi}^*\mathbf{K}'\mathbf{B}' + \boldsymbol{\Theta}^2,$$

where $\boldsymbol{\Phi}^*$ is the variance-covariance matrix of $u^*_{\nu 1}$, $u^*_{\nu 2}$, $u^*_{\nu 3}$, and $u^*_{\nu 4}$.

The estimated variances in $\hat{\boldsymbol{\Phi}}^*$ give useful information for the interpretation of individual differences. If there are no effects due to test classification and specificity, i.e., if $\sigma_b^2 = \sigma_c^2 = \sigma_d^2 = 0$, $u_{\nu 1}$ estimates separately the level of general ability of individual ν with respect to the tests used. In the absence of specific effects, $u_{\nu 2}$ and $u_{\nu 3}$ estimate separately contrasts in abilities defined by the 'Stimulus' and 'Response' dichotomies of the test classification. Finally, $u_{\nu 4}$ estimates separately a function of the specific effects, namely, the interaction effects

$$\tfrac{1}{2}(d_{\nu 11} - d_{\nu 12}) - \tfrac{1}{2}(d_{\nu 21} - d_{\nu 22}).$$

A positive $u_{\nu 4}$ indicates that specific effects cause test scores in the first categories of the 'Stimulus' classification to differ more than those in the second. Covariances between the various *u*s are also useful. For example, the covariance between $u_{\nu 1}$ and $u_{\nu 2}$ gives an indication of the extent to which $\sigma_{b_1}^2 = \sigma_{b_2}^2$.

The classical formulation of the mixed model with the assumptions of homogenous error variances and uncorrelated latent random variables is not realistic in most applications in the behavioral sciences. Bock, Dicken, and van Pelt (1969) demonstrated heuristically the inadequacy of the specification of uncorrelated latent variables, i.e., the inadequacy of $\boldsymbol{\Phi}$ being specified as a diagonal matrix, which is the case considered by Bock (1960) and by Bock and Bargmann (1966). As already pointed out, in our method of covariance structure analysis, the assumption that $\boldsymbol{\Phi}$ be diagonal is not necessary. If the model provides information enough, so that all the variances and covariances of the latent variables are identified, these may also be estimated, and the assumption of zero covariances may be examined empirically.

A General Class of Components of Covariance Models

Wiley, Schmidt, and Bramble (1973) suggested the study of a general class of components of covariance models. This class of models is a special case of Equation 1, namely, when $\mathbf{B}$ is diagonal, $\boldsymbol{\Lambda}$ is known a priori, $\boldsymbol{\Phi}$ is symmetric and positive definite, and $\boldsymbol{\Psi}$ or $\boldsymbol{\Theta}$ are either zero or diagonal. The covariance matrix $\boldsymbol{\Sigma}$ will then be of the form

$$\boldsymbol{\Sigma} = \boldsymbol{\Delta}\mathbf{A}\boldsymbol{\Phi}\mathbf{A}'\boldsymbol{\Delta} + \boldsymbol{\Theta}^2 \quad \text{or} \quad \boldsymbol{\Sigma} = \boldsymbol{\Delta}(\mathbf{A}\boldsymbol{\Phi}\mathbf{A}' + \boldsymbol{\Psi}^2)\boldsymbol{\Delta}. \qquad (25a, 25b)$$

The matrix $\mathbf{A}(p \times k)$ is assumed to be known and gives the coefficient of the linear functions connecting the manifest and latent variables, $\boldsymbol{\Delta}$ is a $p \times p$ diagonal matrix of unknown scale factors, $\boldsymbol{\Phi}$ is the $k \times k$ symmetric and positive definite covariance matrix of the latent variables and $\boldsymbol{\Psi}^2$ and $\boldsymbol{\Theta}^2$ are $p \times p$ diagonal matrices of error variances.

Within this class of models eight different special cases are of interest. These are generated by the combination of the following set of conditions:

$$\text{on } \boldsymbol{\Delta}: \quad \begin{Bmatrix} \boldsymbol{\Delta} = \mathbf{I} \\ \boldsymbol{\Delta} \neq \mathbf{I} \end{Bmatrix};$$

$$\text{on } \boldsymbol{\Phi}: \quad \left\{\begin{matrix} \boldsymbol{\Phi} \text{ is diagonal} \\ \boldsymbol{\Phi} \text{ is not diagonal} \end{matrix}\right\};$$

$$\text{on } \boldsymbol{\Psi}^2 \text{ or } \boldsymbol{\Theta}^2: \quad \left\{\begin{matrix} \boldsymbol{\Psi}^2 \text{ or } \boldsymbol{\Theta}^2 = \sigma^2\mathbf{I} \\ \boldsymbol{\Psi}^2 \text{ or } \boldsymbol{\Theta}^2 \text{ general diagonal} \end{matrix}\right\}.$$

The classical formulation of the mixed model and its generalizations assume that $\boldsymbol{\Delta} = \mathbf{I}$. This is appropriate if the observed variables are in the same metric as, for example, when the observed variables represent physical measurements, time to criterion measures, reaction times, or items similarly scaled such as semantic differential responses. However, if the observed variables are measured in different metrics, then the classical model will not fit. In such cases the inclusion of $\boldsymbol{\Delta}$ in the model as a general diagonal matrix of scaling factors will provide a useful alternative specification. It should be pointed out that the elements of $\boldsymbol{\Delta}$ do not have to be related to the variances of the variables.

The classical components of the variance model assume that $\boldsymbol{\Phi}$ is diagonal. However, as has already been pointed out, there are no substantive reasons for assuming this.

The two conditions on $\boldsymbol{\Psi}^2$ or $\boldsymbol{\Theta}^2$ correspond to homogeneous and heterogeneous error variances. If the variables are in the same metric and if the measurement situation is sufficiently similar from variable to variable, then it seems reasonable to hypothesize that the variances of the errors of measurement ought to be homogeneous, i.e., in Equation 25a we take $\boldsymbol{\Delta} = \mathbf{I}$ and $\boldsymbol{\Theta}^2 = \sigma^2\mathbf{I}$.

If, on the other hand, the scale of measurement is the same, but the measurement situation from variable to variable is different enough so as to generate different kinds of error structures, then the variances of the errors of measurement might differ systematically from variable to variable. For this situation it seems best to take $\boldsymbol{\Delta} = \mathbf{I}$ but to leave $\boldsymbol{\Theta}^2$ free in Equation 25a. This situation could arise where the variable time to task completion is measured under different treatment conditions on the same individuals. Suppose that under the first set of conditions, time to completion takes on the average of 1 minute while in the second treatment group the average time required is 10 minutes. It is reasonable to hypothesize that some recording procedures will vary in their degree of accuracy, since in the first treatment condition the observations will most likely be more accurate than under the second set of conditions. It is also possible that individuals are inherently more variable under some conditions than others. This will result in different error variances despite the constant metric of time.

If the manifest variables are in different metrics, then clearly the error variances in the observed metric will most likely be heterogeneous. One useful hypothesis to test in this context is that the standard deviations of the errors of measurement are proportional to the rescaling factors. This corresponds to taking $\boldsymbol{\Psi}^2 = \sigma^2\mathbf{I}$ in Equation 25b. When both $\boldsymbol{\Delta}$ and $\boldsymbol{\Psi}^2$ are free, Equations 25a and 25b are equivalent.

Analysis of 2^3 Factorial Design

This example is taken from Wiley, Schmidt, and Bramble (1973). The original data are from a study by Miller and Lutz (1966) and consist of the scores of 51 education students on a test designed to assess teachers' judgments about the effects of situation and instruction factors on the facilitation of pupil learning. The items used were designed according to three factors that were hypothesized to influence classroom learning situations and teaching practices. The three factors and their levels are given as follows (see Miller & Lutz, 1966).

> Grade Level (G). The levels of this factor, the first grade (G_1) and the sixth grade (G_2), were chosen to represent extremes of the elementary grades. In this way it was possible to maximize the opportunity for observing any differences in teachers' judgments that might occur as a result of variations due to grade level.
>
> Teacher Approach (T). The teacher-centered approach (T_1) and the pupil-centered approach (T_2) were distinguished as levels of this factor on the basis of the locus of described activity control, and the direction of described pupils' attention. In the case of the teacher-centered approach, the locus and direction were oriented to the teacher; in the case of the pupil-centered approach, the locus and direction were oriented to the pupil.
>
> Teaching Method (M). Level one of this factor was drill (M_1) which was used strictly to refer to rote learning activities; discovery (M_2) was used to refer to an approach in which the teacher attempts to develop pupil understanding through procedures aimed at stimulating insight without recourse to rote memorization or rigid learning routines.

The eight subtest scores are each based on eight items. The eight subtests conform to a 2^3 factorial arrangement, which is given by

$$\boldsymbol{A} = \begin{bmatrix} 1 & 1 & 1 & 1 \\ 1 & 1 & 1 & -1 \\ 1 & 1 & -1 & 1 \\ 1 & 1 & -1 & -1 \\ 1 & -1 & 1 & 1 \\ 1 & -1 & 1 & -1 \\ 1 & -1 & -1 & 1 \\ 1 & -1 & -1 & -1 \end{bmatrix}.$$

The sample covariance matrix $\mathbf{S}$ for these data is given in Table 10.

For this data the use of $\boldsymbol{\Delta} = \boldsymbol{I}$ was considered appropriate, because the same scale (1–7) was employed for each item; the numbers of items in the subtests were also equal.

The data were analyzed under each of the four remaining cases in the general class of models considered in the previous section. Note also the χ^2 values and their degrees of freedom as shown in Table 10. In the margin of this table, χ^2 values are given for testing the hypotheses of uncorrelated components and of homogeneous error variances. It appears that these hypotheses were both rejected by the tests. The only model that seems reasonable is the one

TABLE 10
Miller-Lutz data

Sample Covariance Matrix

$$\mathbf{S} = \begin{bmatrix} 18.74 & & & & & & & \\ 9.28 & 18.80 & & & & & & \\ 15.51 & 7.32 & 21.93 & & \text{(symmetric)} & & & \\ 3.98 & 15.27 & 4.10 & 26.62 & & & & \\ 15.94 & 4.58 & 13.79 & -2.63 & 19.82 & & & \\ 7.15 & 13.63 & 3.86 & 15.33 & 3.65 & 16.81 & & \\ 11.69 & 6.05 & 10.18 & 1.13 & 13.55 & 5.72 & 16.58 & \\ 2.49 & 12.35 & 0.03 & 16.93 & -0.86 & 14.33 & 2.99 & 18.26 \end{bmatrix}$$

χ^2 Values for Testing the Fit of Four Models

	$\boldsymbol{\Theta}^2 = \sigma^2\mathbf{I}$	$\boldsymbol{\Theta}^2 \neq \sigma^2\mathbf{I}$	
$\boldsymbol{\Phi}$ diagonal	$\chi^2_{31} = 68.25$	$\chi^2_{24} = 46.16$	$\chi^2_7 = 22.09$
$\boldsymbol{\Phi}$ not diagonal	$\chi^2_{25} = 51.00$	$\chi^2_{18} = 25.98$	$\chi^2_7 = 25.02$
	$\chi^2_6 = 17.25$	$\chi^2_6 = 20.18$	

Maximum-Likelihood Estimates of $\boldsymbol{\Phi}$ and $\boldsymbol{\Theta}^2$
(Standard Errors in Parentheses)

$$\hat{\boldsymbol{\Phi}} = \begin{bmatrix} 9.16\,(1.95) & & & \\ 0.75\,(0.48) & 0.70\,(0.34) & & \\ 0.63\,(0.43) & -0.05\,(0.33) & 0.43\,(0.91) & \\ -0.62\,(1.10) & -0.51\,(0.81) & 1.13\,(0.51) & 5.21\,(1.58) \end{bmatrix}$$

$$\hat{\boldsymbol{\Theta}}^2 = \text{diag}\,[1.52(0.83),\ 4.95(1.41),\ 8.25(1.88),\ 5.58(1.60),\ 1.95(0.96),\ 5.76(1.21),\ 2.52(0.92)]$$

Note: Miller-Lutz data taken from Wiley, Schmidt, and Bramble (1972).

that assumes both correlated components and heterogenous error variances. The maximum-likelihood estimates of the variances and covariances of the components and the error variances, together with their standard errors, are also given in Table 10.

The relative magnitudes of the estimated variance components for the latent variables indicate the major sources of variation in the performance of the subjects. The estimate of the first variance component ($\hat{\varphi}_{11} = 9.16$) is the largest as would be expected, since this component reflects the variation due to individual differences between the subjects. The estimated values of the other components indicate that another major source of variation in the responses of the subjects is due to the different type of teaching method specified in the content of the item (i.e., $\hat{\varphi}_{44} = 5.21$). Apparently the contrast between the drill and the discovery methods of instruction caused the responses of the education students to vary considerably. Variation contributed by grade level was intermediate in magnitude ($\hat{\varphi}_{22} = 0.70$). The estimate of the variance component for the teacher-approach factor and its large standard error (i.e., $\hat{\varphi}_{33} = 0.43$, S.E. ($\hat{\varphi}_{33}$) = 0.91) indicate that this was not an important source of variation in the performance of the subjects. One of the estimated covariances was relatively large—that between the teaching-method factor and the teacher-approach factor—indicating these latent variables to be highly correlated. This would indicate that the responses of the education students to the different types of teaching methods specified in the items were related to their responses to the teacher-approach factor found in the items.

6. SIMPLEX AND CIRCUMPLEX MODELS

Various Types of Simplex Models

Since the fundamental paper of Guttman (1954) on the simplex structure for correlations between ordered tests, many investigators have found data displaying the typical simplex structure. Guttman gave several examples of this structure. His Table 5 is reproduced here as Table 11. In this example all the tests involve verbal ability and are ordered according to increasing complexity.

The typical property in a simplex correlation structure, such as that in Table 11, is that the correlations decrease as one moves away from the main diagonal. Such data will not usually fit a factor analysis model with one common factor for the following reasons. Let the tests be ordered so that the factor loadings decrease. Then if the factor model holds, the correlations in the first row decrease as one moves away from the diagonal, but the correlations in the last row *increase* as one moves away from the diagonal. Also the

TABLE 11
Intercorrelations of six verbal-ability tests for 1046 Bucknell College sophomores

Test	Spelling	Punctuation	Grammar	Vocabulary	Literature	Foreign Literature
	A	C	B	D	E	H
A	—	0.621	0.564	0.476	0.394	0.389
C	0.621	—	0.742	0.503	0.461	0.411
B	0.564	0.742	—	0.577	0.472	0.429
D	0.476	0.503	0.577	—	0.688	0.548
E	0.394	0.461	0.472	0.688	—	0.639
H	0.389	0.411	0.429	0.548	0.639	—
Total	2.444	2.738	2.784	2.792	2.654	2.416

correlations just below the diagonal *decrease markedly* as one moves down, which does not hold in Table 11.

Jöreskog (1970b) considered several statistical models for such simplex structures. Following Anderson (1960), he formulated these models in terms of the well-known Wiener and Markov stochastic processes. A distinction was made between a perfect simplex and a quasi-simplex. A perfect simplex is reasonable only if the measurement errors in the test scores are negligible. A quasi-simplex on the other hand allows for sizable errors of measurement.

Table 12 shows six types of simplex models and some of their character-

TABLE 12
Six types of simplex models

Model	Covariance Structure	Scale Dependence	No. of Parameters
Markov Simplex	$\boldsymbol{\Sigma} = \mathbf{D}_\alpha \mathbf{T} \mathbf{D}_{s*} \mathbf{T}' \mathbf{D}_\alpha$	Scale Free	$2p - 1$
Wiener Simplex	$\boldsymbol{\Sigma} = \mathbf{T} \mathbf{D}_{s*} \mathbf{T}'$	Scale Dependent	p
Quasi-Wiener Simplex	$\boldsymbol{\Sigma} = \mathbf{T} \mathbf{D}_{s*} \mathbf{T}' + \boldsymbol{\Theta}^2$	Scale Dependent	$2p - 1$
Quasi-Wiener Simplex with Equal Error Variances	$\boldsymbol{\Sigma} = \mathbf{T} \mathbf{D}_{s*} \mathbf{T}' + \theta^2 \mathbf{I}$	Scale Dependent	$p + 1$
Quasi-Markov Simplex	$\boldsymbol{\Sigma} = \mathbf{D}_\alpha \mathbf{T} \mathbf{D}_{s*} \mathbf{T}' \mathbf{D}_\alpha + \boldsymbol{\Theta}^2$	Scale Free	$3p - 3$
Restricted Quasi-Markov Simplex	$\boldsymbol{\Sigma} = \mathbf{D}_\alpha (\mathbf{T} \mathbf{D}_{s*} \mathbf{T}' + \psi^2 \mathbf{I}) \mathbf{D}_\alpha$	Scale Free	$2p$

istics. The Wiener simplexes are scale-dependent models and are appropriate only when the units of measurement are the same for all tests. The Markov simplexes, on the other hand, are scale-free models and can therefore be used with data like those in Table 11, where the units of measurements are arbitrary.

The Markov Simplex

Consider a real stochastic process $X(t)$ and arbitrary scale points $t_1 < t_2 < \cdots < t_p$. The Markov simplex is defined by

$$\begin{aligned} E[X(t_i)] &= \mu_i, \\ \mathrm{Var}[X(t_i)] &= \sigma_i^2, \\ \mathrm{Corr}[X(t_i), X(t_j)] &= \rho^{|t_i - t_j|}, \end{aligned}$$

where $0 < \rho < 1$.

Consider the following transformation from $t' = (t_1, t_2, \ldots, t_p)$ to $t^{*\prime} = (t_1^*, t_2^*, \ldots, t_p^*)$, where

$$\left.\begin{aligned} t_1^* &= t_1 \\ t_2^* &= t_2 - t_1 \\ t_3^* &= t_3 - t_2 \\ &\vdots \\ t_p^* &= t_p - t_{p-1} \end{aligned}\right\}.$$

This transformation is one-to-one and the ts may be obtained from the t^*s as follows

$$\left.\begin{aligned} t_1 &= t_1^* \\ t_2 &= t_1^* + t_2^* \\ t_3 &= t_1^* + t_2^* + t_3^* \\ &\vdots \\ t_p &= t_1^* + t_2^* + \cdots + t_p^* \end{aligned}\right\}.$$

In matrix form these transformations are

$$\begin{aligned} \mathbf{t}^* &= \mathbf{T}^{-1}\mathbf{t}, \\ \mathbf{t} &= \mathbf{T}\mathbf{t}^*, \end{aligned}$$

where

$$\mathbf{T} = \begin{bmatrix} 1 & 0 & 0 & \cdots & 0 \\ 1 & 1 & 0 & \cdots & 0 \\ 1 & 1 & 1 & \cdots & 0 \\ \vdots & \vdots & \vdots & & \vdots \\ 1 & 1 & 1 & & 1 \end{bmatrix}; \quad \mathbf{T}^{-1} = \begin{bmatrix} 1 & 0 & 0 & \cdots & 0 & 0 \\ -1 & 1 & 0 & \cdots & 0 & 0 \\ 0 & -1 & 1 & \cdots & 0 & 0 \\ \vdots & \vdots & \vdots & & \vdots & \vdots \\ 0 & 0 & 0 & \cdots & -1 & 1 \end{bmatrix}. \qquad (26a, 26b)$$

In terms of $t_2^*, t_3^*, \ldots, t_p^*$, the correlation matrix of $x' = [X(t_1), X(t_2), \ldots, X(t_p)]$ is

$$\mathbf{P} = \begin{bmatrix} 1 & & & \text{symmetric} \\ \rho^{t_2^*} & 1 & & \\ \rho^{t_2^*+t_3^*} & \rho^{t_3^*} & & \\ \vdots & \vdots & & \\ \rho^{t_2^*+\cdots+t_p^*} & \rho^{t_3^*+\cdots+t_p^*} & \cdots & 1 \end{bmatrix}. \tag{27}$$

It is seen that there are only $p - 1$ independent correlations, namely, those just below (or above) the main diagonal and the other correlations are products of these. For example,

$$\rho_{ji} = \prod_{k=i}^{k=j} \rho^{t_k^*} = \prod_{k=i}^{k=j} \rho_{k+1,k}, \quad i < j.$$

It is also seen that the correlations fall off as one moves away from the main diagonal. Thus, the model of a Markov simplex may be consistent with the simplex pattern referred to in the previous section. The variance-covariance matrix of $\mathbf{x}' = [X(t_1), X(t_2), \ldots, X(t_p)]$ is

$$\boldsymbol{\Sigma} = \mathbf{D}_\sigma \mathbf{P} \mathbf{D}_\sigma, \tag{28}$$

where $\mathbf{P}$ is defined by Equation 27 and $\mathbf{D}_\sigma = \text{diag}(\sigma_1, \sigma_2, \ldots, \sigma_p)$.

The values $t_2^*, t_3^*, \ldots, t_p^*$ are determined only up to a multiplicative constant, and since only differences between scale points enter into $t_2^*, t_3^*, \ldots, t_p^*$, it is evident that the origin and the unit of the scale are arbitrary. One may, for example, choose the origin at $t_1 = 0$ and define the unit so that $t_2 = 1$.

The Wiener Simplex

The Wiener process $X(s)$, with scale points $s_1 < s_2 \cdots < s_{p-1} < s_p$, is defined by

$$E[X(s_i)] = \mu_i,$$
$$\text{Var}[X(s_i)] = s_i,$$
$$\text{Cov}[X(s_i), X(s_j)] = s_i, \quad i < j.$$

The correlation between $X(s_i)$ and $X(s_j)$ is

$$\rho[X(s_i), X(s_j)] = \frac{s_i}{\sqrt{s_i s_j}} = \sqrt{\frac{s_i}{s_j}}, \quad i < j,$$

so that the correlation matrix of $\mathbf{x}' = [X(s_1), X(s_2), \ldots, X(s_p)]$ is of the form

$$\mathbf{P} = \begin{bmatrix} 1 & & & & \\ \sqrt{\frac{s_1}{s_2}} & 1 & & & \\ \sqrt{\frac{s_1}{s_3}} & \sqrt{\frac{s_2}{s_3}} & 1 & & \\ \vdots & \vdots & & & \\ \sqrt{\frac{s_1}{s_p}} & \sqrt{\frac{s_2}{s_p}} & \cdots & 1 \end{bmatrix}. \tag{29}$$

It is seen that **P** in Equation 29 has the same properties as the **P** in Equation 27. In fact, the correlation structures of the Markov simplex and the Wiener simplex are equivalent, for if $i < j$, then

$$\sqrt{\frac{s_i}{s_j}} = e^{-1/2(\log s_j - \log s_i)} = e^{-\alpha(t_j - t_i)} = \rho^{t_j - t_i},$$

with $\rho = e^{-\alpha}$, $\alpha t_i = \frac{1}{2} \log s_i$, and $\alpha > 0$ arbitrary. Thus, the scale for the Markov simplex is just a logarithmic transformation of the scale for the Wiener simplex. In the Markov simplex, correlations correspond to distances between scale points, whereas in the Wiener simplex, correlations correspond to square roots of ratios between scale points.

Although the correlation matrix **P** for the Markov simplex and the Wiener simplex are identical, it does not hold for the corresponding dispersion matrices. Writing $\mathbf{s}^* = \mathbf{T}^{-1}\mathbf{s}$, the dispersion matrix $\boldsymbol{\Sigma}$ for the Wiener simplex has the form

$$\boldsymbol{\Sigma} = \begin{bmatrix} s_1 & & & \\ s_1 & s_2 & & \\ \vdots & \vdots & \cdots & \\ s_1 & s_2 & \cdots & s_p \end{bmatrix} = \begin{bmatrix} s_1^* & & & \\ s_1^* & s_1^* + s_2^* & & \\ \vdots & \vdots & \cdots & \\ s_1^* & s_1^* + s_2^* & \cdots & s_1^* + s_2^* + \cdots + s_p^* \end{bmatrix} \tag{30}$$
$$= \mathbf{T}\mathbf{D}_{s^*}\mathbf{T}',$$

where **T** is defined by Equation 26 and $\mathbf{D}_{s^*} = \operatorname{diag}(s_1^*, s_2^*, \ldots, s_p^*)$. It is seen that $\boldsymbol{\Sigma}$ has p independent parameters, whereas the $\boldsymbol{\Sigma}$ for the Markov simplex has $2p - 1$ independent parameters. Thus the Wiener simplex is a more restricted model than the Markov simplex. Since **P** in Equations 27 and 29 are the same, namely, $\mathbf{P} = \mathbf{D}_s^{-1/2}\mathbf{T}\mathbf{D}_{s^*}\mathbf{T}'\mathbf{D}_s^{-1/2}$, this may be substituted into Equation 28 to give the variance-covariance matrix for the Markov simplex as

$$\boldsymbol{\Sigma} = \mathbf{D}_\sigma \mathbf{D}_s^{-1/2}\mathbf{T}\mathbf{D}_{s^*}\mathbf{T}'\mathbf{D}_s^{-1/2}\mathbf{D}_\sigma,$$

which, with $\mathbf{D}_\alpha = \mathbf{D}_\sigma \mathbf{D}_s^{-1/2} = \operatorname{diag}(\alpha_1, \alpha_2, \ldots, \alpha_p)$, becomes

$$\boldsymbol{\Sigma} = \mathbf{D}_\alpha \mathbf{T}\mathbf{D}_{s^*}\mathbf{T}'\mathbf{D}_\alpha. \tag{31}$$

It is clear that one α may be fixed at unity, so that there are $2p - 1$ independent parameters. Equations 30 and 31 are in the form of Equation 1. Although simpler methods are available (see, e.g., Jöreskog, 1970b), the Wiener and Markov simplexes may be estimated by means of the ACOVS program, which also provides for the testing of the goodness of fit of the models.

The Wiener simplex is most suitable when all variates $x_1, x_2, \ldots, x_p$ are measured in the same units. If they are not measured in the same units, the results will depend on the various units of measurement used. In contrast to the Markov simplex, the Wiener simplex is scale dependent. Analyses of **S** and **DSD**, where **D** is an arbitrary diagonal matrix of scale factors, do not yield results that are properly related. Whereas the correlation matrix **R** may be used to estimate a Markov simplex, the dispersion matrix **S** must be used to estimate a Wiener simplex.

The Quasi-Wiener Simplex

In a perfect simplex it is assumed that the observed variates are infallible, containing no errors of measurement, and that the simplex structure holds for the observed variates themselves. In a quasi-simplex it is assumed that the observed variates contain errors of measurement and that the simplex structure holds for the true variates. Since most measurements in the behavioral sciences contain sizable errors, quasi-simplex models are more realistic for such data.

The Markov model and the Wiener model of the preceding subsections are both perfect simplexes. While the estimation of these models is a simple matter, this is not so for most quasi-simplexes. Here iterative techniques have to be used to obtain the estimates.

When the observed variates $x_i = X(s_i)$ contain errors of measurement, we write

$$x_i = \tau_i + e_i, \quad i = 1, 2, \ldots, p,$$

where τ_i is the true measurement and e_i is the error. About the errors we assume that $E(e_i) = 0$ and $E(e_i e_j) = 0$ for $i \neq j$ and that $\mathrm{Cov}(e_i, \tau_i) = 0$. Let $\boldsymbol{T}$ be the dispersion matrix of $\tau' = (\tau_1, \tau_2, \ldots, \tau_p)$ and let $\boldsymbol{\Theta}^2$ be the diagonal dispersion matrix of $\mathbf{e}' = (e_1, e_2, \ldots, e_p)$. Then the dispersion matrix $\boldsymbol{\Sigma}$ of the observed variates is

$$\boldsymbol{\Sigma} = \boldsymbol{T} + \boldsymbol{\Theta}^2. \tag{32}$$

We shall first consider the case where $\boldsymbol{T}$ has the Wiener simplex structure $\mathbf{T D_{s^*} T'}$ and then the case where $\boldsymbol{T}$ has the Markov simplex structure $\mathbf{D_\alpha P D_\sigma} = \mathbf{D_\alpha T D_{s^*} T' D_\alpha}$.

The dispersion matrix of the quasi-Wiener simplex is

$$\boldsymbol{\Sigma} = \begin{bmatrix} s_1 + \theta_1^2 & & & \\ s_1 & s_2 + \theta_2^2 & & \\ \vdots & \vdots & \cdots & \\ s_1 & s_2 & \cdots & s_p + \theta_p^2 \end{bmatrix} = \mathbf{T}\mathbf{D}_{s^*}\mathbf{T}' + \boldsymbol{\Theta}^2. \qquad (33)$$

It is seen that s_p and θ_p are only involved in σ_{pp} so that they are not separately identified; only their sum is identified. The other $2p - 1$ parameters in $\boldsymbol{\Sigma}$ are independently identified.

Equation 33 is in the form of Equation 1 and may therefore be estimated using the ACOVS program. There are many different ways in which this may be done. The simplest is probably to choose $\mathbf{B} = \mathbf{T}$, $\boldsymbol{\Lambda} = \mathbf{0}$, $\boldsymbol{\Phi} = \mathbf{0}$, $\boldsymbol{\Psi}^2 = \mathbf{D}_{s^*}$, and the last element of $\boldsymbol{\Theta}$ equal to zero. By specifying the model in this way, the likelihood function may be maximized with respect to $\boldsymbol{\Psi}$ and $\boldsymbol{\Theta}$, and as a consequence, the maximum-likelihood estimates $\hat{\mathbf{D}}_{s^*} = \hat{\boldsymbol{\Psi}}^2$ and $\hat{\boldsymbol{\Theta}}^2$ will be nonnegative. Standard errors may be obtained for each estimated parameter.

The Quasi-Wiener Simplex with Equal Error Variances

Since the quasi-Wiener simplex is applied to data where the units of measurements are the same for all variates, it is sometimes useful to consider a more restrictive model where the error variances are assumed to be equal. The variance-covariance matrix $\boldsymbol{\Sigma}$ for this model is

$$\boldsymbol{\Sigma} = \mathbf{T}\mathbf{D}_{s^*}\mathbf{T}' + \theta^2\mathbf{I}, \qquad (34)$$

where $\mathbf{D}_{s^*}$ is as before and θ^2 is the common error variance. It has $p + 1$ independent parameters.

The maximum-likelihood estimates of this model can be obtained as before with the ACOVS program by specifying the equality of all the elements of $\boldsymbol{\Theta}$. An illustrative example is given in Table 13. The variance-covariance matrix of a proficiency measure in six trials is shown. The data were obtained in a study of a two-hand coordination task that was conducted by Bilodeau (1957). The task requires the subject to move a pin around a clover-shaped runway by the coordinated turning of two control handles. The subjects were 152 basic airmen, and the trials were 60 seconds long with 30-second rest intervals between trials. The data have previously been analyzed by Bock and Bargmann (1966).

The data from the six trials were analyzed under the three models listed under Summary of Analyses. It is seen that a perfect Wiener simplex does not fit the data, whereas the quasi-Wiener simplex with equal error variances has a very good fit. In general, when such a good fit has been obtained there is seldom reason to relax the model further by introducing more parameters. However, for the sake of completeness we have also analyzed the data with-

TABLE 13
Bilodeau's data

Variance-Covariance Matrix of Proficiency Measures in Six Trials
$N = 152$

Trial						
1	521					
2	477	576				
3	484	536	601			
4	510	575	593	755		
5	523	580	598	718	797	
6	528	584	613	722	751	802

Summary of Analyses

	Model	χ^2	d.f.	P
1	Perfect Wiener Simplex	119.05	15	0.00
2	Quasi-Wiener Simplex with Equal Error Variances	9.39	14	0.81
3	Quasi-Wiener Simplex	8.36	10	0.59

Solution for Model 2

i	s_i^*	s.e.(s_i^*)	
1	482.6	58.7	
2	54.6	14.6	$\hat{\theta} = 6.73$
3	16.0	10.2	$\hat{\theta}^2 = 45.3$
4	81.4	14.9	s.e.(θ) = 0.35
5	21.6	9.6	
6	1.61	10.3	

out the assumption of equal error variances. The drop in χ^2 is 1.03 with a drop in degrees of freedom equal to 4. Thus the hypothesis of equal error variances cannot be rejected. The maximum-likelihood solution for the model with equal error variances is given with standard errors for each parameter.

The Quasi-Markov Simplex

In the preceding example the data were obtained from learning trials where there is an a priori given ordering of the variables (trials) and where the unit of measurement in each variable is the same. For the data of Table 11, neither of these conditions is necessarily true, and it is therefore not appropriate to analyze the variance-covariance matrix according to the quasi-Wiener simplex model. In this case a scale-free model such as the quasi-Markov simplex is needed. The variance-covariance matrix of a quasi-Markov simplex is

$$\boldsymbol{\Sigma} = \mathbf{D}_{\alpha}\mathbf{T}\mathbf{D}_{s^*}\mathbf{T}'\mathbf{D}_{\alpha} + \boldsymbol{\Theta}^2. \tag{35}$$

In a four-variate case, $\boldsymbol{\Sigma}$ is explicitly given by

$$\Sigma = \begin{bmatrix} \alpha_1^2 s_1^* + \theta_1^2 & & & \\ \alpha_1\alpha_2 s_1^* & \alpha_2^2(s_1^* + s_2^*) + \theta_2^2 & & \\ \alpha_1\alpha_3 s_1^* & \alpha_2\alpha_3(s_1^* + s_2^*) & \alpha_3^2(s_1^* + s_2^* + s_3^*) + \theta^2 & \\ \alpha_1\alpha_4 s_1^* & \alpha_2\alpha_4(s_1^* + s_2^*) & \alpha_3\alpha_4(s_1^* + s_2^* + s_3^*) & \alpha_4^2(s_1^* + s_2^* + s_3^* + s_4^*) + \theta_4^2 \end{bmatrix}.$$

As pointed out by Anderson (1960; see also Jöreskog, 1970b), there are three kinds of indeterminacies in this model in that arbitrary numbers may be added to the two first and the last diagonal element of $\boldsymbol{\Theta}^2$, the effect of which may be counterbalanced by certain changes in $\mathbf{D}_{\alpha}$ and $\mathbf{D}_{s^*}$. To eliminate the indeterminacies, we must impose three independent conditions. The most convenient conditions seem to be to set s_1^*, s_2^*, and s_p^* at unity. It is clear that in this model one cannot determine the distance between s_1 and s_2 and the distance between s_{p-1} and s_p. Only distances between points $s_2 < s_3 < \cdots < s_{p-1}$ are estimable.

For the quasi-Markov simplex the maximum-likelihood estimates may be computed using the program ACOVS. To do so one specifies $\mathbf{B}$ as the diagonal matrix $\mathbf{D}_{\alpha}$, $\boldsymbol{\Lambda}$ the fixed matrix $\mathbf{T}$, $\boldsymbol{\Phi}$ the diagonal matrix $\mathbf{D}_{s^*}$ with $\varphi_{11} = 1$, $\varphi_{22} = 1$ and $\varphi_{pp} = 1$ and $\boldsymbol{\Theta}^2$ free. In the resulting solution the estimated error variances will be nonnegative. In some cases it may happen that one or more of the estimates $\hat{\varphi}_{33}, \hat{\varphi}_{44}, \ldots, \hat{\varphi}_{p-1,p-1}$ are negative. This is usually an indication that the variables are in the wrong order or that the model is otherwise wrong. The computer program also gives a large sample χ^2 for testing the goodness of fit of the model.

The maximum-likelihood method for the quasi-Markov simplex is scale free in the following sense. If, on the basis of a sample dispersion matrix $\mathbf{S}$, the estimates $\hat{\mathbf{D}}_{\alpha}$, $\hat{\mathbf{D}}_{s^*}$, and $\hat{\boldsymbol{\Delta}}$ are obtained with $s_1^* = 1$, $s_2^* = 1$, and $s_p^* = 1$, then if $\mathbf{DSD}$ is used instead of $\mathbf{S}$, where $\mathbf{D}$ is a diagonal matrix of arbitrary scale factors, the estimates will be $\mathbf{D}\hat{\mathbf{D}}_{\alpha}$, $\hat{\mathbf{D}}_{s^*}$, and $\hat{\mathbf{D}}^2\boldsymbol{\Theta}^2$. Hence one is free to use the correlation matrix $\mathbf{R}$ instead of $\mathbf{S}$. Whichever scaling is used, the estimate of $\mathbf{D}_{s^*}$ is the same. The maximum-likelihood method may be used when the units of measurement differ from one variate to another, and when they are arbitrary or irrelevant.

A Restricted Quasi-Markov Simplex

Equation 35 for the quasi-Markov simplex may also be written

$$\boldsymbol{\Sigma} = \mathbf{D}_\alpha(\mathbf{T}\mathbf{D}_{s^*}\mathbf{T}' + \mathbf{D}_\alpha^{-1}\boldsymbol{\Theta}^2\mathbf{D}_\alpha^{-1})\mathbf{D}_\alpha.$$

If one makes the assumption that $\mathbf{D}_\alpha^{-1}\boldsymbol{\Theta}^2\mathbf{D}_\alpha^{-1}$ is a scalar matrix, one obtains a certain restricted quasi-Markov simplex. The variance-covariance matrix for this model thus is

$$\boldsymbol{\Sigma} = \mathbf{D}_\alpha(\mathbf{T}\mathbf{D}_{s^*}\mathbf{T}' + \psi^2\mathbf{I})\mathbf{D}_\alpha, \tag{36}$$

where ψ^2 is a scalar. This may also be viewed as a scaling of the quasi-Wiener simplex with equal error variances. The advantage of the model of Equation 36 compared with that of Equation 35 is that the three indeterminacies with their interpretational difficulties are eliminated. The $\boldsymbol{\Sigma}$ in Equation 36 has $2p$ independent parameters. One element in $\mathbf{D}_\alpha$ or $\mathbf{D}_{s^*}$ or ψ may be fixed. The most meaningful choice is $s_1^* = 1$, which fixes the unit of the s scale or equivalently the zero point of the t scale.

The model of Equation 36 may also be estimated by the ACOVS method. One chooses $\mathbf{B}$ as the diagonal matrix $\mathbf{D}_\alpha$, $\boldsymbol{\Lambda}$ as $\mathbf{T}$, $\boldsymbol{\Phi}$ as the diagonal matrix $\mathbf{D}$ with $\varphi_{11} = 1$, $\boldsymbol{\Psi}$ of the form $\psi\mathbf{I}$ and $\boldsymbol{\Theta} = \mathbf{0}$. When the estimates $\hat{\mathbf{D}}_\alpha$, $\hat{\mathbf{D}}_{s^*}$, and $\hat{\psi}$ have been obtained, one can interpret $\hat{\mathbf{D}}_\alpha\mathbf{T}\hat{\mathbf{D}}_{s^*}\mathbf{T}\,\hat{\mathbf{D}}_\alpha$ as an estimate of the true variance-covariance matrix $\mathbf{T}$ and $\hat{\psi}^2\hat{\mathbf{D}}_\alpha^2$ as an estimate of the matrix of error variances. A test of goodness of fit of the model may also be obtained by the ACOVS method.

The model of Equation 36 is also scale free. The estimate $\hat{\mathbf{D}}_{s^*}$ with $s_1^* = 1$ is invariant under scale changes in the original variables.

Consider again the data of Table 11. This represents six verbal ability tests and the sample size is very large ($N = 1046$). Some results of analyses are shown in Table 14. It is seen that although the value 43.81 of χ^2 is significant, the fit appears to be very good as judged by the small residuals. Only two residuals 0.056 and 0.090 are larger than 0.030 in absolute value. It is clear that the fit is acceptable for most practical purposes. The estimated scale values on the t scale are given in the last column under Solution for Model 3.

Circumplex Models

Simplex models, as considered in the previous sections, are models for tests that may be conceived of as having a linear ordering. The circumplex is another model considered by Guttman (1954) and this yields a circular instead of a linear ordering. The circular order has no beginning and no end, but there is still a law of neighboring that holds.

TABLE 14
Guttman's simplex data

Summary of Analyses

Model	No. Par.	χ^2	d.f.	P
1 Perfect Markov Simplex	11	202.64	10	0.000
2 Restricted Quasi-Markov Simplex	12	70.36	9	0.000
3 Quasi-Markov Simplex	15	43.81	6	0.000

Solution for Model 3

i	$\hat{\alpha}_i$	$\hat{s}_i^*$	$\hat{\theta}_i^2$	$\hat{s}_i$	$\hat{t}_i = \log \hat{s}_i$
1	0.989	1.000*	0.022	1.000	0.00
2	0.628	1.000*	0.212	2.000	0.69
3	0.586	0.290	0.212	2.290	0.83
4	0.425	2.065	0.216	4.355	1.47
5	0.370	1.357	0.218	5.712	1.74
6	0.302	1.000*	0.386	6.712	1.90

Residuals: $\mathbf{S} - \boldsymbol{\Sigma}$

Spelling	0.000					
Punctuation	−0.000	−0.000				
Grammar	−0.016	0.005	−0.000			
Vocabulary	0.056	−0.030	0.007	−0.000		
Literature	0.028	−0.004	−0.025	0.004	−0.000	
Foreign literature	0.090	0.031	0.023	−0.011	−0.000	0.000

The circumplex model suggested by Guttman (1954) is a circular moving average process. Let $\xi_1, \xi_2, \ldots, \xi_p$ be uncorrelated, random latent variables. Then the perfect circumplex of order m with p variables is defined by

$$x_i = \xi_i + \xi_{i+1} + \cdots + \xi_{i+m-1},$$

where $x_{p+i} = x_i$. In matrix form we may write this as $\mathbf{x} = \mathbf{C}\boldsymbol{\xi}$, where $\mathbf{C}$ is a matrix of order $p \times p$ with zeros and ones. In the case of $p = 6$ and $m = 3$,

$$\mathbf{C} = \begin{bmatrix} 1 & 1 & 1 & 0 & 0 & 0 \\ 0 & 1 & 1 & 1 & 0 & 0 \\ 0 & 0 & 1 & 1 & 1 & 0 \\ 0 & 0 & 0 & 1 & 1 & 1 \\ 1 & 0 & 0 & 0 & 1 & 1 \\ 1 & 1 & 0 & 0 & 0 & 1 \end{bmatrix}$$

Let $\varphi_1, \varphi_2, \ldots, \varphi_p$ be the variances of $\xi_1, \xi_2, \ldots, \xi_p$, respectively. Then the variance-covariance matrix of $\mathbf{x}$ is

$$\boldsymbol{\Sigma} = \mathbf{C}\mathbf{D}_\varphi\mathbf{C}', \tag{37}$$

where $\mathbf{D}_\varphi = \text{diag}\,(\varphi_1, \varphi_2, \ldots, \varphi_p)$. The variance of x_i is $\sum_{k=i}^{i+m-1} \varphi_k$, the covariance between x_i and x_j, for $i < j$, is $\sum_{k=j}^{i+m-1} \varphi_k$ for $j = i + 1, i + 2, \ldots, i + m - 1$ and 0 otherwise, and the correlation between x_i and x_j, $i < j$, is

$$\rho_{ij} = \frac{\sum_{k=j}^{i+m-1} \varphi_k}{\sqrt{\left(\sum_{k=i}^{i+m-1} \varphi_k\right)\left(\sum_{k=j}^{j+m-1} \varphi_k\right)}}.$$

Here φ_{p+k} should be interpreted as φ_k.

For a given i, the correlations ρ_{ij} decrease as j increases beyond i, reach a minimum, and then increase as j approaches $p + i$. It is convenient to think of $1, 2, \ldots, p$ as points on a circle. Then for adjacent points the correlation tends to be high; for points far apart, the correlation tends to be low or 0. If $m < p/2$, then $\rho_{ij} = 0$ if $j - i > m$ (modulus p). Since zero correlations are not expected in practice, we assume that m is chosen to be greater than or equal to $p/2$.

Guttman gives several examples of sets of tests showing this property of the correlations. His Table 19 is reproduced here as Table 15. Note that in each row the correlations fall away from the main diagonal, reach a minimum, and then increase. Guttman argues that the tests form a circular order. In

TABLE 15
Intercorrelations among tests of six different kinds of abilities for 710 Chicago school children

	Association	Incomplete Words	Multi-plication	Dot Patterns	ABC	Directions
Test	6	32	37	17	1	12
6	1.	0.446	0.321	0.213	0.234	0.442
32	0.446	1.	0.388	0.313	0.208	0.330
37	0.321	0.388	1.	0.396	0.325	0.328
17	0.213	0.313	0.396	1.	0.352	0.247
1	0.234	0.208	0.325	0.352	1.	0.347
12	0.442	0.330	0.328	0.247	0.347	1.

the example the Association test is about equally related to the Incomplete Words test and the Directions test. It is this circular ordering that is the significant feature of the correlation matrix.

The perfect circumplex is too restrictive a model. First, it cannot account for random measurement error in the test scores, and second, since the model is not scale free, it cannot be used to analyze correlations such as those in Table 15. However, these difficulties are easily remedied by considering the quasi-circumplex

$$\mathbf{x} = \mathbf{D}_\alpha \mathbf{C}\boldsymbol{\xi} + \mathbf{e}$$

with dispersion matrix

$$\boldsymbol{\Sigma} = \mathbf{D}_\alpha \mathbf{C}\mathbf{D}_\varphi \mathbf{C}'\mathbf{D}_\alpha + \boldsymbol{\Theta}^2, \tag{38}$$

where $\mathbf{e}$ is the vector of error scores with variances in the diagonal matrix $\boldsymbol{\Theta}^2$, and $\mathbf{D}_\alpha$ is a diagonal matrix of scale factors. One element in $\mathbf{D}_\alpha$ or $\mathbf{D}_\varphi$ must be fixed at unity. It seems most natural to fix $\varphi_1 = 1$. Then the model is scale free and may be estimated by the ACOVS program. The results of an analysis of the data in Table 15 are given in Table 16.

TABLE 16
Quasi-circumplex solution for the data of Table 15 ($m = 4$)

i	$\hat{\alpha}_i$	$\hat{\varphi}_i$	$\hat{\theta}_i$
1	0.268	1.000	0.625
2	0.219	2.012	0.709
3	0.210	3.133	0.714
4	0.235	2.129	0.711
5	0.220	2.793	0.758
6	0.254	2.767	0.639

Note: $\chi^2 = 16.47$ with 4 d.f.

7. PATH ANALYSIS MODELS

Introduction

Path analysis, due to Wright (1918), is a technique sometimes used to assess the direct causal contribution of one variable to another in a nonexperimental situation. The problem in general is that of estimating the parameters of a set of linear structural equations representing the cause and effect relationships hypothesized by the investigator. Recently, several models have been studied that involve hypothetical constructs, i.e., latent variables that, while not directly observed, have operational implications for relationships among observable variables (see, e.g., Werts & Linn, 1970; Hauser & Goldberger, 1971). In some models, the observed variables appear only as effects (indicators) of the hypothetical constructs, while in others, the observed variables appear as causes (components) or as both causes and effects of latent variables. We give one simple example of each kind of model to indicate how such models may be handled within the framework of covariance structure analysis.

A Model with Correlated Measurement Errors

Consider the model discussed by Costner (1969) shown in Figure 6. Note that the errors δ_3 and ϵ_3 are assumed to be correlated as might be the case, for

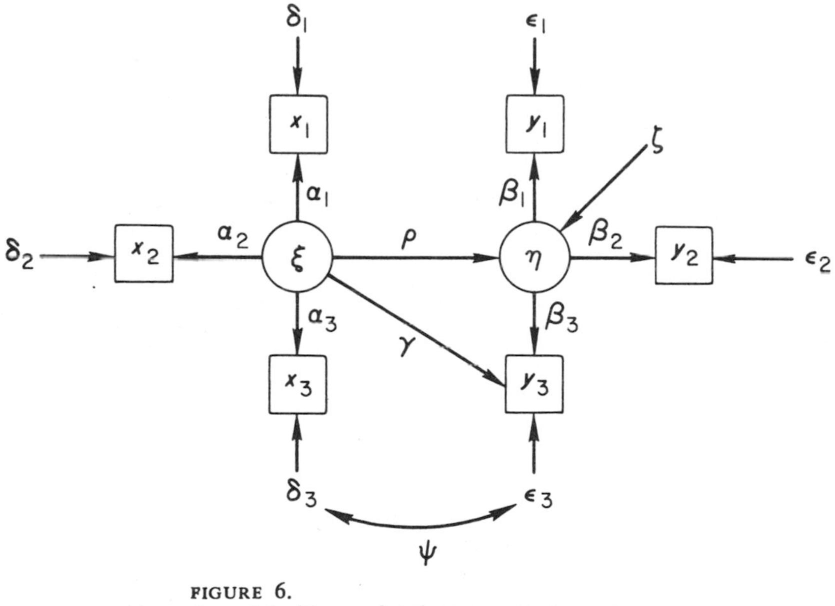

FIGURE 6.
A model with correlated measurement errors.

example, if x_3 and y_3 are scores from the same measuring instrument used at two different occasions. In algebraic form the model may be written, ignoring the means of the observed variables,

$$\begin{pmatrix} x_1 \\ x_2 \\ x_3 \\ y_1 \\ y_2 \\ y_3 \end{pmatrix} = \begin{bmatrix} \alpha_1 & 0 & 1 & 0 & 0 & 0 & 0 & 0 \\ \alpha_2 & 0 & 0 & 1 & 0 & 0 & 0 & 0 \\ \alpha_3 & 0 & 0 & 0 & 1 & 0 & 0 & 0 \\ \gamma & \beta_1 & 0 & 0 & 0 & 1 & 0 & 0 \\ 0 & \beta_2 & 0 & 0 & 0 & 0 & 1 & 0 \\ 0 & \beta_3 & 0 & 0 & 0 & 0 & 0 & 1 \end{bmatrix} \begin{pmatrix} \xi \\ \eta \\ \delta_1 \\ \delta_2 \\ \delta_3 \\ \epsilon_1 \\ \epsilon_2 \\ \epsilon_3 \end{pmatrix}. \tag{39}$$

Let $\boldsymbol{\Lambda}$ be the matrix in Equation 39 and $\boldsymbol{\Phi}$ the variance-covariance matrix of the vector on the right side. Then $\boldsymbol{\Phi}$ is of the form

$$\boldsymbol{\Phi} = \begin{bmatrix} 1 & & & & & & & \\ \rho & 1 & & & & & & \\ 0 & 0 & \theta_1^2 & & & & & \\ 0 & 0 & 0 & \theta_2^2 & & & & \\ 0 & 0 & 0 & 0 & \theta_3^2 & & & \\ 0 & 0 & 0 & 0 & 0 & \theta_4^2 & & \\ 0 & 0 & 0 & 0 & 0 & 0 & \theta_5^2 & \\ 0 & 0 & 0 & 0 & \varphi & 0 & 0 & \theta_6^2 \end{bmatrix},$$

where ρ is the correlation between the latent variables ξ and η, φ the covariance between δ_3 and ϵ_3, and $\theta_1^2, \theta_2^2, \ldots, \theta_6^2$ the variances of the errors $\delta_1, \delta_2, \delta_3, \epsilon_1, \epsilon_2, \epsilon_3$. The variance-covariance matrix of the observed variables is

$$\boldsymbol{\Sigma} = \boldsymbol{\Lambda}\boldsymbol{\Phi}\boldsymbol{\Lambda}'. \tag{40}$$

Note that in this example $\boldsymbol{\Lambda}$ has more columns than rows and includes also the error part of the model. This representation is necessary since the covariance matrix of the errors is not diagonal.

This model has 15 parameters to be estimated and the covariance matrix in Equation 40 has six degrees of freedom. The investigator may be interested in testing the specific hypothesis $\gamma = 0$, i.e., that ξ affects y_3 only via η. This may be done in large samples, assuming that the rest of the model holds, by a χ^2 test with one degree of freedom.

A Model with Multiple Causes and Multiple Indicators of a Single Latent Variable

Consider the model discussed by Hauser and Goldberger (1971) shown in Figure 7. This model involves a single hypothetical variable ξ, which appears as both cause and effect variable. The equations are

$$\xi = \boldsymbol{\alpha}'\mathbf{x} + v,$$
$$\mathbf{y} = \boldsymbol{\beta}\xi + \mathbf{u}.$$

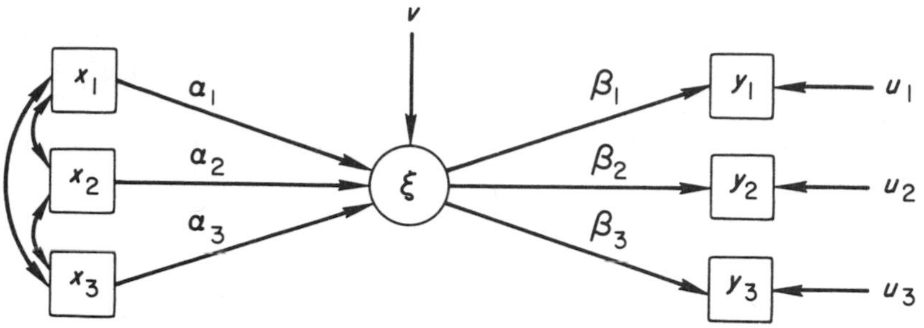

FIGURE 7.
A model with multiple causes and indicators of a latent variable.

The case where the residuals u_1, u_2, and u_3 are mutually correlated and $v = 0$ was considered by Hauser and Goldberger (1971). The case shown in Figure 7, where u_1, u_2, and u_3 are mutually uncorrelated, will be considered in detail in a forthcoming paper by Jöreskog and Goldberger (1973). In this case the structure of the variance-covariance matrix of the observed variables is

$$\begin{aligned} \boldsymbol{\Sigma}_{yy} &= \boldsymbol{\beta\alpha}'\boldsymbol{\Sigma}_{xx}\boldsymbol{\alpha\beta}' + \boldsymbol{\beta\beta}' + \boldsymbol{\Theta}^2, \\ \boldsymbol{\Sigma}_{yx} &= \boldsymbol{\beta\alpha}'\boldsymbol{\Sigma}_{xx}, \\ \boldsymbol{\Sigma}_{xx} & \quad \text{unconstrained.} \end{aligned}$$

The residual v may be scaled to unit variance, as assumed here. Alternatively, the latent variable ξ may be scaled to unit variance or one of the αs fixed at some nonzero value. It is readily verified that this model may be represented in terms of Equation 1 by specifying

$$\mathbf{B} = \begin{bmatrix} \beta_1 & 0 & 0 & 0 \\ \beta_2 & 0 & 0 & 0 \\ \beta_3 & 0 & 0 & 0 \\ 0 & 1 & 0 & 0 \\ 0 & 0 & 1 & 0 \\ 0 & 0 & 0 & 1 \end{bmatrix}, \quad \boldsymbol{\Lambda} = \begin{bmatrix} \alpha_1 & \alpha_2 & \alpha_3 \\ 1 & 0 & 0 \\ 0 & 1 & 0 \\ 0 & 0 & 1 \end{bmatrix}, \quad \boldsymbol{\Phi} = \boldsymbol{\Sigma}_{xx},$$

$$\boldsymbol{\Psi} = \operatorname{diag}(1, 0, 0, 0), \ \boldsymbol{\Theta} = \operatorname{diag}(\sigma_{u_1}, \sigma_{u_2}, \sigma_{u_3}, 0, 0, 0).$$

The model has 15 independent parameters and six degrees of freedom.

Models for Longitudinal Data

Many studies in the behavioral sciences involve longitudinal data, i.e., measurements that have been repeated on the same individuals over time or under different conditions. Jöreskog (1970c) developed a general model for analyzing multitest-multioccasion correlation matrices and Werts, Jöreskog, and Linn (1973) discussed the implications of this model for the study of growth. Here we confine ourselves to a path analysis formulation of the quasi-Markov simplex.

A path diagram of the model is shown in Figure 8. The corresponding equations are

$$x_i = \alpha_i \tau_i + \epsilon_i, \quad i = 1, 2, \ldots, p, \tag{41}$$

$$\tau_i = \beta_i \tau_{i-1} + \zeta_i, \quad i = 2, 3, \ldots, p, \tag{42}$$

where it is assumed, without loss of generality, that $\tau_1, \tau_2, \ldots, \tau_p$ are scaled to unit variance. With this scaling, α_i is the covariance between x_i and τ_i, and β_i is the correlation between τ_i and τ_{i-1}; the covariance between any two variates is the product of the path coefficients of the lines going from one variate to the other. For example, $\text{cov}(x_1, \tau_3) = \alpha_1\beta_2\beta_3$ and generally $\text{cov}(x_i, x_j) = \alpha_i\beta_{i+1}, \beta_{i+2} \ldots \beta_j\alpha_j$ for $i < j$.

When $p = 4$, the variance-covariance matrix is

$$\Sigma = \begin{bmatrix} \alpha_1^2 + \theta_1^2 & & & \\ \alpha_2\alpha_1\beta_2 & \alpha_2^2 + \theta_2^2 & & \\ \alpha_3\alpha_1\beta_2\beta_3 & \alpha_3\alpha_2\beta_3 & \alpha_3^2 + \theta_3^2 & \\ \alpha_4\alpha_1\beta_2\beta_3\beta_4 & \alpha_4\alpha_2\beta_3\beta_4 & \alpha_4\alpha_3\beta_4 & \alpha_4^2 + \theta_4^2 \end{bmatrix}.$$

If α_1 is replaced by $a\alpha_1$ and β_2 by (β_2/a), the off-diagonal elements in the first row and column are unchanged; if θ_1^2 is properly adjusted the first-diagonal element is also unchanged. Thus, α_1 and β_2 are not identified, but their product is, since

$$(\alpha_1\beta_2)^2 = \frac{\sigma_{21}\sigma_{13}}{\sigma_{23}} = \frac{\sigma_{21}\sigma_{14}}{\sigma_{24}}.$$

The same kind of argument can be used to show that α_4 and β_4 are not identified, but that $\alpha_4\beta_4$ is. The parameters α_2, α_3, and β_3 are identified since

$$\alpha_2^2 = \frac{\sigma_{12}\sigma_{23}}{\sigma_{13}} = \frac{\sigma_{12}\sigma_{24}}{\sigma_{14}},$$

$$\alpha_3^2 = \frac{\sigma_{13}\sigma_{34}}{\sigma_{14}} = \frac{\sigma_{23}\sigma_{34}}{\sigma_{24}},$$

$$\beta_3^2 = \frac{\sigma_{32}\sigma_{24}}{\sigma_{34}\alpha_2^2} = \frac{\sigma_{13}\sigma_{32}}{\sigma_{12}\alpha_3^2}.$$

Since α_2 and α_3 are identified, θ_2 and θ_3 are also identified.

The above reasoning generalizes to an arbitrary number of variables, p, in an obvious way. For the 'inner' variables the parameters $\alpha_2, \alpha_3, \ldots, \alpha_{p-1}$, $\beta_3, \beta_4, \ldots, \beta_{p-1}, \theta_2, \theta_3, \ldots, \theta_{p-1}$ are all identified, but for the 'outer' variates only the products $\alpha_1\beta_2$ and $\alpha_p\beta_p$ are identified. The indeterminacy in α_1 and β_2

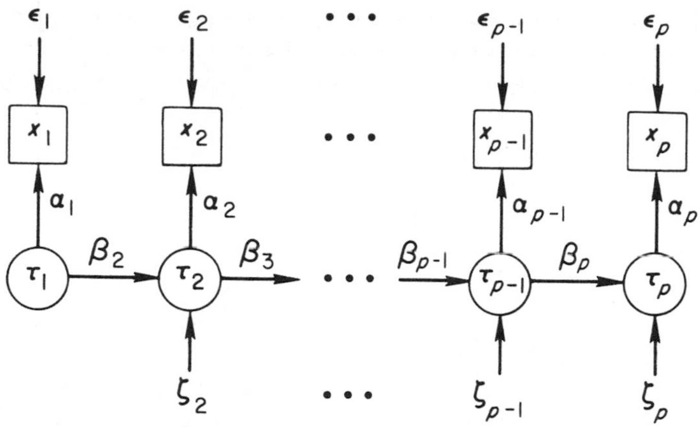

FIGURE 8.
Path analysis diagram of the quasi-Markov simplex.

and in α_p and β_p may be eliminated by choosing $\beta_2 = \beta_p = 1$ and Figure 8 may, for estimation purposes, be replaced by Figure 9. The condition $\beta_2 = \beta_p = 1$ is assumed in what follows.

The special cases $p = 3$ and $p = 4$ may now be examined more closely and shown to be equivalent to certain factor analysis models. In the case $p = 3$, Equation 41 may be written

$$\begin{pmatrix} x_1 \\ x_2 \\ x_3 \end{pmatrix} = \begin{pmatrix} \alpha_1 \\ \alpha_2 \\ \alpha_3 \end{pmatrix} \tau_2 + \begin{pmatrix} \epsilon_1 \\ \epsilon_2 \\ \epsilon_3 \end{pmatrix},$$

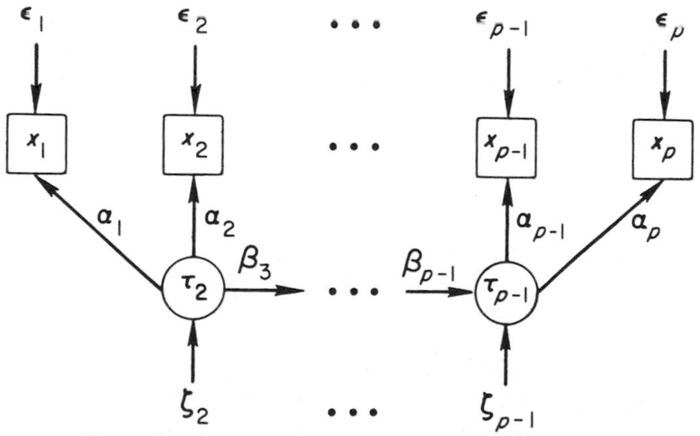

FIGURE 9.
Path analysis diagram of the quasi-Markov simplex after elimination of indeterminacies.

which is formally equivalent to a factor analysis model with one common factor. This model, however, has zero degrees of freedom, and if the data are consistent with the model the fit will be perfect. In the case $p = 4$, the equation may be written

$$\begin{pmatrix} x_1 \\ x_2 \\ x_3 \\ x_4 \end{pmatrix} = \begin{pmatrix} \alpha_1 & 0 \\ \alpha_2 & 0 \\ 0 & \alpha_3 \\ 0 & \alpha_4 \end{pmatrix} \begin{pmatrix} \tau_2 \\ \tau_3 \end{pmatrix} + \begin{pmatrix} \epsilon_1 \\ \epsilon_2 \\ \epsilon_3 \\ \epsilon_4 \end{pmatrix},$$

with β_3 equal to the correlation between τ_2 and τ_3. The corresponding $\boldsymbol{\Sigma}$ is

$$\Sigma = \begin{bmatrix} \alpha_1 & 0 \\ \alpha_2 & 0 \\ 0 & \alpha_3 \\ 0 & \alpha_4 \end{bmatrix} \begin{bmatrix} 1 & \beta_3 \\ \beta_3 & 1 \end{bmatrix} \begin{bmatrix} \alpha_1 & \alpha_2 & 0 & 0 \\ 0 & 0 & \alpha_3 & \alpha_4 \end{bmatrix} + \begin{bmatrix} \theta_1^2 & 0 & 0 & 0 \\ 0 & \theta_2^2 & 0 & 0 \\ 0 & 0 & \theta_3^2 & 0 \\ 0 & 0 & 0 & \theta_4^2 \end{bmatrix}. \qquad (43)$$

$$= \boldsymbol{\Lambda\Phi\Lambda}' + \boldsymbol{\Theta}^2.$$

This model is equivalent to a factor analysis model with two nonoverlapping oblique factors. The model has one degree of freedom. Given this model, one can test each of the hypotheses $\beta_3 = 1$, $\alpha_2 = \alpha_3$, and $\theta_2^2 = \theta_3^2$ using a large sample χ^2 with one degree of freedom. Assumptions of this kind have traditionally been thought necessary in growth studies, but a test of their validity has not been available earlier (see, e.g., Harris, 1963, Chs. 1-3).

A Growth Data Example

To illustrate the model (Eq. 43), we use some data from a longitudinal growth study conducted at Educational Testing Service (Anderson & Maier, 1963; Hilton, 1969). A nationwide sample of fifth graders was tested in 1961 and again in 1963, 1965, and 1967 as seventh, ninth, and eleventh graders, respectively. Table 17 shows the variance-covariance matrix of the scores on the quantitative part of SCAT (Scholastic Aptitude Test) at the four different occasions for a random sample of 799 girls from all the girls that took all tests at all occasions. The tests have been scaled so that the unit of measurement is approximately the same at all occasions.

The matrix in Table 17 has been analyzed under five different hypotheses based on the model of Equation 43:

(1) Equation 43;
(2) Equation 43 with $\beta_3 = 1$;
(3) Equation 43 with $\alpha_2 = \alpha_3$;
(4) Equation 43 with $\theta_2 = \theta_3$;
(5) Equation 43 with $\alpha_2^* = \alpha_3^*$.

TABLE 17
SCATQ growth data

Variance-Covariance Matrix
$N = 799$

Q_5	67.951			
Q_7	71.071	141.578		
Q_9	85.966	134.655	249.748	
Q_{11}	97.153	151.068	218.757	300.669

Summary of Analysis

Hypothesis	χ^2	d.f.	P
1	0.092	1	0.761
2	50.020	2	0.000
3	65.543	2	0.000
4	22.289	2	0.000
5	0.100	2	0.951

Solution for Model 1

$\hat{\alpha}_i$		$\hat{\psi}_i^2$
6.75		22.39
10.53	$\hat{\beta}_3 = 0.92$	30.71
13.95		55.28
15.69		54.60

In model 5 it is assumed that the standardized regression coefficients, i.e., the reliabilities, are equal. This condition may be specified by letting the true scores τ_2 and τ_3 be in the same metric as x_1 and x_2; the resulting regression coefficients α_2 and α_3 will then be standardized. The values of χ^2 for each hypothesis are given under Summary of Analyses. Each hypothesis has been tested against the general alternative that Σ is unconstrained. It is seen that models 1 and 5 have good fits, but the other models have very poor fits. Testing hypothesis 5 against hypothesis 1, one obtains $\chi^2 = 0.100 - 0.092 = 0.008$ with one degree of freedom. Thus, the hypothesis of equal reliabilities cannot be rejected. It should be noted that this hypothesis is equivalent to the assertion that the error variances are in a fixed proportion of the total variances. The maximum-likelihood solution under hypothesis 1 is given under Solution for Model 1.

REFERENCES

Anderson, S. B., & Maier, M. H. 34,000 pupils and how they grow. *Journal of Teacher Education*, 1963, **14**, 212–216.

Anderson, T. W. Some stochastic process models for intelligence test scores. In K. J. Arrow, S. Karlin, and P. Suppes (Eds.), *Mathematical methods in the social sciences*, 1959. Stanford, California: Stanford University Press, 1960.

Bilodeau, E. A. The relationship between a relatively complex motor skill and its components. *American Journal of Psychology*, 1957, **70**, 49–55.

Bock, R. D. Components of variance analysis as a structural and discriminal analysis for psychological tests. *British Journal of Statistical Psychology*, 1960, **13**, 151–163.

Bock, R. D., & Bargmann, R. E. Analysis of covariance structures. *Psychometrika*, 1966, **31**, 507–534.

Bock, R. D., Dicken, D., & van Pelt, J. Methodological implications of content-acquiescence correlation in the MMPI. *Psychological Bulletin*, 1969, **71**, 127–139.

Burt, C. Factor analysis and analysis of variance. *British Journal of Psychology*, Statistical Section, 1947, **1**, 3–26.

Campbell, D. T., & Fiske, D. W. Convergent and discriminant validation by the multitrait-multimethod matrix. *Psychological Bulletin*, 1959, **56**, 81–105.

Costner, H. L. Theory, deduction and rules of correspondence. *American Journal of Sociology*, 1969, **75**, 245–263.

Fletcher, R., & Powell, M. J. D. A rapidly convergent descent method for minimization. *The Computer Journal*, 1963, **6**, 163–168.

Graybill, F. A. *An introduction to linear statistical models*. Vol. 1. New York: McGraw-Hill, 1961.

Gruvaeus, G. T., & Jöreskog, K. G. *A computer program for minimizing a function of several variables*. Research Bulletin 70–14. Princeton, N.J.: Educational Testing Service, 1970.

Guilford, J. P. The structure of intellect. *Psychological Bulletin*, 1956, **53**, 267–293.

Guttman, L. A new approach to factor analysis: The Radex. In P. F. Lazarsfeld (Ed.), *Mathematical thinking in the social sciences*. New York: Columbia University Press, 1954.

Harris, C. W. (Ed.) *Problems in measuring change*. Madison: University of Wisconsin Press, 1963.

Hauser, R. M., & Goldberger, A. S. The treatment of unobservable variables in path analysis. In H. L. Costner (Ed.), *Sociological methodology*. London: Jossey-Bass, 1971.

Hilton, T. L. *Growth study annotated bibliography*. Progress Report 69–11. Princeton, N.J.: Educational Testing Service, 1969.

Jöreskog, K. G. Some contributions to maximum likelihood factor analysis. *Psychometrika*, 1967, **32**, 443–482.

Jöreskog, K. G. A general approach to confirmatory maximum likelihood factor analysis. *Psychometrika*, 1969, **34**, 183–202.

Jöreskog, K. G. A general method for analysis of covariance structures. *Biometrika*, 1970, **57**, 239–251. (a)

Jöreskog, K. G. Estimation and testing of simplex models. *British Journal of Mathematical and Statistical Psychology*, 1970, **23**, 121–145. (b)

Jöreskog, K. G. Factoring the multitest-multioccasion correlation matrix. In C. E. Lunneborg (Ed.), *Current problems and techniques in multivariate psychology*. (Proceedings of a Conference Honoring Professor Paul Horst.) Seattle: University of Washington, 1970. (c)

Jöreskog, K. G. Statistical analysis of sets of congeneric tests. *Psychometrika*, 1971, **36**, 109–133.

Jöreskog, K. G., & Goldberger, A. S. Factor analysis by generalized least squares. *Psychometrika*, 1972, **37**, 243–260.

Jöreskog, K. G., & Goldberger, A. S. *Estimation of a model with multiple indicators and multiple causes of a single latent variable*. Research Report 73-14. Uppsala: Dept. of Statistics, University of Uppsala, 1973. Also issued as Report 7328. Madison: Social Systems Res. Inst. Workshop Series, University of Wisconsin, 1973.

Jöreskog, K. G., Gruvaeus, G. T., & van Thillo, M. *ACOVS—A general computer program for analysis of covariance structures*. Research Bulletin 70–15. Princeton, N.J.: Educational Testing Service, 1970.

Jöreskog, K. G., & Lawley, D. N. New methods in maximum likelihood factor analysis. *British Journal of Mathematical and Statistical Psychology*, 1968, **21**, 85–96.

Kelley, E. L., & Fiske, D. W. *The prediction of performance in clinical psychology*. Ann Arbor: University of Michigan Press, 1951.

Kristof, W. On the theory of a set of tests which differ only in length. *Psychometrika*, 1971, **36**, 207–225.

Language modalities survey: A test for aphasia. Speech Clinic, University of Chicago and Psychometric Laboratory, University of North Carolina, Chapel Hill, 1959.

Lawley, D. N., & Maxwell, A. E. *Factor analysis as a statistical method*. London: Butterworths, 1963.

Lord, F. M. A study of speed factors in tests and academic grades. *Psychometrika*, 1956, **21**, 31–50.

Lord, F. M. A significance test for the hypothesis that two variables measure the same trait except for errors of measurement. *Psychometrika*, 1957, **22**, 207–220.

Lord, F. M., & Novick, M. R. *Statistical theories of mental test scores* (with contributions by A. Birnbaum). Reading, Massachusetts: Addison-Wesley, 1968.

McNemar, Q. Attenuation and interaction. *Psychometrika*, 1958, **23**, 259–265.

Miller, D. M., & Lutz, M. V. Item design for an inventory of teaching practices and learning situations. *Journal of Educational Measurement*, 1966, **3**, 53–61.

Werts, C. E., Jöreskog, K. G., & Linn, R. L. A multitrait-multimethod model for studying growth. *Educational and Psychological Measurement*, 1973, **33**, in press.

Werts, C. E., & Linn, R. L. Path analysis: Psychological examples. *Psychological Bulletin*, 1970, **67**, 193–212.

Wiley, D. E., Schmidt, W. H., & Bramble, W. J. Studies of a class of covariance structure models. *Journal of the American Statistical Association*, 1973, **68,** 317–323.

Wright, S. On the nature of size factors. *Genetics*, 1918, **3,** 367–374.

Reprinted from: *Contemporary Developments in Mathematical Psychology*, Vol. II, edited by David H. Krantz, Richard C. Atkinson, R. Duncan Luce, and Patrick Suppes. W. H. Freeman and Company.

Part II
Structural Equation Models

Introduction to Part II

In part II the factor analysis measurement model developed in the preceding chapters is merged with a structural model specifying causal linkages among the factors. Measurement models associated with independent variables and dependent variables are formulated separately and serve to define the factors within the structural model. Some factors may be assumed to be perfectly measured by individual variables; others are measured using multiple indicators. The structural model and the system of equations that make up the measurement models form the Linear Structural Relations (LISREL) model.

In chapter 4 the problem of identification for the LISREL model is discussed, and the approaches to estimation and hypothesis testing are described and illustrated. A detailed example applies these methods to a sociological model for analyzing the way in which one's peers influence one's decisions.

A model is said to be underidentified if one or more parameters cannot be estimated because of lack of information. In such cases additional restrictions must be imposed on the parameters. A model is said to be overidentified if all its parameters can be estimated and restrictions on the parameters are more than sufficient to estimate at least one of the model parameters. In the case of overidentification a goodness-of-fit test is available and reflects whether the data are consistent with all the restrictions. Various subsets of restrictions can also be tested by subtracting the goodness-of-fit statistic for models that include the subset of restrictions from the corresponding chi-square statistic for the similar model that does not impose this subset of restrictions.

Chapter 5 further describes the LISREL approach to longitudinal models and presents detailed examples for one- and two-variable models. It also describes the general approach for multiwave, multivariable models.

Sometimes a model does not fit the data well because an additional source of correlation between two or more variables is not taken into account in the model. Although the explanation of such correlation may not appear to be important to the theory being tested, it nevertheless must be taken into account to avoid misspecification and bias in the estimates. For example, a hypothetical Socio-Economic Status (SES) factor may be measured by mother's education, father's education, occupational status, and income in a single-factor model. However, for reasons unrelated to SES, persons tend to marry another who has a similar level of education. Thus additional correlation is likely to be present between mother's and father's education that cannot be attributed to SES. One way to modify the measurement model without explicitly postulating an additional factor is to allow the residuals in the measurement model associated with mother's education and father's education to be correlated. Chapter 6 describes a general approach for detecting the presence of such contamination due to "nuisance" factors within the context of the LISREL model. Applications for longitudinal data are emphasized.

Chapter 4

Structural Equation Models in the Social Sciences: Specification, Estimation and Testing

Karl G. Jöreskog

1. INTRODUCTION

One of the most difficult problems for a social scientist, when it comes to the formulation of a causal model, arises because many of the concepts and constructs that he/she wants to work with are not directly measurable (see e.g. Torgerson, [22], Chapter 1 and Goldberger, [11]). A sociologist works with concepts like social (or socioeconomic) status, occupational aspiration, motivations and attitudes. A psychologist uses concepts like intelligence, verbal ability, spatial ability, etc. An educational researcher studying the effect of home factors on school achievements may be using a construct such as "preschool verbal stimulation in the home." Also in economics one uses hypothetical constructs like price level, permanent income, economic expectations, etc. Although such hypothetical concepts and constructs, or latent variables, as we shall call them, cannot be directly measured, a number of variables can be used to measure various aspects of these latent variables more or less accurately. Thus, while the latent variables cannot be directly observed, they have operational implications for relationships among observed variables. We may regard the observed variables as indicators of the latent variables. Each indicator has a relationship with the latent variable, but if we take one indicator alone to measure the latent variable we would obtain a poor measurement. By using several indicators of each latent variable we get a better measurement of these latent variables.

Another reason for using latent variables in socioeconomic studies is that most of the measurements employed contain sizeable errors of measurement (observational errors). Errors of measurement arise because of imperfection in the various measurement instruments (questionnaires, tests, etc), that are used to measure such abstractions as people's behavior, attitudes, feelings and motivations. Even if we could construct valid measurement of these it is usually impossible to obtain perfectly reliable variables.

In recent years there has been a strong interest in more complex models in-

---

The research reported in this paper has been supported by the Bank of Sweden Tercentenary Foundation under project *Structural Equation Models in the Social Sciences*, Karl G. Jöreskog, project director. I thank Dag Sörbom for helpful assistance.

volving specified causal structures among the latent variables. These have usually been formulated in terms of path analysis, a term and a technique introduced by Wright [27] and further developed by him in a series of papers (see e.g. Wright, [28] and references therein). For sociological applications of path analysis, see Duncan [4] and for psychological applications, see Werts and Linn [23]. More recent studies are Goldberger [10, 11], Hauser and Goldberger [13], Jöreskog and Goldberger [18], and Werts, Jöreskog and Linn [24].

In this paper we shall describe a general method for estimating the unknown coefficients in a set of linear structural equations. The variables in the equation system may be latent variables and there may be multiple indicators or causes of each latent variable. Also, the method allows for both errors in equations (residuals, disturbances) and errors in the observed variables (errors of measurement, observational errors) and yields estimates of the residual covariance matrix and the measurement error covariance matrix as well as estimates of the unknown coefficients in the structural equations, provided that all these parameters are identified. The method covers a wide range of models useful in the social and behavioral sciences. It is particularly designed to handle models with latent variables, measurement errors and reciprocal causation (simultaneity, interdependence). The general model on which the method is based is a generalization of that considered in a previous paper [17].

2. GENERAL THEORY

2.1. *The General Model*

Consider random vectors $\eta' = (\eta_1, \eta_2, \ldots, \eta_m)$ and $\xi' = (\xi_1, \xi_2, \ldots, \xi_n)$ of latent dependent and independent variables, respectively, and the following system of linear structural relations

$$B\eta = \Gamma\xi + \zeta \qquad (1)$$

where $B(m \times m)$ and $\Gamma(m \times n)$ are coefficient matrices and $\zeta' = (\zeta_1, \zeta_2, \ldots, \zeta_m)$ is a random vector of residuals (errors in equations, random disturbance terms). Without loss of generality it may be assumed that $E(\eta) = E(\zeta) = 0$ and $E(\xi) = 0$. It is furthermore assumed that ζ is uncorrelated with ξ and that B is nonsingular.

The vectors η and ξ are not observed but instead vectors $y' = (y_1, y_2, \ldots, y_p)$ and $x' = (x_1, x_2, \ldots, x_q)$ are observed, such that

$$y = \Lambda_y\eta + \varepsilon , \qquad (2)$$

$$x = \Lambda_x\xi + \delta , \qquad (3)$$

where ε and δ are vectors of errors of measurement in y and x, respectively. We take $\mathbf{y}$ and $\mathbf{x}$ to be measured as deviations from their means. The matrices $\Lambda_y (p \times m)$ and $\Lambda_x (q \times n)$ are regression matrices of $\mathbf{y}$ on $\boldsymbol{\eta}$ and of $\mathbf{x}$ on $\boldsymbol{\xi}$, respectively. It is convenient to refer to $\mathbf{y}$ and $\mathbf{x}$ as the observed variables and $\boldsymbol{\eta}$ and $\boldsymbol{\xi}$ as the latent variables. The errors of measurement $\boldsymbol{\varepsilon}$ and $\boldsymbol{\delta}$ are assumed to be uncorrelated and uncorrelated with the latent variables.

Let $\Phi\ (n \times n)$ and $\Psi\ (m \times m)$ be the covariance matrices of $\boldsymbol{\xi}$ and $\boldsymbol{\zeta}$, respectively, and let Θ_ε and Θ_δ be the covariance matrices of $\boldsymbol{\varepsilon}$ and $\boldsymbol{\delta}$, respectively. Then it follows, from the above assumptions that the covariance matrix $\Sigma[(p + q) \times (p + q)]$ of $\mathbf{z} = (\mathbf{y}', \mathbf{x}')'$ is

$$\Sigma = \begin{bmatrix} \Lambda_y (B^{-1}\Gamma\Phi\Gamma'B'^{-1} + B^{-1}\Psi B'^{-1})\Lambda_y' + \Theta_\varepsilon & \Lambda_y B^{-1}\Gamma\Phi\Lambda_x' \\ \Lambda_x \Phi\Gamma'B'^{-1}\Lambda_y' & \Lambda_x\Phi\Lambda_x' + \Theta_\delta \end{bmatrix} \tag{4}$$

The elements of Σ are functions of the elements of Λ_y, Λ_x, B, Γ, Φ, Ψ, Θ_δ and Θ_ε. In applications some of these elements are fixed and equal to assigned values. In particular, this is so for elements Λ_y, Λ_x, B and Γ, but we shall allow for fixed values in the other matrices also. For the remaining non-fixed elements of the six parameter matrices one or more subsets may have identical but unknown values. Thus elements in Λ_y, Λ_x, B, Γ, Φ, Ψ, Θ_δ and Θ_ε are of three kinds:

(i) *fixed parameters* that have been assigned given values,

(ii) *constrained parameters* that are unknown but equal to one or more other parameters and

(iii) *free parameters* that are unknown and not constrained to be equal to any other parameter.

Equations (1), (2) and (3), with the accompanying assumptions, define the general model. Equations (2) and (3) constitute the measurement model and equation (1) constitutes the structural equation model. The measurement model can be written more compactly as

$$\mathbf{z} = \Lambda \mathbf{f} + \mathbf{e}\ , \tag{5}$$

where

$$\mathbf{z} = (\mathbf{y}', \mathbf{x}')', \qquad \mathbf{f} = (\boldsymbol{\eta}', \boldsymbol{\xi}')', \qquad \mathbf{e} = (\boldsymbol{\varepsilon}', \boldsymbol{\delta}')'$$

and

$$\Lambda = \begin{bmatrix} \Lambda_y & 0 \\ 0 & \Lambda_x \end{bmatrix}$$

From (5) it can be seen that the model is a restricted factor analysis model [16] in which the factors η and ξ satisfy a linear structural equation system of the form (1). It should be noted, however, that there is no requirement that $m < p$, $n < q$ and that Θ_ε and Θ_δ be diagonal as in traditional factor analysis. The only requirement is that Σ in (4) is nonsingular and that the model is identified (see next section).

2.2. Identification of Parameters

Before an attempt is made to estimate a model of this kind the identification problem must be examined. Identifiability depends on the choice of model and on the specification of fixed, constrained, and free parameters. Under a given specification, a given structure Λ_y, Λ_x, B, Γ, Φ, Ψ, Θ_ε, Θ_δ generates one and only one Σ but there may be several structures generating the same Σ. If two or more structures generate the same Σ, the structures are said to be equivalent. If a parameter has the same value in all equivalent structures, the parameter is said to be identified. If all parameters of the model are identified, the whole model is said to be identified.

We assume that the distribution of the observed variables is sufficiently well described by the moments of first and second order, so that information contained in moments of higher order may be ignored. In particular, this will hold if the distribution is multivariate normal. Since the mean vector is unconstrained, the distribution of $z = (y', x')'$ is described by the independent parameters in Λ_y, Λ_x, B, Γ, Φ, Ψ, Θ_ε, Θ_δ. Let θ be a vector of all the independent, free and constrained parameters (i.e. counting each distinct constrained parameter once only) and let s be the order of θ. The identification problem then is the problem of whether or not θ is determined by Σ.

If a parameter is not identified it will not be possible to find a consistent estimator of it.

To examine the identification problem for a particular model consider the equations in (4) of the form

$$\sigma_{ij} = f_{ij}(\theta) \; , \; i \leq j \; . \tag{6}$$

There are (1/2) (p + q) (p + q + 1) equations and s unknown elements in θ. Hence a necessary condition for identification of all parameters is that

$$s \leq (1/2)\,(p + q)\,(p + q + 1) \tag{7}$$

If a parameter θ can be determined from Σ, this parameter is identified, otherwise it is not. Often some parameters can be determined from Σ in different ways. This gives rise to overidentifying conditions on Σ which must hold if the model is true. The solution of (6) is often complicated and tedious and explicit solutions for all θ's seldom exist. Examples of how the identification problem can be resolved are given in section 3.

2.3. *Estimation of the Model*

Let $\underset{\sim}{z}_1, \underset{\sim}{z}_2, \ldots, \underset{\sim}{z}_M$ be independent observations of $\underset{\sim}{z} = (\underset{\sim}{y}', \underset{\sim}{x}')'$ and let $\bar{\underset{\sim}{z}} = (\bar{\underset{\sim}{y}}', \bar{\underset{\sim}{x}}')$be the sample mean vector, and

$$\underset{\sim}{S} = \frac{1}{N} \sum_{\alpha=1}^{N} (\underset{\sim}{z}_\alpha - \bar{\underset{\sim}{z}}) (\bar{\underset{\sim}{z}}_\alpha - \bar{\underset{\sim}{z}})', \tag{8}$$

with N = M - 1, the sample covariance matrix.

We assume that the distribution of the observed variables can be described by the mean vector and the covariance matrix. The mean vector is unconstrained, so the estimation problem is essentially that of fitting the $\underset{\sim}{\Sigma}$ imposed by the model to the sample covariance matrix $\underset{\sim}{S}$. As a fitting function we use

$$F = \log|\underset{\sim}{\Sigma}| + \mathrm{tr}(\underset{\sim}{S}\, \underset{\sim}{\Sigma}^{-1}) - \log|\underset{\sim}{S}| - (p + q) \ , \tag{9}$$

which is to be minimized with respect to $\underset{\sim}{\theta}$. If the distribution of $\underset{\sim}{z} = (\underset{\sim}{y}, \underset{\sim}{x}')'$ is multimormal this yields maximum likelihood estimates which are efficient in large samples. If the distribution deviates far from the multimormal it is probably wise to "robustify" the variances and covariances in $\underset{\sim}{S}$ before the analysis: see Andrews, Gnanadesikan and Warner [1] for methods for assessing multivariate normality, Gnanadesikan and Kettenring [9] for robust estimates of variances and covariances and Devlin, Gnanadesikan and Kettenring [3] for robust estimates of correlations. The minimization of F with respect to the independent parameters is done by an iterative method which makes use of the first derivatives of F and the information matrix. It converges rapidly from an arbitrary starting point to a local minimum of F. If there are several minima of F there is no guarantee that the method will converge to the absolute minimum. At the minimum of F, the information matrix may be computed and used to compute standard errors for all estimated parameters.

The derivatives of F with respect to the elements of $\underset{\sim}{\Lambda}_y$, $\underset{\sim}{\Lambda}_x$, $\underset{\sim}{B}$, $\underset{\sim}{\Gamma}$, $\underset{\sim}{\Phi}$, $\underset{\sim}{\Psi}$, $\underset{\sim}{\Theta}_\delta$, and $\underset{\sim}{\Theta}_\varepsilon$ may be obtained by matrix differentiation. Writing $\underset{\sim}{A} = \underset{\sim}{B}^{-1}$, $\underset{\sim}{D} = \underset{\sim}{B}^{-1}\underset{\sim}{\Gamma}$, $\underset{\sim}{C} = \underset{\sim}{D}\underset{\sim}{\Phi}\underset{\sim}{D}' + \underset{\sim}{A}\underset{\sim}{\Psi}\underset{\sim}{A}'$ and

$$\underset{\sim}{\Omega} = \begin{bmatrix} \underset{\sim}{\Omega}_{yy} & \underset{\sim}{\Omega}_{yx} \\ & \\ \underset{\sim}{\Omega}_{xy} & \underset{\sim}{\Omega}_{xx} \end{bmatrix} = \underset{\sim}{\Sigma}^{-1}(\underset{\sim}{\Sigma} - \underset{\sim}{S})\underset{\sim}{\Sigma}^{-1} \ , \tag{10}$$

the derivatives are, except for a scale factor of 2 in the first four derivatives,

$$\partial F/\partial \Lambda_y = \Omega_{yy}\Lambda_y C + \Omega'_{xy}\Lambda_x \Phi D' \quad , \tag{11}$$

$$\partial F/\partial \Lambda_x = \Omega_{xy}\Lambda_y D\Phi + \Omega_{xx}\Lambda_x \Phi \quad , \tag{12}$$

$$\partial F/\partial B = -A'\Lambda'_y(\Omega_{yy}\Lambda_y C + \Omega'_{xy}\Lambda_x \Phi D') \quad , \tag{13}$$

$$\partial F/\partial \Gamma = A'\Lambda'_y(\Omega_{yy}\Lambda_y D + \Omega'_{xy}\Lambda_x)\Phi \quad , \tag{14}$$

$$\partial F/\partial \Phi = D'\Lambda_y \Omega_{yy}\Lambda_y D + \Lambda'_x \Omega_{xy}\Lambda_y D + D'\Lambda_y \Omega'_{xy}\Lambda_x + \Lambda'_x \Omega_{xx}\Lambda_x \quad , \tag{15}$$

$$\partial F/\partial \Psi = A'\Lambda_y \Omega_{yy}\Lambda_y A \quad , \tag{16}$$

$$\partial F/\partial \Theta_\varepsilon = \Omega_{yy} \quad , \tag{17}$$

$$\partial F/\partial \Theta_\delta = \Omega_{xx} \quad . \tag{18}$$

In these expressions the symmetry of Φ, Ψ, Θ_ε and Θ_δ has not been taken into account. Symmetry may be handled by equality constraints on the off-diagonal elements.

2.4. Information Matrix

The elements of S have an asymptotic multinormal distribution with mean matrix Σ . Let ε_{ij} be a typical element of $S - \Sigma$. Then, asymptotically

$$NE(\varepsilon_{gh}\varepsilon_{ij}) = \sigma_{gi}\sigma_{hj} + \sigma_{gj}\sigma_{hi} \, . \tag{19}$$

The asymptotic distribution of the elements of $\Omega = \Sigma^{-1}(\Sigma - S)\Sigma^{-1}$ is multivariate normal with means zero and variances and covariances given by

$$NE(\omega_{\alpha\beta}\omega_{\mu\nu}) = \sigma^{\alpha\mu}\sigma^{\beta\nu} + \sigma^{\alpha\nu}\,\sigma^{\beta\mu}.$$

This follows immediately by multiplying

$$\omega_{\alpha\beta} = \sum_{gh}\sum \sigma^{\alpha g}(\sigma_{gh} - s_{gh})\sigma^{h\beta} \quad \text{and} \quad \omega_{\mu\nu} = \sum_{ij}\sum \sigma^{\mu i}(\sigma_{ij} - s_{ij})\sigma^{j\nu}$$

and using (19).

The information matrix may be obtained by using the following lemma.

LEMMA. Let the element of Σ be functions of two parameter matrices $M = (\mu_{gh})$ and $N = (\nu_{ij})$ and let $F(M, N) = \frac{1}{2}[\log|\Sigma| + \mathrm{tr}(S\Sigma^{-1})]$ with $\partial F/\partial M = A\Omega B$ and $\partial F/\partial N = C\Omega D$, where A, B, C, D are independent of S. Then we have asymptotically

$$E(\partial^2 F/\partial\mu_{gh}\partial\nu_{ij}) = (\underset{\sim}{A}\underset{\sim}{\Sigma}^{-1}\underset{\sim}{C}')_{gi}(\underset{\sim}{B}'\underset{\sim}{\Sigma}^{-1}\underset{\sim}{D})_{hj} + (\underset{\sim}{A}\underset{\sim}{\Sigma}^{-1}\underset{\sim}{D})_{gj}(\underset{\sim}{B}'\underset{\sim}{\Sigma}^{-1}\underset{\sim}{C}')_{hi}.$$

Proof. Writing $\partial f/\partial\mu_{gh} = a_{g\alpha}\omega_{\alpha\beta}b_{\beta h}$ and $\partial F/\partial\nu_{ij} = c_{i\mu}\omega_{\mu\nu}d_{\nu j}$, where it is assumed that every repeated subscript is to be summed over, we have (cf. Kendall and Stuart, [21] Eq. 18.57)

$$E(\partial^2 F/\partial\mu_{gh}\partial\nu_{ij}) = NE(\partial F/\partial\mu_{gh}\partial F/\partial\nu_{ij}) = NE(a_{g\alpha}\omega_{\alpha\beta}b_{\beta h}c_{i\mu}\omega_{\mu\nu}d_{\nu j})$$

$$= Na_{g\alpha}b_{\beta h}c_{i\mu}d_{\nu j}E(\omega_{\alpha\beta}\omega_{\mu\nu}) = a_{g\alpha}b_{\beta h}c_{i\mu}d_{\nu j}\ (\sigma^{\alpha\mu}\sigma^{\beta\nu} + \sigma^{\alpha\nu}\sigma^{\beta\mu})$$

$$= (a_{g\alpha}\sigma^{\alpha\mu}c_{i\mu})\ (b_{\beta h}\sigma^{\beta\nu}d_{\nu j}) + (a_{g\alpha}\sigma^{\alpha\nu}d_{\nu j})\ (b_{\beta h}\sigma^{\beta\mu}c_{i\mu})$$

$$= (\underset{\sim}{A}\underset{\sim}{\Sigma}^{-1}\underset{\sim}{C}')_{gi}(\underset{\sim}{B}'\underset{\sim}{\Sigma}^{-1}\underset{\sim}{D})_{hj} + (\underset{\sim}{A}\underset{\sim}{\Sigma}^{-1}\underset{\sim}{D})_{gj}(\underset{\sim}{B}'\underset{\sim}{\Sigma}^{-1}\underset{\sim}{C}')_{hi}.$$

It should be noted that the lemma is quite general in that both $\underset{\sim}{M}$ and $\underset{\sim}{N}$ may be row or column vectors or scalars and $\underset{\sim}{M}$ and $\underset{\sim}{N}$ may be identical in which case, of course, $\underset{\sim}{A} \equiv \underset{\sim}{C}$ and $\underset{\sim}{B} \equiv \underset{\sim}{D}$.

The above lemma can be applied repeatedly to compute all the elements of the information matrix. For example, with $\underset{\sim}{\Omega}$ and $\underset{\sim}{\Lambda}$ as before and with

$$\underset{\sim}{T}'[m \times (m + n)] = [\underset{\sim}{A}' \quad \underset{\sim}{0}]$$

$$\underset{\sim}{P}'[m \times (m + n)] = [\underset{\sim}{C} \quad \underset{\sim}{D}\underset{\sim}{\Phi}] ,$$

$$\partial F/\partial\underset{\sim}{B} = -2\underset{\sim}{T}'\underset{\sim}{\Lambda}'\underset{\sim}{\Omega}\underset{\sim}{\Lambda}\underset{\sim}{P}.$$

2.5. *The Handling of Fixed and Constrained Parameters*

Let $\underset{\sim}{\mu}' = (\mu_1, \mu_2,..., \mu_k)$ be a vector of *all* elements in all eight parameter matrices $\underset{\sim}{\Lambda}_y$, $\underset{\sim}{\Lambda}_x$, $\underset{\sim}{B}$, $\underset{\sim}{\Gamma}$, $\underset{\sim}{\Phi}$, $\underset{\sim}{\Psi}$, $\underset{\sim}{\Theta}_\varepsilon$, and $\underset{\sim}{\Theta}_\delta$, and consider F as a function of $\underset{\sim}{\mu}$. This function is continuous and has continuous derivatives of first and second order except where $\underset{\sim}{\Sigma}$ is singular. The totality of these derivatives is represented by a gradient vector $\partial F/\partial\underset{\sim}{\mu}$ and a symmetric Hessian matrix $\partial^2 F/\partial\underset{\sim}{\mu}\partial\underset{\sim}{\mu}'$. Some $k - r$ of the μ's are fixed. Let the remaining r μ's form a vector $\underset{\sim}{\nu}' = (\nu_1, \nu_2,..., \nu_r)$. Derivatives $\partial F/\partial\underset{\sim}{\nu}$ and $\partial^2 F/\partial\underset{\sim}{\nu}\partial\underset{\sim}{\nu}'$ are obtained from $\partial F/\partial\underset{\sim}{\mu}$ and $\partial^2 F/\partial\underset{\sim}{\mu}\partial\underset{\sim}{\mu}'$ by elimination of the rows and columns corresponding to the fixed $\underset{\sim}{\mu}$'s. Among $\nu_1, \nu_2,..., \nu_r$ there are some distinct parameters $\theta_1, \theta_2,..., \theta_s$, assumed to be identifiable. Let $k_{ig} = 1$ if $\nu_i = \theta_g$ and $k_{ig} = 0$ otherwise and let $\underset{\sim}{K} = (k_{ig})$, $i = 1, 2,..., r$, $g = 1, 2,..., s$. Then we have

$$\partial F/\partial\underset{\sim}{\theta} = \underset{\sim}{K}'\ \partial F/\partial\underset{\sim}{\nu} \qquad \partial^2 F/\partial\underset{\sim}{\theta}\ \partial\underset{\sim}{\theta}' = \underset{\sim}{K}'\ \partial^2 F/\partial\underset{\sim}{\nu}\partial\underset{\sim}{\nu}'\underset{\sim}{K}$$

and

$$E(\partial^2 F/\partial\underset{\sim}{\theta}\partial\underset{\sim}{\theta}') = \underset{\sim}{K}'E(\partial^2 F/\partial\underset{\sim}{\nu}\ \partial\nu')\underset{\sim}{K} \tag{20}$$

The elements of the information matrix on the right-hand side of (20) are obtained as described in the previous section.

2.6. Basic Minimization Algorithm

The function $F(\theta)$ may be minimized numerically by Fisher's scoring method or the method of Fletcher and Powell [8] (see also Gruvaeus and Jöreskog [12]).

The minimization starts at an arbitrary starting point $\theta^{(1)}$ and generates successively new points $\theta^{(2)}$, $\theta^{(3)}, \ldots$, such that $F(\theta^{(s+1)}) < F(\theta^{(s)})$ until convergence is obtained. Let $g^{(s)}$ be the gradient vector $\partial F/\partial \theta$ at $\theta = \theta^{(s)}$ and let $E^{(s)}$ be the information matrix $E(\partial^2 F/\partial\theta\partial\theta')$ evaluated at $\theta = \theta^{(s)}$. Then Fisher's scoring method computes a correction vector by solving the equation system

$$E^{(s)}\delta^{(s)} = g^{(s)}$$

and then computes the new point as

$$\theta^{(s+1)} = \theta^{(s)} - \delta^{(s)}$$

This requires the computation of $E^{(s)}$ and the solution of a symmetric equation system in each iteration and this is often quite time consuming. An alternative is to use the method of Fletcher and Powell, which evaluates only the inverse of $E^{(1)}$ and in subsequent iterations E is improved, using information built up about the function, so that ultimately E converges to an approximation of the inverse of $\partial^2 F/\partial\theta\, \partial\theta'$ at the minimum.

A computer program LISREL based on the Fletcher and Powell method has been written by Jöreskog and Sörbom [20]*.

2.7. Tests of Hypotheses

When the maximum likelihood estimates of the parameters have been obtained, the goodness of fit of the model may be tested, in large samples, by the likelihood ratio technique. Let H_0 be the null hypothesis of the model under the given specifications of fixed, constrained, and free parameters. First consider the case when the alternative hypothesis H_1 is that Σ is any positive definite matrix. Then minus twice the logarithm of the likelihood ratio is $(N/2)F_0$ where F_0 is the minimum value of F. If the model holds, this is distributed, in large samples, as χ^2 with

$$d = \frac{1}{2}(p + q)\ (p + q + 1) - s$$

degrees of freedom, where, as before, s is the total number of independent

* See appendix.

parameters estimated under H_0.

Let H_0 be any specified hypothesis concerning the parametric structure of the general model which is more restrictive than an alternative hypothesis, H_1, say. In large samples one can then test H_0 against H_1. Let F_0 be the minimum of F under H_0 and let F_1 be the minimum of F under H_1. Then $F_1 < F_0$ and minus twice the logarithm of the likelihood ratio becomes $(N/2)(F_0 - F_1)$. Under H_0 this is distributed approximately as χ^2 with degrees of freedom equal to $s_1 - s_0$, the difference in number of independent parameters estimated under H_1 and H_0. In many situation, it is possible to set up a sequence of hypotheses such that each one is a special case of the preceding and to test these hypotheses sequentially.

In a more exploratory situation the χ^2-goodness-of-fit-values can be used as follows. If a value of χ^2 is obtained, which is large compared to the number of degrees of freedom, the fit may be examined by an inspection of the magnitudes of the first derivatives of F with respect to the fixed parameters. Often such an inspection or the results of analysis will suggest ways to relax the model somewhat by introducing more parameters. The new model usually yields a smaller χ^2. A large drop in χ^2, compared to the difference in degrees of freedom, indicates that the changes made in the model represent a real improvement. On the other hand, a drop in χ^2 close to the difference in number of degrees of freedom indicates that the improvement in fit is obtained by "capitalizing on chance", and the added parameters may not have real significance and meaning.

3. APPLICATIONS

3.1. Models for Longitudinal Data

The characteristic feature of a longitudinal research design is that the same measurements are obtained from the same people at two or more occasions. The purpose of a longitudinal or panel study is to assess the changes that occur between the occasions and to attribute these changes to certain background characteristics and events existing or occurring before the first occasion and/or to various "treatments" that occur after the first occasion. Wiley and Harnischfeger [26] have given an account of the conceptual issues in the attribution of change in educational studies. In the sociological literature there has been a number of articles concerned with the specification of models incorporating causation and measurement errors, and analysis of data from panel studies; see Bohrnstedt [2], Heise [14, 15] , Duncan [5, 6]. Jöreskog and Sörbom [19] have given an outline of statistical models and methods for longitudinal data.

An Example: The stability of alienation

As an example of a longitudinal study, we use some data analyzed in more detail by Wheaton, Muthen, Alwin and Summers [25]. This study was concerned with the stability over time of attitudes such as alienation and the relation to background variables such as education and occupation. Data on attitude scales were collected from 932 persons in two rural regions in Illinois at three points in time: 1966, 1967 and 1971. The variables we use for the present illustration are the *Anomia* subscale and the *Powerlessness* subscale, taken to be indicators of *Alienation*. We use these subscales from 1967 and 1971 only. The background variables are respondent's education (years of schooling completed) and Duncan's Socioeconomic Index (SEI). These are taken to be indicators of respondent's socioeconomic status (SES). We analyze these variables under two different models as shown in Figures 1 A - B, none of which correspond to Wheaton et al's final model. In these figures, observed variables are enclosed in rectangles whereas latent variables are enclosed in ellipses. Residuals (errors in equations) and errors of measurements are included in the diagram but are not enclosed. A one-way arrow pointing from one variable x to another variable y indicates a possible direct causal influence of x on y, whereas a curved two-way arrow between x and y indicates that x and y may correlate without any causal interpretation of this correlation being given. A coefficient associated with an arrow may also be included in the path diagram.

Specification

Let

y_1 = ANOMIA 67 $\quad$ y_3 = ANOMIA 71

y_2 = POWERLESSNESS 67 $\quad$ y_4 = POWERLESSNESS 71

x_1 = EDUCATION $\quad$ x_2 = SEI

ξ = SES $\quad$ η_1 = ALIENATION 67 $\quad$ η_2 = ALIENATION 71

The LISREL specification of model 1A is

$$\begin{pmatrix} y_1 \\ y_2 \\ y_3 \\ y_4 \end{pmatrix} = \begin{bmatrix} 1 & 0 \\ \lambda_1 & 0 \\ 0 & 1 \\ 0 & \lambda_2 \end{bmatrix} \begin{pmatrix} \eta_1 \\ \eta_2 \end{pmatrix} + \begin{pmatrix} \varepsilon_1 \\ \varepsilon_2 \\ \varepsilon_3 \\ \varepsilon_4 \end{pmatrix} \tag{21}$$

$$\begin{pmatrix} x_1 \\ x_2 \end{pmatrix} = \begin{pmatrix} 1 \\ \lambda_3 \end{pmatrix} \xi \begin{pmatrix} \delta_1 \\ \delta_2 \end{pmatrix} \tag{22}$$

$$\begin{pmatrix} 1 & 0 \\ \beta & 1 \end{pmatrix} \begin{pmatrix} \eta_1 \\ \eta_2 \end{pmatrix} = \begin{pmatrix} \gamma_1 \\ \gamma_2 \end{pmatrix} \xi + \begin{pmatrix} \zeta_1 \\ \zeta_2 \end{pmatrix} \tag{23}$$

FIG 1 A

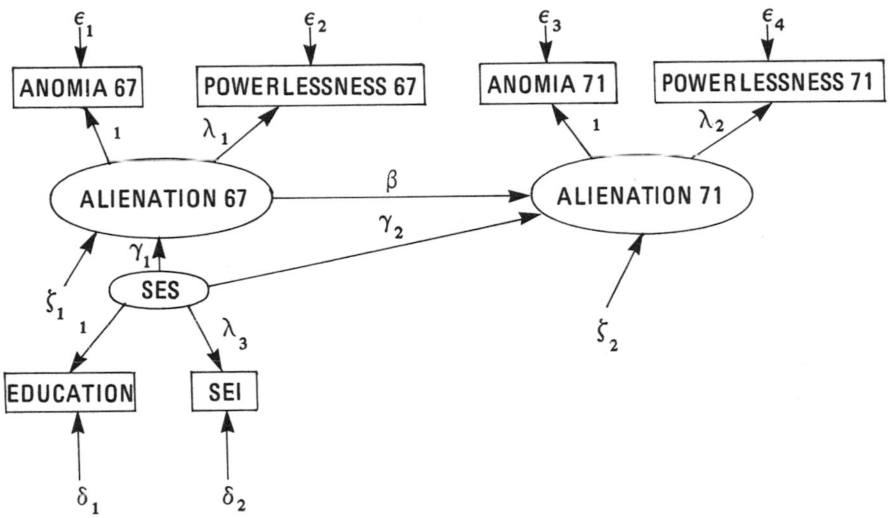

FIG 1 B

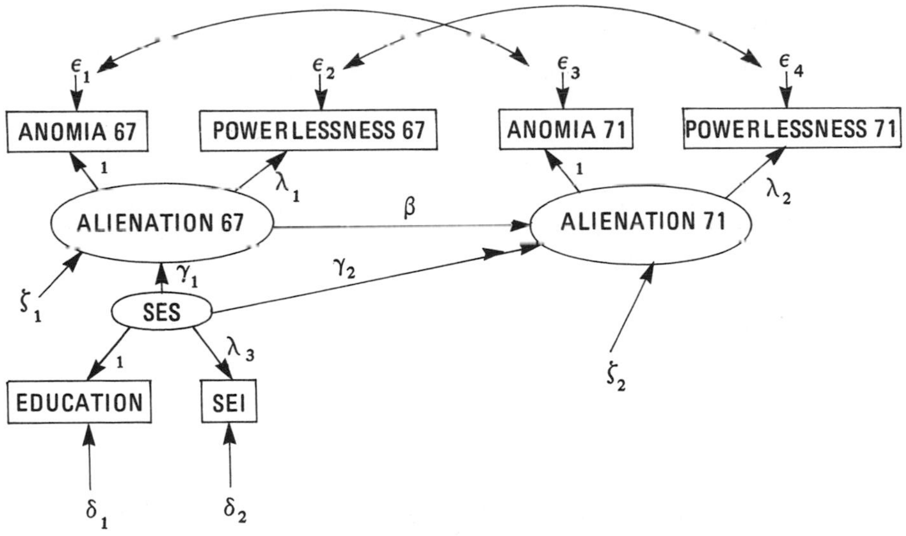

Fig. 1 A – B. Models for the study of Stability of Alienation.

with $\underset{\sim}{\Phi}(1 \times 1) = \phi = \text{Var}(\xi)$, $\underset{\sim}{\Psi}(2 \times 2) = \text{diag}\,(\psi_{11}, \psi_{22})$, where $\psi_{ii} = \text{Var}(\zeta_i)$ i = 1,2, and $\underset{\sim}{\Theta}_{\varepsilon}(4 \times 4)$ and $\underset{\sim}{\Theta}_{\delta}(2 \times 2)$ diagonal. Here we have assumed that the scales for η_1, η_2 and ξ have been chosen to be the same as for y_1, y_3 and x_1, respectively.

In model 1B the only difference is that the $\underset{\sim}{\Theta}_{\varepsilon}$ is not diagonal but

$$\underset{\sim}{\Theta}_{\varepsilon} = \begin{bmatrix} \theta_{11}^{(\varepsilon)} & & & \\ 0 & \theta_{22}^{(\varepsilon)} & & \\ \theta_{31}^{(\varepsilon)} & 0 & \theta_{33}^{(\varepsilon)} & \\ 0 & \theta_{42}^{(\varepsilon)} & 0 & \theta_{44}^{(\varepsilon)} \end{bmatrix},$$

with $\theta_{ii}^{(\varepsilon)} = \text{Var}(\varepsilon_i)$ and $\theta_{ij}^{(\varepsilon)} = \text{Cov}(\varepsilon_i, \varepsilon_j)$.

Identification

We have six observed variables with 21 variances and covariances. Model 1A has 15 parameters (3 λ's, 1 β, 2 γ's, 1 ϕ, 2 ψ's and 6 θ's) so that if all these are identified the model will have 6 degrees of freedom.

The reduced form of (23) is

$$\eta_1 = \gamma_1\xi + \zeta_1$$

$$\eta_2 = (\gamma_2 - \beta\,\gamma_1)\xi + (\zeta_2 - \beta\,\zeta_1)$$

$$= \pi\,\xi + \nu \quad \text{, say.}$$

We have

$$\text{Cov}(y_1, x_1) = \text{Cov}(\eta_1, x_1) = \gamma_1\phi \tag{24}$$

$$\text{Cov}(y_2, x_1) = \lambda_1\text{Cov}(\eta_1, x_1) = \lambda_1\gamma_1\phi \tag{25}$$

$$\text{Cov}(y_3, x_1) = \text{Cov}(\eta_2, x_1) = \pi\,\phi \tag{26}$$

$$\text{Cov}(y_4, x_1) = \lambda_2\text{Cov}(\eta_2, x_1) = \lambda_2\pi\,\phi\ . \tag{27}$$

If we use x_2 instead of x_1 in these equations all four right sides will be multiplied by λ_3. Hence λ_3 is overdetermined since

$$\lambda_3 = \text{Cov}(y_i, x_2) \,/\, \text{Cov}(y_i, x_1)\ ,\ i = 1, 2, 3, 4.$$

With λ_3 determined, ϕ is determined by

$$Cov(x_1, x_2) = \lambda_3 \phi \ .$$

With ϕ determined, equations (24) - (27) determine γ_1, λ_1, π and λ_2, respectively. Furthermore

$$Cov(y_1, y_2) = \lambda_1 \, Var(\eta_1) = \lambda_1(\gamma_1^2 \phi + \psi_{11}) \ ,$$

which determines ψ_{11}, and

$$Cov(y_3, y_4) = \lambda_2 \, Var(\eta_2) = \lambda_2[\pi^2 \phi + Var(\nu)]$$

which determines

$$Var(\nu) = \psi_{22} + \beta^2 \psi_{11} \ .$$

For given λ_1, λ_2, γ_1, π , ϕ and ψ_{11} the four equations

$$Cov(y_1, y_3) = \gamma_1 \pi \phi - \beta \psi_{11} \ , \qquad (28)$$

$$Cov(y_1, y_4) = \lambda_2(\gamma_1 \pi \phi - \beta \psi_{11}) \ , \qquad (29)$$

$$Cov(y_2, y_3) = \lambda_1(\gamma_1 \pi \phi - \beta \psi_{11}) \ , \qquad (30)$$

$$Cov(y_2, y_4) = \lambda_1 \lambda_2(\gamma_1 \pi \phi - \beta \psi_{11}) \ , \qquad (31)$$

show that β is overdetermined. Then, with β determined, $\gamma_2 = \pi + \beta \gamma_1$ and ψ_{22} are obtained. The errors variances $\theta_{ii}^{(\varepsilon)}$ are determined from $Var(y_i)$, $i = 1, 2, 3, 4$, and $\theta_{ii}^{(\delta)}$ from $Var(x_i)$, $i = 1, 2$. Hence it is clear that model 1A is identified and has six independent restrictions on Σ .

In model 1B we have two more parameters namely $\theta_{31}^{(\varepsilon)}$ and $\theta_{42}^{(\varepsilon)}$. These will be added to the right sides of (28) and (31), respectively. However, since (29) and/or (30) can be used to determine β it is clear that also $\theta_{31}^{(\varepsilon)}$ and $\theta_{42}^{(\varepsilon)}$ are determined by (28) and (31) respectively. Hence model 1B is also identified and has four degrees of freedom.

Estimation and testing

To estimate models 1A - B we use the data in TABLE I. The maximum likelihood estimates of the parameters with standard errors in parenthesis are given in Table II. The stability of alienation over time is reflected in the parameter β . The influence of SES on Alienation at the two occasions is significant in model 1A. The coefficient for 1967, γ_1, is -0.614 with a standard error of 0.056 and for 1971, γ_2, it is -0.174 with a standard error equal to 0.054. The negative sign of the γ-coefficients indicates that for high socioeconomic status the alienation is low

TABLE I

Covariance Matrix for the Model of Fig. 1 (n = 932)

y_1	11.834					
y_2	6.947	9.364				
y_3	6.819	5.091	12.532			
y_4	4.783	5.028	7.495	9.986		
x_1	- 3.839	- 3.889	- 3.841	- 3.625	9.610	
x_2	-21.899	-18.831	-21.748	-18.775	35.522	450.288

and vice versa. However, the overall fit of the model 1A is not acceptable; χ^2 with six degrees of freedom equals 71.544. Since the same scales are used at the two occasions, it seems plausible that errors may be correlated for the same variable over time, see e.g. Jöreskog and Sörbom [19]. Thus the model 1B is intuitively more plausible. As can be seen from TABLE II the inclusion of these correlations results in a model with an acceptable overall fit. A test of the hypothesis that both $\theta_{31}^{(\varepsilon)}$ and $\theta_{42}^{(\varepsilon)}$ are zero yields $\chi^2 = 66.774$ with 2 degrees of freedom so that this hypothesis must be rejected.

3.2. Models with Reciprocal Causation

The models in section 3.1 were such that the structural equation system was recursive, i.e. with $\underset{\sim}{B}$ lower triangular. Another characteristic feature was that the latent variables were acting as underlying causes of observed variables. In this section we consider a different type of model namely with reciprocal causation (simultaneity or interdependence). Another feature of this type of models is that latent variables may appear not only as underlying causes but also as effects of observed variables [18].

A model of Duncan, Haller and Portes

Sociologists have often called attention to the way in which one's peers -- e.g. best friends -- influence one's decisions -- e.g. choice of occupation. They have recognized that the relation must be reciprocal -- if my best friend influences my choice, I must influence his. Duncan, Haller and Portes [7] presented a simultaneous equation model of peer influences on occupational choice, using a sample of Michigan high-school students paired with their best friends. A striking result was that the direct path from respondent's choice to friend's choice turned out to be very similar to the direct path from friend's choice to respondent's choice. The authors interpret educational and occupational choice as two indicators of a single latent variable "ambition", and specify reciprocal causation between the ambitions rather than between the choices. This model with simultaneity and errors of measurement is displayed in Figure 2.

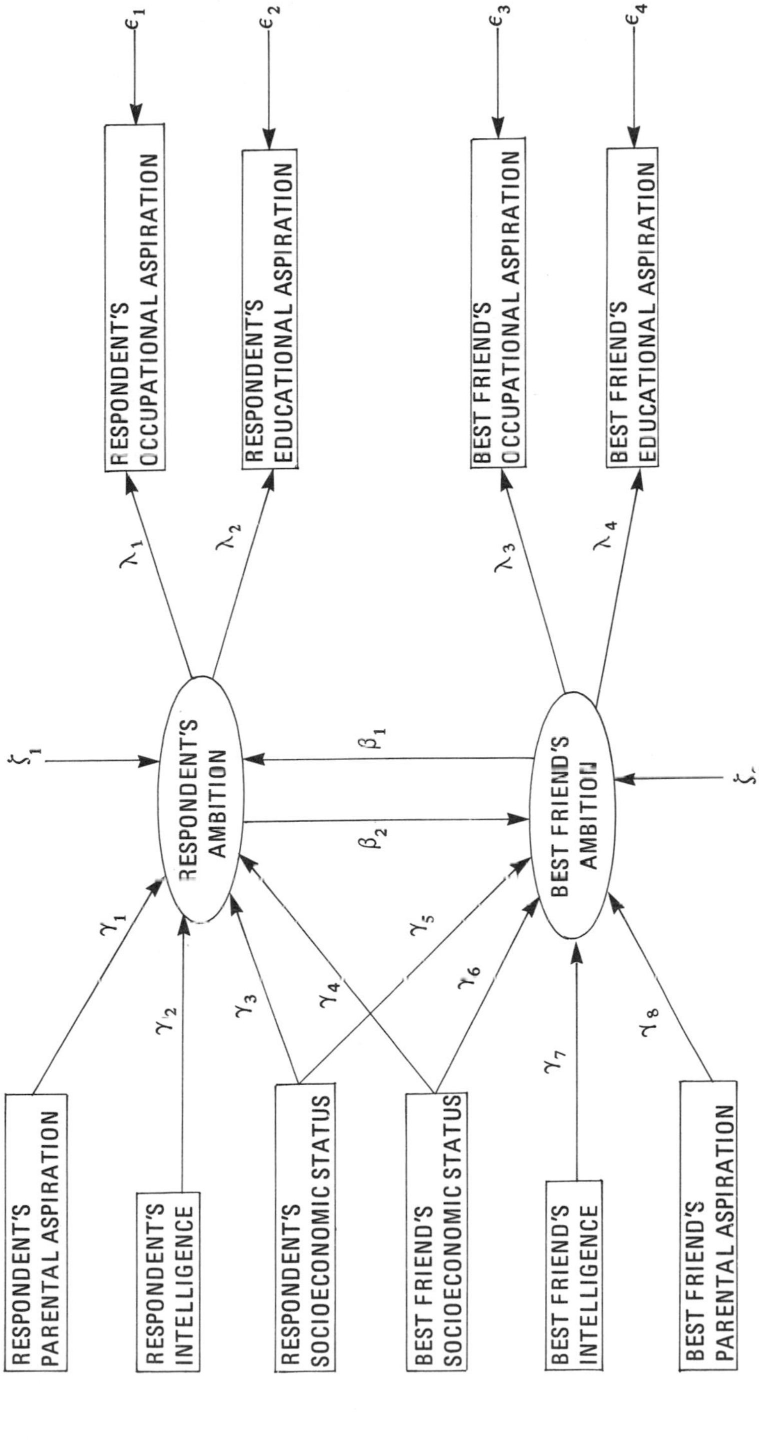

Fig. 2. A model of Duncan, Haller and Portes (1971).

TABLE II

Maximum Likelihood Estimates for the Models in Fig. 1A - B

The standard errors of the estimates are given within parenthesis.

Parameter	Model in Figure 1A	Model in Figure 1B
λ_1	0.888 (.041)	0.979 (.062)
λ_2	0.849 (.040)	0.922 (.059)
λ_3	5.331 (.430)	5.221 (.422)
β	0.705 (.054)	0.607 (.051)
γ_1	- 0.614 (.056)	- 0.575 (.056)
γ_2	- 0.174 (.054)	- 0.227 (.052)
ψ_{11}	5.307 (.473)	4.847 (.468)
ψ_{22}	3.742 (.388)	4.089 (.405)
ϕ	6.663 (.641)	6.803 (.650)
$\theta_{11}^{(\delta)}$	2.947 (.500)	2.807 (.508)
$\theta_{22}^{(\delta)}$	260.911 (18.241)	264.818 (18.155)
$\theta_{11}^{(\varepsilon)}$	4.014 (.343)	4.734 (.454)
$\theta_{22}^{(\varepsilon)}$	3.199 (.271)	2.566 (.404)
$\theta_{33}^{(\varepsilon)}$	3.699 (.373)	4.401 (.516)
$\theta_{44}^{(\varepsilon)}$	3.626 (.292)	3.075 (.435)
$\theta_{31}^{(\varepsilon)}$	- - -	1.624 (.314)
$\theta_{42}^{(\varepsilon)}$	- - -	0.340 (.261)
χ^2	71.544	4.770
d.f.	6	4

Specification

Let

x_1 = RESPONDENT'S PARENTAL ASPIRATION

x_2 = RESPONDENT'S INTELLIGENCE

x_3 = RESPONDENT'S SOCIOECONOMIC STATUS

x_4 = BEST FRIEND'S SOCIOECONOMIC STATUS

x_5 = BEST FRIEND'S INTELLIGENCE

x_6 = BEST FRIEND'S PARENTAL ASPIRATION

y_1 = RESPONDENT'S OCCUPATIONAL ASPIRATION

y_2 = RESPONDENT'S EDUCATIONAL ASPIRATION

y_3 = BEST FRIEND'S OCCUPATIONAL ASPIRATION

y_4 = BEST FRIEND'S EDUCATIONAL ASPIRATION

η_1 = RESPONDENT'S AMBITION

η_2 = BEST FRIEND'S AMBITION

In terms of our general model we take $\xi_i \equiv x_i$, i.e. in equation (3) we take $\Lambda_x(6 \times 6) = I$ and $\delta = 0$.

The structural equations are

$$\begin{pmatrix} 1 & \beta_1 \\ \beta_2 & 1 \end{pmatrix} \begin{pmatrix} \eta_1 \\ \eta_2 \end{pmatrix} = \begin{pmatrix} \gamma_1 & \gamma_2 & \gamma_3 & \gamma_4 & 0 & 0 \\ 0 & 0 & \gamma_5 & \gamma_6 & \gamma_7 & \gamma_8 \end{pmatrix} \begin{pmatrix} \xi_1 \\ \xi_2 \\ \xi_3 \\ \xi_4 \\ \xi_5 \\ \xi_6 \end{pmatrix} + \begin{pmatrix} \zeta_1 \\ \zeta_2 \end{pmatrix}, \tag{32}$$

and the equations relating the η's to the y's are

$$\begin{pmatrix} y_1 \\ y_2 \\ y_3 \\ y_4 \end{pmatrix} = \begin{bmatrix} 1 & 0 \\ \lambda_1 & 0 \\ 0 & 1 \\ 0 & \lambda_2 \end{bmatrix} \begin{pmatrix} \eta_1 \\ \eta_2 \end{pmatrix} + \begin{pmatrix} \varepsilon_1 \\ \varepsilon_2 \\ \varepsilon_3 \\ \varepsilon_4 \end{pmatrix} \tag{33}$$

In Λ_y we have fixed the scales for η_1 and η_2 to be the same as in y_1 and y_3, respectively. Since $\xi \equiv x$ and there are no constraints on Φ, $\Phi = \Sigma_{xx}$. The matrix $\Psi(2 \times 2)$ is

$$\underset{\sim}{\Psi} = \begin{bmatrix} \psi_{11} & \\ \psi_{21} & \psi_{22} \end{bmatrix}$$

with $\psi_{ii} = \mathrm{Var}(\zeta_i)$, i = 1, 2 and $\psi_{21} = \mathrm{Cov}(\zeta_1, \zeta_2)$ and the matrix $\underset{\sim}{\Theta}_\varepsilon$ is diagonal with diagonal elements $\theta_{ii}^{(\varepsilon)} = \mathrm{Var}(\varepsilon_i)$.

Since $\xi \equiv x$, the structural equations are equivalent to

$$\underset{\sim}{B}\underset{\sim}{\eta} = \underset{\sim}{\Gamma}\underset{\sim}{x} + \underset{\sim}{\zeta}$$

with reduced form

$$\underset{\sim}{\eta} = \underset{\sim}{B}^{-1}\underset{\sim}{\Gamma}\underset{\sim}{x} + \underset{\sim}{B}^{-1}\underset{\sim}{\zeta}. \tag{34}$$

Furthermore we have

$$\underset{\sim}{y} = \underset{\sim}{\Lambda}_y\underset{\sim}{\eta} + \underset{\sim}{\varepsilon} = \underset{\sim}{\Lambda}_y\underset{\sim}{B}^{-1}\underset{\sim}{\Gamma}\underset{\sim}{x} + \underset{\sim}{\Lambda}_y\underset{\sim}{B}^{-1}\underset{\sim}{\zeta} + \underset{\sim}{\varepsilon} = \underset{\sim}{\Pi}\underset{\sim}{x} + \underset{\sim}{z}\ , \tag{35}$$

where

$$\underset{\sim}{\Pi} = \underset{\sim}{\Sigma}_{yx}\underset{\sim}{\Sigma}_{xx}^{-1} = \underset{\sim}{\Lambda}_y\underset{\sim}{B}^{-1}\underset{\sim}{\Gamma}$$

and

$$\underset{\sim}{z} = \underset{\sim}{\Lambda}_y\underset{\sim}{B}^{-1}\underset{\sim}{\zeta} + \underset{\sim}{\varepsilon}$$

Identification

Since $\underset{\sim}{\Psi}$ is unconstrained, it follows from (34) that there is a one-to-one correspondence between $\underset{\sim}{\Psi}$ and the covariance matrix of $\underset{\sim}{\eta}$, $\underset{\sim}{\Omega}$ say, where

$$\underset{\sim}{\Omega} = \begin{bmatrix} \omega_{11} & \\ \omega_{21} & \omega_{22} \end{bmatrix}$$

From the first part of (35) we have

$$\underset{\sim}{\Sigma}_{yy} = \underset{\sim}{\Lambda}_y\underset{\sim}{\Omega}\underset{\sim}{\Lambda}_y' + \underset{\sim}{\Theta}_\varepsilon = \begin{bmatrix} \omega_{11} + \theta_{11}^{(\varepsilon)} & & & \\ \lambda_1\ \omega_{11} & \lambda_1^2\ \omega_{11} + \theta_{22}^{(\varepsilon)} & & \\ \omega_{21} & \lambda_1\ \omega_{21} & \omega_{22} + \theta_{33}^{(\varepsilon)} & \\ \lambda_2\ \omega_{21} & \lambda_1\ \lambda_2\omega_{21} & \lambda_2\omega_{22} & \lambda_2\omega_{22} + \theta_{44}^{(\varepsilon)} \end{bmatrix} \tag{36}$$

Furthermore we have

$$\underset{\sim}{B}^{-1} = (1 - \beta_1\beta_2)^{-1} \begin{pmatrix} 1 & -\beta_1 \\ -\beta_2 & 1 \end{pmatrix}$$

and

$$\underset{\sim}{B}^{-1}\underset{\sim}{\Gamma} = (1-\beta_1\beta_2)^{-1} \begin{pmatrix} \gamma_1 & \gamma_2 & \gamma_3-\beta_1\gamma_5 & \gamma_4-\beta_1\gamma_6 & -\beta_1\gamma_7 & -\beta_1\gamma_8 \\ -\beta_2\gamma_1 & -\beta_2\gamma_2 & \gamma_5-\beta_2\gamma_3 & \gamma_6-\beta_2\gamma_4 & \gamma_7 & \gamma_8 \end{pmatrix}$$

The first and third rows of $\underset{\sim}{\Pi}$ are identical to the first and second row of $\underset{\sim}{B}^{-1}\underset{\sim}{\Gamma} = \underset{\sim}{D}$, say, respectively. The second row of $\underset{\sim}{\Pi}$ is λ_1 times the first row of $\underset{\sim}{D}$ and the fourth row of $\underset{\sim}{\Pi}$ is λ_2 times the second row of $\underset{\sim}{D}$. Hence it is clear that λ_1 and λ_2 are identified and that

$$\lambda_1 = \frac{\pi_{2i}}{\pi_{1i}} \quad \text{and} \quad \lambda_2 = \frac{\pi_{4i}}{\pi_{3i}}, \quad i = 1, 2, \ldots, 6 .$$

Since D are two rows of $\underset{\sim}{\Pi}$, $\underset{\sim}{D}$ is identified. From $\underset{\sim}{D}$ it follows that γ_1, γ_2, γ_7, γ_8, β_1 and β_2 are determined as

$$\gamma_1 = d_{11}, \quad \gamma_2 = d_{12}, \; \gamma_7 = d_{25}, \; \gamma_8 = d_{26}$$

$$\beta_1 = -(d_{15}/d_{25}) = -(d_{16}/d_{26}) \quad ,$$

$$\beta_2 = -(d_{21}/d_{11}) = -(d_{22}/d_{12}) \; .$$

γ_3, γ_4, γ_5 and γ_6 are then determined by d_{13}, d_{14}, d_{23}, d_{24}. With λ_1 and λ_2 determined we can now determine ω_{11}, ω_{12} and ω_{22} from the off-diagonal elements of $\underset{\sim}{\Sigma}_{yy}$. Finally, the $\theta_{ii}^{(\varepsilon)}$, i = 1, 2, 3, 4, can be determined from the diagonal elements of $\underset{\sim}{\Sigma}_{yy}$.

This analysis shows that all parameters are identified. All together there are 40 parameters (2 β's, 8 γ's, 3 ω's, 2 λ's, 4 $\theta^{(\varepsilon)}$'s and 21 ϕ's). The degrees of freedom will be (1/2) 10 · 11 - 40 = 15.

Analysis of data

To illustrate the analysis we use the correlation matrix published by Duncan, Haller and Portes [7]. This is reproduced here in TABLE III. An overall goodness of fit test gives χ^2 = 26.70 with 15 degrees of freedom which is significant at the 5% level but not at the 2.5% level. A test of hypothesis $\psi_{21} = 0$ gives χ^2 = 0.19 with 1 degree of freedom and a test of $\beta_1 = \beta_2$, given $\psi_{21} = 0$, gives χ^2 = 0.06 with 1 degree of freedom. Hence it is clear that these hypotheses can-

TABLE III

Correlation Matrix for Duncan-Haller-Portes Model (N = 329)

y_1	1.000									
y_2	0.625	1.000								
y_3	0.327	0.367	1.000							
y_4	0.422	0.327	0.640	1.000						
x_1	0.214	0.274	0.112	0.084	1.000					
x_2	0.411	0.404	0.290	0.260	0.184	1.000				
x_3	0.324	0.405	0.305	0.279	0.049	0.222	1.000			
x_4	0.293	0.241	0.411	0.361	0.019	0.186	0.271	1.000		
x_5	0.300	0.286	0.519	0.501	0.078	0.336	0.230	0.295	1.000	
x_6	0.076	0.070	0.278	0.199	0.115	0.102	0.093	-0.044	0.209	1.000

not be rejected. An overall test of goodness of fit of the model with $\psi_{21} = 0$ and $\beta_1 = \beta_2$ gives $\chi^2 = 26.95$ with 17 degrees of freedom. This is not significant at the 5% level. The maximum likelihood estimates with their standard errors are given in TABLE IV. The standardized solution in which η_1 and η_2 are scaled to unit variance is also given in TABLE IV.

4. CONCLUSION

We have described a general approach to handle many models involving latent variables, measurement error and simultaneity. By allowing the parameters of the model to be fixed, free or constrained, great flexibility is obtained in that the general model contains a wide range of specific models useful in the social and behavioral sciences. Using the LISREL program, each model may be estimated by the maximum-likelihood method on the assumption that the observed variables have a multinormal distribution. When the estimates have been obtained, the information matrix may be computed and used to determine standard errors for the estimated parameters. The over-all fit of the model may be assessed by means of a large sample χ^2-test and specific parametric hypotheses may be tested by large sample χ^2-tests. The general approach has been illustrated by means of two examples, one, in which the latent variables are acting as underlying causes of observed variables and which includes a recursive structural equation system and one, in which the latent variables appear both as underlying causes and effects of observed variables and which also includes reciprocal causation.

TABLE IV

Maximum Likelihood Estimates for the Model in Fig. 2 with $\psi_{21} = 0$ and $\beta_1 = \beta_2$.

The standard errors of the estimates are given in parenthesis.

Parameter	Unscaled Solution	Standardized Solution
λ_1	1.000*	0.768
λ_2	1.059 (0.089)	0.813
λ_3	1.000*	0.827
λ_4	0.934 (0.070)	0.772
β_1	- 0.180 (0.039)	- 0.194
β_2	- 0.180 (0.039)	- 0.167
γ_1	0.163 (0.039)	0.213
γ_2	0.253 (0.042)	0.329
γ_3	0.219 (0.042)	0.285
γ_4	0.074 (0.042)	0.096
γ_5	0.077 (0.041)	0.093
γ_6	0.236 (0.041)	0.285
γ_7	0.356 (0.043)	0.431
γ_8	0.163 (0.039)	0.197
ψ_{11}	0.281 (0.046)	0.476
ψ_{22}	0.265 (0.045)	0.387
$\theta_{11}^{(\varepsilon)}$	0.411 (0.051)	0.411
$\theta_{22}^{(\varepsilon)}$	0.339 (0.051)	0.339
$\theta_{33}^{(\varepsilon)}$	0.314 (0.046)	0.314
$\theta_{44}^{(\varepsilon)}$	0.403 (0.046)	0.403

Asterisks denote parameter values fixed by scaling.

REFERENCES

[1] ANDREWS, D. F., GNANADESIKAN, R. and WARNER, J. L. (1973). Methods for assessing multivariate normality. *Multivariate Analysis - III*, (P. R. Krishnaiah, ed.). Academic Press, New York. 95-116.

[2] BOHRNSTEDT, G. W. (1969). Observations on the measurement of change. *Sociological Methodology*. (E. F. Borgatta, ed.). Jossey-Bass, San Francisco.

[3] DEVLIN, S. J., GNANADESIKAN, R. and KETTENRING, J. R. (1975). Robust estimation and outlier detection with correlation coefficients. *Biometrika*, 62, 531-545.

[4] DUNCAN, O. D. (1966). Path analysis: Sociological examples. *American Journal of Sociology*, 72, 1-16.

[5] DUNCAN, O. D. (1969). Some linear models for two-wave, two-variable panel analysis. *Psychological Bulletin*, 72, 177-182.

[6] DUNCAN, O. D. (1972). Unmeasured variables in linear models for panel analysis. *Sociological Methodology*. (H. L. Costner, ed.). Jossey-Bass, San Francisco.

[7] DUNCAN, O. D., HALLER, A. O. and PORTES, A. (1971). Peer influence on aspirations, a reinterpretation. *Causal Models in the Social Sciences*. (H. M. Blalock, Jr., ed.). Aldine-Atherton, Inc.

[8] FLETCHER, R. and POWELL, M. J. D. (1963). A rapidly convergent descent method for minimization. *Comput. J.*, 163-168.

[9] GNANADESIKAN, R. and KETTENRING, J. R. (1972). Robust estimates, residuals, and outlier detection with multiresponse data. *Biometrics*, 28, 81-124.

[10] GOLDBERGER, A. S. (1971). Econometrics and psychometrics: a survey of communalities. *Psychometrika*, 36, 83-107.

[11] GOLDBERGER, A. S. (1972). Structural equation methods in the social sciences. *Econometrica*, 40, 929-1001.

[12] GRUVAEUS, G. T. and JÖRESKOG, K. G. (1970). A computer program for minimizing a function of several variables. *Research Bulletin* 70-14. Educational Testing Service, Princeton, N. J.

[13] HAUSER, R. M. and GOLDBERGER, A. S. (1971). The treatment of unobservable variables in path analysis. *Sociological Methodology*. (H. S. Costner, ed.). Jossey-Bass, San Francisco. 81-117.

[14] HEISE, D. R. (1969). Separation reliability and stability in test-retest correlation. *American Sociological Review*, 34, 93-101.

[15] HEISE, D. R. (1970). Causal inference from panel data. *Sociological Methodology*. (E. F. Borgatta and G. W. Bohrnstedt, eds.). Jossey-Bass, San Francisco. 3-27.

[16] JÖRESKOG, K. G. (1969). A general approach to confirmatory maximum likelihood analysis. *Psychometrika*, 34, 183-202.

[17] JÖRESKOG, K. G. (1973). A general method for estimating a linear structural equation system. *Structural Equation Models in the Social Sciences*. (A. S. Goldberger and O. D. Duncan, eds.). Seminar Press, New York. 85-112.

[18] JÖRESKOG, K. G. and GOLDBERGER, A. S. (1975). Estimation of a model with multiple indicators and multiple causes of a single latent variable. *J. Amer. Statist. Assoc.*, 10, 631-639.

[19] JÖRESKOG, K. G. and SÖRBOM, D. (1977). Statistical models and methods for analysis of longitudinal data. *Latent variables in Socioeconomic models*. (D. J. Aigner and A. S. Goldberger, eds.). North-Holland Publishing Co., Amsterdam.

[20] JÖRESKOG, K. G. and SÖRBOM, D. (1976). LISREL III-Estimation of linear structural equation systems by maximum likelihood methods. International Educational Services, Chicago.

[21] KENDALL, M. G. and STUART, A. (1961). *The Advanced Theory of Statistics. Inference and Relationship.* Griffin, London. Vol. 2.

[22] TORGERSON, W. S. (1958). *Theory and Methods of Scaling.* Wiley, New York.

[23] WERTS, C. E. and LINN, R. L. (1970). Path Analysis: Psychological examples. *Psychological Bulletin*, **74**, 193-212.

[24] WERTS, C. E., JÖRESKOG, K. G. and LINN, R. L. (1973). Identification and estimation in path analysis with unmeasured variables. *American Journal of Sociology*, **78**, 1469-1484.

[25] WHEATON, B., MUTHÉN, B., ALWIN, D. and SUMMERS, G. (1977). Assessing reliability and stability in panel models with multiple indicators. *Sociological Methodology*, (D. R. Heise, ed.).

[26] WILEY, D. E. and HARNISCHFEGER, A. (1973). Post hoc, ergo propter hoc. Problems in the attribution of change. Studies of Educative Processes, Chicago, Report No. 7.

[27] WRIGHT, S. (1934). The method of path coefficients. *Ann. Math. Statist.*, **5**, 161-215.

[28] WRIGHT, S. (1960). Path coefficients and regressions: Alternative or complementary concepts? *Biometrics*, **16**, 189-202.

Reprinted from: P. R. Krishnaiah, *Applications of Statistics* (Amsterdam: No. Holland Publishing Co., 1977), pp. 265-287.

Chapter 5

Statistical Models and Methods for Analysis of Longitudinal Data

Karl G. Jöreskog

1. Introduction

In the psychometric literature various models for the analysis of longitudinal data have been proposed and discussed. For example, Jöreskog (1970a), Werts, Jöreskog and Linn (1972), Corballis (1973), Schmidt and Wiley (1974), and Frederiksen (1974) have considered complex models which involve multiple measurements at several occasions. In the recent sociological literature a number of articles treat the analysis of data from panel studies [e.g., Bohrnstedt (1969), Heise (1969, 1970), Wiley and Wiley (1970), Duncan (1969, 1972)]. In this chapter we consider some of these models and various others from the point of view of formulation and statistical specification, estimation and testing.

The general setup is that of a longitudinal study where the same or similar quantitative measurements have been obtained at two or more occasions. Our models and examples are relevant for psychological and educational studies involving latent traits as well as for social and socio-economic studies involving unobserved variables or hypothetical constructs. In both types of studies there are several features that need to be considered.

(i) Most measurements employed in the behavioral and social sciences contain sizeable errors of measurement and any adequate model for the measurement of change must take this fact into account.

(ii) It is often difficult to obtain repeated measurements using the same measuring instrument (test) because of retest effects. When the measurements at the different occasions are not in the same units, scaling devices may be used to obtain approximately equal units.

(iii) When there are several measurements employed at each occasion, the question arises as to what traits are actually measured at each

Research reported in this paper has been supported by the Swedish Council for Social Science Research under project 'Statistical Methods for Analysis of Longitudinal Data', project director Karl G. Jöreskog.

occasion and how these are related. Special care must be taken to obtain measurements that actually measure the relevant latent traits or hypothetical constructs.

(iv) It may not be possible, nor even desirable, to specify the model completely. There may be other models which are equally plausible and a technique for choosing the most reasonable one is required. Further, if there is sufficient evidence to reject a given model the technique should suggest which part of the model is causing the poor fit.

In presenting our models it is convenient to use a path diagram. In the path diagram observed variables are enclosed in squares whereas latent (unobserved) variables are enclosed in circles. Residuals (errors in equations) and errors of measurements are included in the diagram but are not enclosed. A one-way arrow pointing from one variable x to another variable y indicates a possible direct causal influence of x on y, whereas a curved two-way arrow between x and y indicates that x and y may correlate without any causal interpretation of this correlation being given. The coefficient associated with each arrow is also displayed in the path diagram. For one-way arrows such coefficients will be (partial) regression coefficients. For two-way arrows they will be covariances, or if all variables are standardized, they will be correlations. With these conventions it is possible to write down the model equations from the path diagram.

2. The LISREL approach

2.1. Specification

Before proceeding to discuss various models suitable for longitudinal data, we first consider the general structure of all the models and methods of this paper. The models to be considered may all be formulated, in one way or another, in the following fashion. There are two sets of observed variables $\mathbf{y} = (y_1, y_2, \ldots, y_p)'$ and $\mathbf{x} = (x_1, x_2, \ldots, x_q)'$, here taken to be measured as deviations from their means. In some longitudinal models these are the post- and pre-tests, respectively. In other cases it is more convenient to regard all the observed variables as belonging to $\mathbf{y}$, say, in which case the $\mathbf{x}$-set is empty and we shall say that there is no $\mathbf{x}$. It is assumed that the sets $\mathbf{y}$ and $\mathbf{x}$ satisfy factor analysis models with common factors $\boldsymbol{\eta} = (\eta_1, \eta_2, \ldots, \eta_m)'$ and $\boldsymbol{\xi} = (\xi_1, \xi_2, \ldots, \xi_n)'$ and unique factors $\boldsymbol{\varepsilon} = (\varepsilon_1, \varepsilon_2, \ldots, \varepsilon_p)'$ and $\boldsymbol{\delta} = (\delta_1, \delta_2, \ldots, \delta_q)'$, respectively, so that

$$y = \boldsymbol{\Lambda}_y \boldsymbol{\eta} + \boldsymbol{\varepsilon}, \tag{1}$$

and

$$x = \boldsymbol{\Lambda}_x \boldsymbol{\xi} + \boldsymbol{\delta}, \tag{2}$$

where $\boldsymbol{\Lambda}_y$ and $\boldsymbol{\Lambda}_x$ are factor loading matrices of order $p \times m$ and $q \times n$, respectively. The usual assumptions of factor analysis are made, i.e.,

$$\begin{array}{lllll} E(\boldsymbol{\eta}) = \boldsymbol{0}, & E(\boldsymbol{\xi}) = \boldsymbol{0}, & E(\boldsymbol{\varepsilon}) = \boldsymbol{0}, & E(\boldsymbol{\delta}) = \boldsymbol{0}, & \\ E(\boldsymbol{\eta}\boldsymbol{\varepsilon}') = \boldsymbol{0}, & E(\boldsymbol{\xi}\boldsymbol{\delta}') = \boldsymbol{0}, & E(\boldsymbol{\varepsilon}\boldsymbol{\varepsilon}') = \boldsymbol{\Theta}^2_{\varepsilon}, & E(\boldsymbol{\delta}\boldsymbol{\delta}') = \boldsymbol{\Theta}^2_{\delta}, & E(\boldsymbol{\varepsilon}\boldsymbol{\delta}') = \boldsymbol{0}, \end{array}$$

where $\boldsymbol{\Theta}_\varepsilon$ and $\boldsymbol{\Theta}_\delta$ are diagonal matrices. The factors $\boldsymbol{\eta}$ and $\boldsymbol{\xi}$ are in general correlated (oblique) not only within sets but also between sets. The distinctive feature of LISREL is that a complete system of causal relationships may be postulated to hold among the factors, where the η's are treated as the jointly dependent (endogenous) variables and the ξ's as the independent (exogenous) variables. This system is specified as a set of linear structural equations of the form

$$\boldsymbol{B}\boldsymbol{\eta} = \boldsymbol{\Gamma}\boldsymbol{\xi} + \boldsymbol{\zeta}, \tag{3}$$

where $\boldsymbol{B}$ and $\boldsymbol{\Gamma}$ are parameter matrices and $\boldsymbol{\zeta}$ is a vector of random residuals (disturbance terms or errors in equations) assumed to be uncorrelated with $\boldsymbol{\xi}$. In some models, $\boldsymbol{B} = \boldsymbol{I}$ and (3) may be regarded as the multivariate regression of $\boldsymbol{\eta}$ on $\boldsymbol{\xi}$. In other models when there is no $\boldsymbol{x}$, there is no $\boldsymbol{\xi}$ either, and (3) reduces to

$$\boldsymbol{B}\boldsymbol{\eta} = \boldsymbol{\zeta}. \tag{4}$$

Let $\boldsymbol{\Phi}$ be the covariance matrix of $\boldsymbol{\xi}$, and $\boldsymbol{\Psi}$ be the covariance matrix of $\boldsymbol{\zeta}$. Then the covariance matrix of the latent variables $(\boldsymbol{\eta}', \boldsymbol{\xi}')'$, namely

$$\boldsymbol{\Omega} = \begin{pmatrix} \boldsymbol{\Omega}_{\eta\eta} & \boldsymbol{\Omega}_{\eta\xi} \\ \boldsymbol{\Omega}_{\xi\eta} & \boldsymbol{\Omega}_{\xi\xi} \end{pmatrix},$$

is a function of the elements of $\boldsymbol{B}$, $\boldsymbol{\Gamma}$, $\boldsymbol{\Phi}$ and $\boldsymbol{\Psi}$:

$$\boldsymbol{\Omega}_{\eta\eta} = \boldsymbol{B}^{-1}\boldsymbol{\Gamma}\boldsymbol{\Phi}\boldsymbol{\Gamma}'\boldsymbol{B}'^{-1} + \boldsymbol{B}^{-1}\boldsymbol{\Psi}\boldsymbol{B}'^{-1}, \tag{5}$$

$$\boldsymbol{\Omega}_{\eta\xi} = \boldsymbol{\Omega}'_{\xi\eta} = \boldsymbol{B}^{-1}\boldsymbol{\Gamma}\boldsymbol{\Phi}, \tag{6}$$

$$\boldsymbol{\Omega}_{\xi\xi} = \boldsymbol{\Phi}. \tag{7}$$

When there is no $\boldsymbol{x}$, the covariance matrix $\boldsymbol{\Omega}$ of $\boldsymbol{\eta}$ reduces to

$$\boldsymbol{\Omega} = \boldsymbol{B}^{-1}\boldsymbol{\Psi}\boldsymbol{B}'^{-1}. \tag{8}$$

In the general case, the covariance matrix of the observed variables

$(\boldsymbol{y}', \boldsymbol{x}')'$ implied by the model is

$$\boldsymbol{\Sigma} = \begin{pmatrix} \boldsymbol{\Sigma}_{yy} & \boldsymbol{\Sigma}_{yx} \\ \boldsymbol{\Sigma}_{xy} & \boldsymbol{\Sigma}_{xx} \end{pmatrix},$$

where

$$\boldsymbol{\Sigma}_{yy} = \boldsymbol{\Lambda}_y \boldsymbol{\Omega}_{\eta\eta} \boldsymbol{\Lambda}'_y + \boldsymbol{\Theta}^2_{\varepsilon}, \tag{9}$$

$$\boldsymbol{\Sigma}_{yx} = \boldsymbol{\Sigma}'_{xy} = \boldsymbol{\Lambda}_y \boldsymbol{\Omega}_{\eta\xi} \boldsymbol{\Lambda}'_x, \tag{10}$$

$$\boldsymbol{\Sigma}_{xx} = \boldsymbol{\Lambda}_x \boldsymbol{\Omega}_{\xi\xi} \boldsymbol{\Lambda}'_x + \boldsymbol{\Theta}^2_{\delta}. \tag{11}$$

A model of the form of (1), (2) and (3), with the accompanying assumptions, has been called a LISREL model [Jöreskog and van Thillo (1973)]. The LISREL model consists of two parts, the *measurement model* represented by (1) and (2) and the *structural equation model* represented by (3). The measurement model specifies how the latent variables or hypothetical constructs are measured in terms of the observed variables. Its parameters $\boldsymbol{\Lambda}_x$, $\boldsymbol{\Lambda}_y$, $\boldsymbol{\Theta}_\delta$, $\boldsymbol{\Theta}_\varepsilon$ are used to describe the measurement properties (validities and reliabilities) of the observed variables. The structural equation model specifies the causal relationships among the latent variables. Its parameters $\boldsymbol{B}$, $\boldsymbol{\Gamma}$ and $\boldsymbol{\Psi}$ describe the causal effects and the amount of unexplained variance. A particular model is defined by specifying the nature of the elements of each of the parameter matrices $\boldsymbol{\Lambda}_y$ $(p \times m)$, $\boldsymbol{\Lambda}_x$ $(q \times n)$, $\boldsymbol{B}$ $(m \times m)$, $\boldsymbol{\Gamma}$ $(m \times n)$, $\boldsymbol{\Phi}$ $(n \times n)$, $\boldsymbol{\Psi}$ $(m \times m)$, $\boldsymbol{\Theta}_\varepsilon$ $(p \times p)$, and $\boldsymbol{\Theta}_\delta$ $(q \times q)$. Such an element may be specified as a

(i) fixed parameter, i.e., specified to have a given value,
(ii) free parameter, i.e., unknown to be estimated from data, or
(iii) constrained parameter, i.e., unknown but specified to be equal to one or more other parameters.

In the above presentation of the LISREL model it has been assumed that $\operatorname{cov}(\delta_i, \delta_j) = 0$, $i \neq j$, $\operatorname{cov}(\varepsilon_i, \varepsilon_j) = 0$, $i \neq j$, and $\operatorname{cov}(\delta_i, \varepsilon_j) = 0$ for all i, j. These assumptions are not critical, however, since one can incorporate the δ's and ε's as factors along with the ξ's and η's, respectively. For example, to allow for correlated δ's one simply writes (2) as

$$\boldsymbol{x} = (\boldsymbol{\Lambda}_x \, \boldsymbol{I}_q) \begin{pmatrix} \boldsymbol{\xi} \\ \boldsymbol{\delta} \end{pmatrix} + \boldsymbol{0},$$

where $\boldsymbol{I}_q$ is the identity matrix of order q. Then $\boldsymbol{\Theta}_\delta = \boldsymbol{0}$ and

$$\boldsymbol{\Phi} = \begin{pmatrix} \boldsymbol{\Phi}_{\xi\xi} & \boldsymbol{0} \\ \boldsymbol{0} & \boldsymbol{\Phi}_{\delta\delta} \end{pmatrix},$$

where $\boldsymbol{\Phi}_{\xi\xi}$ is the covariance matrix of $\boldsymbol{\xi}$, and $\boldsymbol{\Phi}_{\delta\delta}$ is the covariance matrix of $\boldsymbol{\delta}$. A similar trick can be used to let the ε's be correlated. To allow the ε's to correlate with the δ's one simply uses the no-x option and treats all observed variables as y variables. Examples of such devices will be given later on in the chapter.

It should be noted that there is no requirement that $m < p$ and $n < q$ as in traditional factor analysis. The only requirement is that $\boldsymbol{\Sigma}$ is nonsingular and that all parameters are identified.

2.2. Identification

We assume that the distribution of the observed variables is sufficiently well described by the moments of first and second order, so that information contained in moments of higher order may be ignored. In particular, this will hold if the distribution is multivariate normal. The distribution of $(\boldsymbol{y}', \boldsymbol{x}')'$ is therefore generated by the parameters in $\boldsymbol{\Lambda}_y$, $\boldsymbol{\Lambda}_x$, $\boldsymbol{B}$, $\boldsymbol{\Gamma}$, $\boldsymbol{\Phi}$, $\boldsymbol{\Psi}$, $\boldsymbol{\Theta}_\varepsilon$, $\boldsymbol{\Theta}_\delta$. Let $\boldsymbol{\theta}$ be a vector of all the independent free and constrained parameters (i.e., counting each distinct constrained parameter only once) and let s be the order of $\boldsymbol{\theta}$. The identification problem then is the problem of whether or not $\boldsymbol{\theta}$ is uniquely determined by $\boldsymbol{\Sigma}$. Every $\boldsymbol{\theta}$ generates one and only one $\boldsymbol{\Sigma}$ but two or more $\boldsymbol{\theta}$'s could possibly generate the same $\boldsymbol{\Sigma}$. If within the model there is only one $\boldsymbol{\theta}$ for every $\boldsymbol{\Sigma}$ then $\boldsymbol{\theta}$ is identified and we say that the whole model is identified. If, on the other hand, there are several $\boldsymbol{\theta}$'s generating the same $\boldsymbol{\Sigma}$ we say that all such $\boldsymbol{\theta}$'s are equivalent. If a parameter in $\boldsymbol{\theta}$ has the same value in all equivalent $\boldsymbol{\theta}$'s we say that this parameter is identified. If a parameter is not identified it will not be possible to find a consistent estimator of it.

To examine the identification of a model consider (9)–(11) after substitution of (5)–(7),

$$\sigma_{ij} = f_{ij}(\boldsymbol{\theta}). \tag{12}$$

There are $\frac{1}{2}(p+q)(p+q+1)$ equations and s unknown elements in $\boldsymbol{\theta}$. Hence a necessary condition for identification of all parameters is

$$s \leq \tfrac{1}{2}(p+q)(p+q+1).$$

If a parameter θ can be determined uniquely from $\boldsymbol{\Sigma}$, this parameter is identified, otherwise it is not. Often some parameters can be determined from $\boldsymbol{\Sigma}$ in different ways. This gives rise to overidentifying conditions on $\boldsymbol{\Sigma}$ which must hold if the model is true. The solution of (12) is often complicated and explicit solutions for all θ's seldom exist. For certain special types of LISREL models general rules for the identification problem have been given by Fisher (1966), Wiley (1973), Geraci (1974), and Werts, Jöreskog and Linn (1973). For most of our models we will consider the identification problem on a case-by-case basis.

2.3. Estimation and testing

Since we have assumed that the distribution of the observed variables is described by the covariance matrix, the estimation problem is essentially that of fitting the $\boldsymbol{\Sigma}$ imposed by the model to the sample covariance matrix $\boldsymbol{S}$. As a fitting function we use

$$F = \log|\boldsymbol{\Sigma}| + \mathrm{tr}\,(\boldsymbol{S}\boldsymbol{\Sigma}^{-1}) - \log|\boldsymbol{S}| - (p+q), \tag{13}$$

which is to be minimized with respect to $\boldsymbol{\theta}$. If the distribution of $(\boldsymbol{y}', \boldsymbol{x}')'$ is multinormal this yields maximum-likelihood estimates which are efficient in large samples. A computer program LISREL-II [Jöreskog and van Thillo (1973)] is available for computing the estimates of the LISREL parameters and their standard errors.

When the maximum-likelihood estimates have been obtained the computer program calculates the information matrix at the estimates. If this matrix is not positive definite the specified model is almost certainly not identified. (This is one way in which the identification problem can be handled if it has not been solved by other means.) For an identified model the information matrix will almost certainly be positive definite and can be inverted to give the standard errors.

The program also gives a χ^2-measure of overall goodness-of-fit of the model, which may be regarded as a test of the specified model against the most general alternative that $\boldsymbol{\Sigma}$ is an unconstrained positive definite matrix. The degrees of freedom for this χ^2-measure are

$$\tfrac{1}{2}(p+q)(p+q+1) - s, \tag{14}$$

where s is the number of identified parameters in the model.

Suppose H_0 represents one model under given specifications of fixed, free, and constrained parameters. To test the model H_0 against any more general model H_1, estimate them separately and compare their χ^2. The difference in χ^2 is asymptotically a χ^2 with degrees of freedom equal to the corresponding difference in degrees of freedom. In many situations, it is possible to set up a sequence of hypotheses such that each one is a special case of the preceding and to test these hypotheses sequentially. We will illustrate this procedure by means of several examples.

In a more exploratory situation the χ^2-goodness-of-fit-values can be used as follows. If a value of χ^2 is obtained which is large compared to the number of degrees of freedom, the fit may be examined by an inspection of the residuals, i.e., the discrepancies between the observed and the reproduced variances and covariances. That examination, in conjunction with subject-matter considerations, may suggest ways to relax the model somewhat by introducing more parameters. The new model usually yields a smaller χ^2. A large drop in χ^2, compared to the difference in degrees of freedom, supports the changes made. On the other hand, a drop in χ^2 which is close to the difference in number of degrees of freedom indicates that the improvement in fit is obtained by capitalizing on chance.

3. Two-wave models

3.1. Two-wave, two-variable models

Consider the situation where we have two observed variables at each of two occasions. Such a two-wave, two-variable model is shown in Figure 1a, where y_1 and y_2 are measured at the first occasion and y_3 and y_4 at the second occasion. To fix the scales for the latent variables η_1 and η_2 we assume that these are measured in the same metric as y_1 and y_3, respectively. The measurement model is

$$\begin{aligned} y_1 &= \eta_1 + \varepsilon_1, \qquad & y_2 &= \lambda_1\eta_1 + \varepsilon_2, \\ y_3 &= \eta_2 + \varepsilon_3, \qquad & y_4 &= \lambda_2\eta_2 + \varepsilon_4. \end{aligned} \tag{15}$$

The implied covariance matrix of $\mathbf{y} = (y_1, y_2, y_3, y_4)'$ is

$$\mathbf{\Sigma} = \begin{bmatrix} \omega_{11} + \theta_{11}^{(\varepsilon)} & & & \\ \lambda_1\omega_{11} & \lambda_1^2\omega_{11} + \theta_{22}^{(\varepsilon)} & & \\ \omega_{21} & \lambda_1\omega_{21} & \omega_{22} + \theta_{33}^{(\varepsilon)} & \\ \lambda_2\omega_{21} & \lambda_1\lambda_2\omega_{21} & \lambda_2\omega_{22} & \lambda_2^2\omega_{22} + \theta_{44}^{(\varepsilon)} \end{bmatrix}, \tag{16}$$

where

$$\mathbf{\Omega} = \begin{pmatrix} \omega_{11} & \omega_{12} \\ \omega_{21} & \omega_{22} \end{pmatrix}$$

is the covariance matrix of η_1 and η_2. The matrix $\mathbf{\Sigma}$ has 10 distinct elements which are expressed in terms of 9 parameters. It is clear that $\omega_{21} = \sigma_{31}$ is identified. With ω_{21} determined, λ_1 is determined from σ_{32} and

λ_2 from σ_{41}. The constraint $\sigma_{42} = \lambda_1\lambda_2\omega_{21}$ represents one overidentifying restriction. The variances ω_{11} and ω_{22} are determined from σ_{21} and σ_{43}, respectively. Finally, $\theta_{ii}^{(\varepsilon)}$ are determined from σ_{ii} $(i = 1, \ldots, 4)$. Hence this model is identified with one overidentifying restriction.

When the same measuring instrument (test or questionnaire) is used at both occasions there is usually a tendency for the errors in each variable to correlate over time because of memory or other re-test effects. In this situation the model should allow for correlations between ε_1 and ε_3 and also between ε_2 and ε_4. Such a model is illustrated in Figure 1b. Duncan (1972) and Kenny (1973) have previously considered models of this type. The covariance matrix $\boldsymbol{\Sigma}$ for the model in Figure 1b is the same as (16) except that $\theta_{31}^{(\varepsilon)} = \text{cov}(\varepsilon_1, \varepsilon_3)$ and $\theta_{42}^{(\varepsilon)} = \text{cov}(\varepsilon_2, \varepsilon_4)$ are added to σ_{31} and σ_{42}, respectively. There are now 11 unknown parameters but only 10 observable moments. Hence the model is not identified.

Indeed none of the 11 parameters are identified without additional conditions. The loadings λ_1 and λ_2 may be multiplied by a constant and the ω's divided by the same constant, without changing σ_{21}, σ_{32}, σ_{41}, and σ_{43}. The changes in the other σ's may be compensated by adjusting the θ's additively. To make the model identified one must fix one λ or one ω at a non-zero value, or one θ at some arbitrary value. There does not seem to be any reasonable way to do so, unless one is willing to make some further assumptions about the nature of the variables. For example, if the variables y_1 and y_2 are tau-equivalent [see, e.g., Lord and Novick (1968, p.

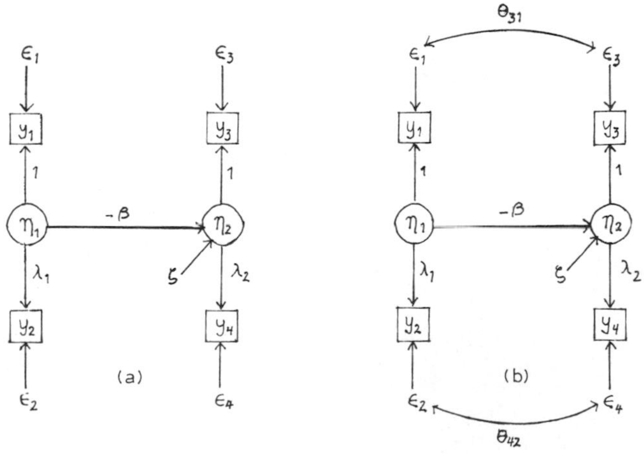

FIGURE 1. Two-wave, two-variable model with (a) uncorrelated and (b) correlated errors.

47)] we set $\lambda_1 = 1$. The model is then just-identified and will fit most data exactly. However, as pointed out by Duncan (1972) and Kenny (1973), the *correlation* between η_1 and η_2 is identified without any restriction, because

$$\rho^2_{\eta_1\eta_2} = \omega^2_{21}/\omega_{11}\omega_{22} = (\sigma_{41}\sigma_{32})/(\sigma_{21}\sigma_{43}).$$

The model may be used to estimate this correlation and to test whether it is unity. When the model is just-identified, the maximum-likelihood estimate of $\rho^2_{\eta_1\eta_2}$ is

$$\hat{\rho}^2_{\eta_1\eta_2} = (s_{41}s_{32})/(s_{21}s_{43}),$$

where s_{ij} is the sample covariance corresponding to σ_{ij}.

3.2. Background variables

In many longitudinal studies the objective is to measure change between two occasions and to relate the change to certain characteristics and events. Such studies include not only pre- and post-measures but also various background variables believed to influence change. The background variables are socio-economic or other characteristics which differentiate the individuals prior to the pre-test occasion.

Consider the model shown in Figure 2. The background variable, denoted x, is the only independent (exogenous) variable. The main purpose of the model is to estimate the direct effect of η_1 on η_2 by eliminating the effect of x. We shall consider this model under two alternative conditions on the background variable, namely,

(i) x is measured without error,
(ii) x is fallible.

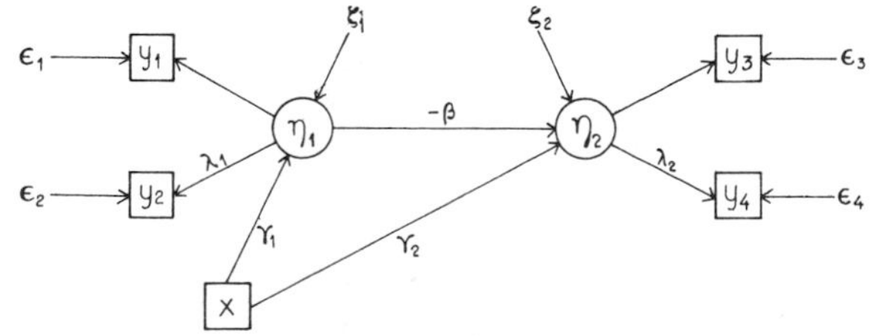

FIGURE 2. Two pre-tests, two post-tests, and an infallible background variable.

3.3. *Accurate background variable*

First suppose x is measured without error. The measurement model for $\mathbf{y}' = (y_1, y_2, y_3, y_4)$ is

$$\begin{pmatrix} y_1 \\ y_2 \\ y_3 \\ y_4 \end{pmatrix} = \begin{bmatrix} 1 & 0 \\ \lambda_1 & 0 \\ 0 & 1 \\ 0 & \lambda_2 \end{bmatrix} \begin{pmatrix} \eta_1 \\ \eta_2 \end{pmatrix} + \begin{pmatrix} \varepsilon_1 \\ \varepsilon_2 \\ \varepsilon_3 \\ \varepsilon_4 \end{pmatrix}, \tag{17}$$

and the structural model for η_1, η_2, and x is

$$\begin{pmatrix} 1 & 0 \\ \beta & 1 \end{pmatrix} \begin{pmatrix} \eta_1 \\ \eta_2 \end{pmatrix} = \begin{pmatrix} \gamma_1 \\ \gamma_2 \end{pmatrix} x + \begin{pmatrix} \zeta_1 \\ \zeta_2 \end{pmatrix}. \tag{18}$$

Solving (18) for η_1 and η_2 we get

$$\begin{aligned} \eta_1 &= \gamma_1 x + \zeta_1, \\ \eta_2 &= (\gamma_2 - \beta\gamma_1)x + (\zeta_2 - \beta\zeta_1) \\ &= \pi x + \upsilon, \end{aligned}$$

say.

Let us first consider the identification problem. We have five observed variables y_1, y_2, y_3, y_4, and x, so that we have fifteen variances and covariances in $\mathbf{\Sigma}$. We have twelve unknown parameters: λ_1, λ_2, β, γ_1, γ_2, $\phi = \text{var}(x)$, $\psi_{11} = \text{var}(\zeta_1)$, $\psi_{22} = \text{var}(\zeta_2)$, and $\theta^2_{\varepsilon_i} = \text{var}(\varepsilon_i)$, $i = 1, 2, 3, 4$. We have

$$\text{cov}(y_1, x) = \text{cov}(\eta_1, x) = \gamma_1\phi, \tag{19}$$

$$\text{cov}(y_2, x) = \lambda_1 \text{cov}(\eta_1, x) = \lambda_1\gamma_1\phi, \tag{20}$$

$$\text{cov}(y_3, x) = \text{cov}(\eta_2, x) = \pi\phi, \tag{21}$$

$$\text{cov}(y_4, x) = \lambda_2 \text{cov}(\eta_2, x) = \lambda_2\pi\phi. \tag{22}$$

Since $\phi = \text{var}(x)$ is identified, these equations determine γ_1, λ_1, π and λ_2, respectively. Furthermore,

$$\text{cov}(y_1, y_2) = \lambda_1 \text{var}(\eta_1) = \lambda_1(\gamma_1^2\phi + \psi_{11}) \tag{23}$$

determines ψ_{11}, and

$$\text{cov}(y_3, y_4) = \lambda_2 \text{var}(\eta_2) = \lambda_2[\pi^2\phi + \text{var}(\upsilon)] \tag{24}$$

determines

$$\text{var}(\upsilon) = \psi_{22} + \beta^2\psi_{11}. \tag{25}$$

With λ_1, λ_2, $\gamma_1\pi$, ϕ and ψ_{11} in hand, any one of the four equations

$$\mathrm{cov}\,(y_1, y_3) = \gamma_1\pi\phi - \beta\psi_{11}, \tag{26}$$
$$\mathrm{cov}\,(y_1, y_4) = \lambda_2(\gamma_1\pi\phi - \beta\psi_{11}), \tag{27}$$
$$\mathrm{cov}\,(y_2, y_3) = \lambda_1(\gamma_1\pi\phi - \beta\psi_{11}), \tag{28}$$
$$\mathrm{cov}\,(y_2, y_4) = \lambda_1\lambda_2(\gamma_1\pi\phi - \beta\psi_{11}), \tag{29}$$

can be used to determine β. Thus there are three overidentifying restrictions. Then, with β determined, $\gamma_2 = \pi + \beta\gamma_1$ and ψ_{22} is obtained from (25). Finally the error variances $\theta^2_{\varepsilon_i}$ are determined from var (y_i), $i = 1, 2, 3, 4$. Hence the whole model is identified and there are three independent restrictions on $\boldsymbol{\Sigma}$.

3.4. Fallible background variable

Now suppose x is fallible,

$$x = \xi + \delta,$$

where ξ is the true score and δ the measurement error, the latter assumed to have zero mean and to be uncorrelated with ξ and all other unobserved variables. We shall consider two cases, namely, (a) x has a known reliability,

$$\rho_{xx} = \mathrm{var}\,(\xi)/\mathrm{var}\,(x),$$

and (b) ξ is measured by two congeneric background variables x_1 and x_2.

The first case is shown in Figure 3 and is similar to the model just analyzed. In (19) through (29) everything is the same except that var (x) is replaced by var (ξ). Since var $(\xi) = \rho_{xx}$ var (x), where ρ_{xx} is known and var (x) is identified, all the other parameters will be determined as before.

An extension of the model in Figure 3 to include correlations between ε_1 and ε_3 and between ε_2 and ε_4 is shown in Figure 4. Then cov $(\varepsilon_1, \varepsilon_3)$ will be added to the right side of (26) and cov $(\varepsilon_2, \varepsilon_4)$ will be added to the right side of (29). For this model, λ_1, λ_2, γ_1, π and ϕ are determined as before. Then (27) and (28) determine β with one overidentifying restriction, and cov $(\varepsilon_1, \varepsilon_3)$ and cov $(\varepsilon_2, \varepsilon_4)$ are then uniquely determined by (26) and (29), respectively. Hence this model is identified and there is one restriction on $\boldsymbol{\Sigma}$.

The case of two congeneric background variables is shown in Figure 5.

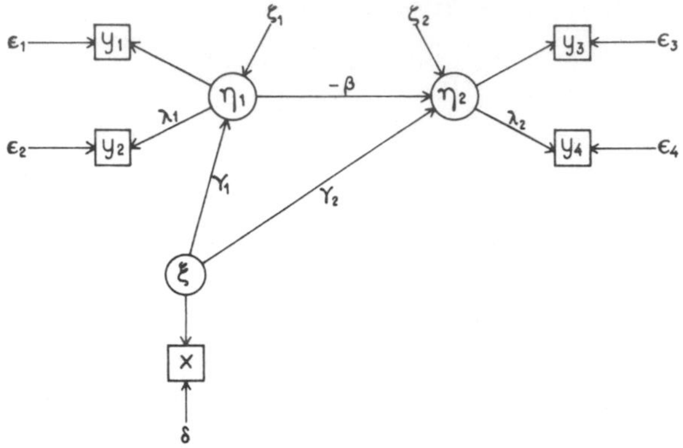

FIGURE 3. Two pre-tests, two post-tests, and a fallible background variable.

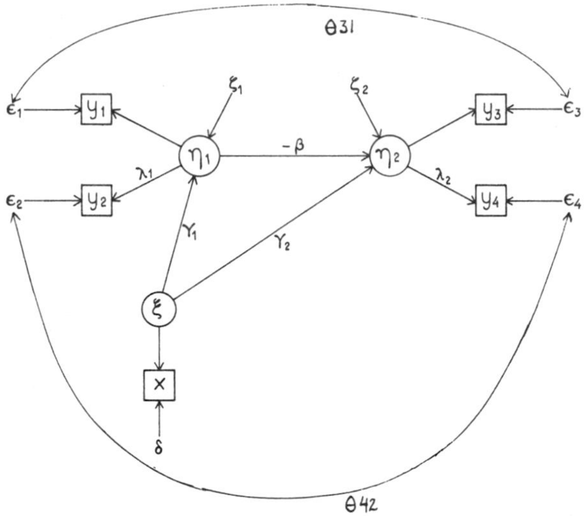

FIGURE 4. Two pre-tests, two post-tests, a fallible background variable, and correlated errors.

Here we write

$$x_1 = \xi + \delta_1, \tag{30}$$

$$x_2 = \lambda_3 \xi + \delta_2, \tag{31}$$

where λ_3 is a parameter to be determined, and δ_1 and δ_2 are uncorrelated measurement errors, uncorrelated with ξ and all other latent variables.

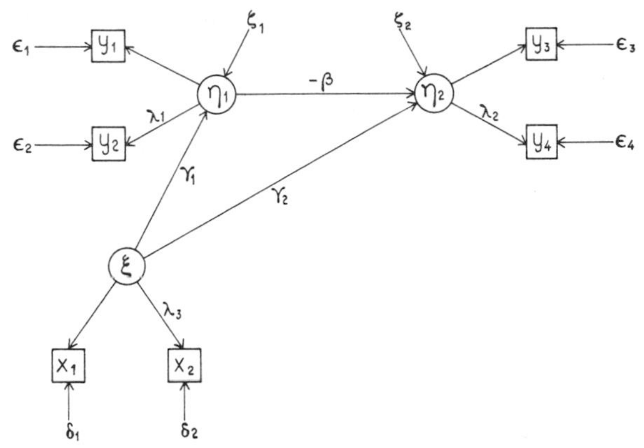

FIGURE 5. Two pre-tests, two post-tests, and two congeneric background variables.

The other equations are as before except that in (18) x is replaced by ξ. We now have three more parameters than before, namely, λ_3, $\theta^2_{\delta_1} = \text{var}(\delta_1)$, and $\theta^2_{\delta_2} = \text{var}(\delta_2)$. The new parameter $\phi = \text{var}(\xi)$ replaces the old $\phi = \text{var}(x)$. On the other hand we now have six more observable moments, so that the model has six degrees of freedom when $\text{cov}(\varepsilon_1, \varepsilon_3) = \text{cov}(\varepsilon_2, \varepsilon_4) = 0$, and four degrees of freedom when these covariances are left free.

The parameter λ_3 is identified with three overidentifying restrictions since

$$\text{cov}(x_2, y_i)/\text{cov}(x_1, y_i) = \lambda_3, \qquad i = 1, 2, 3, 4.$$

All the other parameters are determined as before.

3.5. Estimation

We discuss estimation of the model of Figure 5 explicitly; the other cases can be handled in a similar manner.

When the ε's are all uncorrelated, the model can be directly specified in the LISREL format. We have

$$\begin{pmatrix} y_1 \\ y_2 \\ y_3 \\ y_4 \end{pmatrix} = \begin{pmatrix} 1 & 0 \\ \lambda_1 & 0 \\ 0 & 1 \\ 0 & \lambda_2 \end{pmatrix} \begin{pmatrix} \eta_1 \\ \eta_2 \end{pmatrix} + \begin{pmatrix} \varepsilon_1 \\ \varepsilon_2 \\ \varepsilon_3 \\ \varepsilon_4 \end{pmatrix}, \tag{32}$$

$$\begin{pmatrix} x_1 \\ x_2 \end{pmatrix} = \begin{pmatrix} 1 \\ \lambda_3 \end{pmatrix} \xi + \begin{pmatrix} \delta_1 \\ \delta_2 \end{pmatrix}, \tag{33}$$

$$\begin{pmatrix} 1 & 0 \\ \beta & 1 \end{pmatrix} \begin{pmatrix} \eta_1 \\ \eta_2 \end{pmatrix} = \begin{pmatrix} \gamma_1 \\ \gamma_2 \end{pmatrix} \xi + \begin{pmatrix} \zeta_1 \\ \zeta_2 \end{pmatrix}. \tag{34}$$

The LISREL program gives maximum-likelihood estimates of λ_1, λ_2, λ_3, β, γ_1, γ_2, ψ_{11}, ψ_{22}, ϕ, $\theta^2_{\delta_i}$ $(i = 1, 2)$, and $\theta^2_{\varepsilon_i}$ $(i = 1, 2, 3, 4)$ and standard errors for these estimates.

When cov $(\varepsilon_1, \varepsilon_3)$ and cov $(\varepsilon_2, \varepsilon_4)$ are unrestricted, we incorporate the ε's as ξ's and η's as follows:

$$\begin{pmatrix} y_1 \\ y_2 \\ y_3 \\ y_4 \end{pmatrix} = \begin{bmatrix} 1 & 0 & 1 & 0 & 0 & 0 \\ \lambda_1 & 0 & 0 & 1 & 0 & 0 \\ 0 & 1 & 0 & 0 & 1 & 0 \\ 0 & \lambda_2 & 0 & 0 & 0 & 1 \end{bmatrix} \begin{pmatrix} \eta_1 \\ \eta_2 \\ \varepsilon_1 \\ \varepsilon_2 \\ \varepsilon_3 \\ \varepsilon_4 \end{pmatrix}, \tag{35}$$

$$\begin{pmatrix} x_1 \\ x_2 \end{pmatrix} = \begin{pmatrix} 1 & 0 & 0 & 0 & 0 \\ \lambda_3 & 0 & 0 & 0 & 0 \end{pmatrix} \begin{pmatrix} \xi \\ \varepsilon_1 \\ \varepsilon_2 \\ \varepsilon_3 \\ \varepsilon_4 \end{pmatrix} + \begin{pmatrix} \delta_1 \\ \delta_2 \end{pmatrix}, \tag{36}$$

$$\begin{bmatrix} 1 & 0 & 0 & 0 & 0 & 0 \\ \beta & 1 & 0 & 0 & 0 & 0 \\ 0 & 0 & 1 & 0 & 0 & 0 \\ 0 & 0 & 0 & 1 & 0 & 0 \\ 0 & 0 & 0 & 0 & 1 & 0 \\ 0 & 0 & 0 & 0 & 0 & 1 \end{bmatrix} \begin{pmatrix} \eta_1 \\ \eta_2 \\ \varepsilon_1 \\ \varepsilon_2 \\ \varepsilon_3 \\ \varepsilon_4 \end{pmatrix} = \begin{bmatrix} \gamma_1 & 0 & 0 & 0 & 0 \\ \gamma_2 & 0 & 0 & 0 & 0 \\ 0 & 1 & 0 & 0 & 0 \\ 0 & 0 & 1 & 0 & 0 \\ 0 & 0 & 0 & 1 & 0 \\ 0 & 0 & 0 & 0 & 1 \end{bmatrix} \begin{pmatrix} \xi \\ \varepsilon_1 \\ \varepsilon_2 \\ \varepsilon_3 \\ \varepsilon_4 \end{pmatrix} + \begin{pmatrix} \zeta_1 \\ \zeta_2 \\ 0 \\ 0 \\ 0 \\ 0 \end{pmatrix}. \tag{37}$$

In (35) we represent the ε's as additional η's and in (36) we also represent them as additional ξ's.

So in LISREL we take

$$\boldsymbol{\Phi} = \begin{pmatrix} \phi & & & & \\ 0 & \phi_{11} & & & \\ 0 & 0 & \phi_{22} & & \\ 0 & \phi_{31} & 0 & \phi_{33} & \\ 0 & 0 & \phi_{42} & 0 & \phi_{44} \end{pmatrix},$$

where $\phi = \text{var}(\xi)$, $\phi_{ii} = \text{var}(\varepsilon_i)$ and $\phi_{ij} = \text{cov}(\varepsilon_i, \varepsilon_j)$, $i, j = 1, 2, 3, 4$, and

$$\boldsymbol{\Psi} = \begin{bmatrix} \psi_{11} & & & & & \\ 0 & \psi_{22} & & & & \\ 0 & 0 & 0 & & & \\ 0 & 0 & 0 & 0 & & \\ 0 & 0 & 0 & 0 & 0 & \\ 0 & 0 & 0 & 0 & 0 & 0 \end{bmatrix}$$

An example of this model follows in the next section.

3.6. The stability of alienation

To illustrate the theory we draw on the data in Wheaton et al. (1975). This study was concerned with the stability over time of attitudes such as alienation and the relation of attitudes to background variables such as education and occupation. Data on attitude scales were collected from 932 persons in two rural regions in Illinois at three points in time: 1966, 1967 and 1971 [see Summers et al. (1969) for further description of the research setting]. We use only the data for 1967 and 1971. The variables we use are the *Anomia* subscale and the *Powerlessness* subscale, taken to be indicators of *Alienation*. The background variables are respondent's education (years of schooling completed) and Duncan's Socio-Economic Index (SEI). These are taken to be indicators of respondent's Socio-Economic Status (SES). We analyze the data under the three models shown in Figures 6a–c, none of which correspond to Wheaton's final model.

The maximum-likelihood estimates of the parameters are given in Table 1. The main aim of the Wheaton study was to estimate the stability of alienation over time, which is reflected in the parameter β, or rather in the squared correlation between ALIENATION 71 and ALIENATION 67.

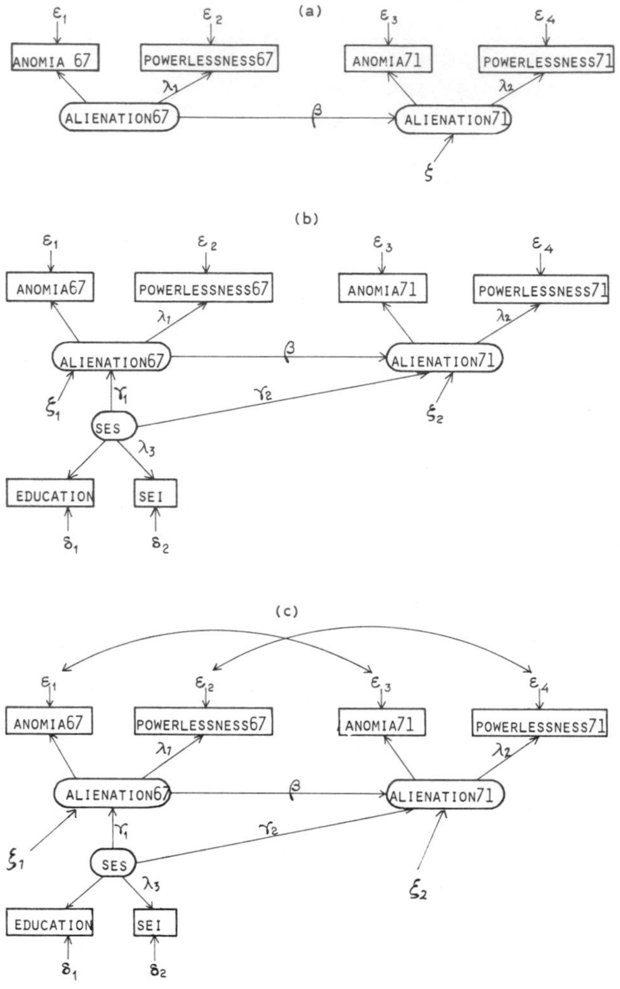

FIGURE 6. Models for study of stability of alienation.

The squared correlation coefficients are 0.56, 0.56 and 0.46 for Models a, b and c, respectively.

As can be seen from Table 1, β is overestimated in Model 6a, which ignores SES. Model 6b indicates that SES has a significant influence on alienation on both occasions. The negative signs of the SES coefficients γ_1 and γ_2 indicate that high socio-economic status reduces alienation. However, the overall fit of Model 6b is not acceptable. Since the same scales are used on both occasions, it seems reasonable to allow error correlations between the same measures at the two occasions. This is

TABLE 1

Parameter estimates for the models in Figure 6a–c (standard errors in parentheses).

Parameter	Model 6a	Model 6b	Model 6c
λ_1	0.815 (0.040)	0.888 (0.041)	0.979 (0.062)
λ_2	0.847 (0.042)	0.849 (0.040)	0.922 (0.059)
λ_3	–	5.331 (0.430)	5.221 (0.422)
β	0.789 (0.044)	0.705 (0.054)	0.607 (0.051)
γ_1	–	−0.614 (0.056)	−0.575 (0.056)
γ_2	–	−0.174 (0.054)	−0.227 (0.052)
ψ_{11}	–	5.307 (0.473)	4.847 (0.468)
ψ_{22}	4.085 (0.432)	3.742 (0.388)	4.089 (0.405)
ϕ	–	6.663 (0.641)	6.803 (0.650)
σ_{δ_1}	–	1.717 (0.145)	1.675 (0.151)
σ_{δ_2}	–	16.153 (0.565)	16.273 (0.558)
σ_{ϵ_1}	1.906 (0.097)	2.004 (0.086)	2.176 (0.104)
σ_{ϵ_2}	1.865 (0.077)	1.786 (0.076)	1.602 (0.126)
σ_{ϵ_3}	1.827 (0.109)	1.923 (0.097)	2.098 (0.123)
σ_{ϵ_4}	1.969 (0.077)	1.904 (0.077)	1.754 (0.124)
corr (ϵ_1, ϵ_3)	–	–	0.356 (0.047)
corr (ϵ_2, ϵ_4)	–	–	0.121 (0.082)
χ^2	61.155	71.544	4.770
d.f.	1	6	4

done in Model 6c, which, as can be seen in Table 1, provides an acceptable fit.

4. Multi-wave, one- and two-variable models

We now turn to the multi-wave case with one or two variables being measured on the subjects on three or more occasions.

4.1. Multi-wave, one-variable models

Suppose one fallible measure y is administered repeatedly to the same group of people. A possible model for this situation is shown in Figure 7. (We confine attention to the four-occasion case here, the extension to the general case being obvious throughout.) Such schemes have been termed simplex models by Guttman (1954) because of the typical pattern of intercorrelations they produce. Anderson (1960) formulated this model in terms of various stochastic processes and treated the identification problem; Jöreskog (1970b) treated the estimation problem. Applications

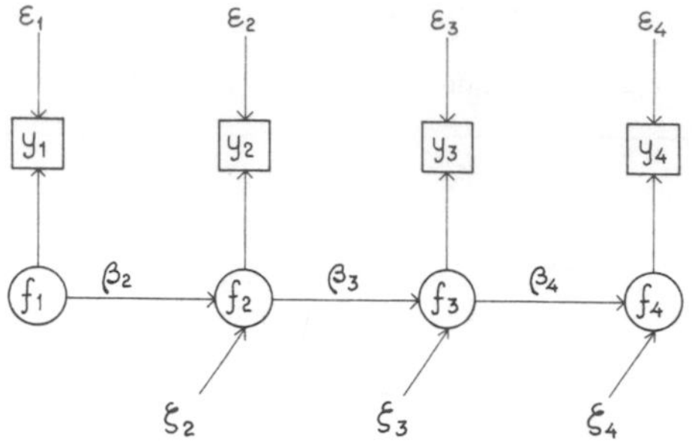

FIGURE 7. A simplex model.

to sociological panel analysis have been discussed by Heise (1969), Wiley and Wiley (1970), and Werts, Jöreskog and Linn (1971), and an application to the measurement of academic growth has been given by Werts, Linn and Jöreskog (1976).

The unit of measurement in each factor f_i is chosen to be the same as in the corresponding y_i. The equations defining the model are

$$y_i = f_i + \varepsilon_i, \qquad i = 1, 2, 3, 4, \tag{38}$$

$$f_{i+1} = \beta_{i+1} f_i + \zeta_{i+1}, \qquad i = 1, 2, 3. \tag{39}$$

Here the ε_i are mutually uncorrelated and uncorrelated with all the f_i, and ζ_{i+1} is uncorrelated with f_i. The unknown parameters of the model are $\phi_i = \text{var}(f_i)$, $\theta_i^2 = \text{var}(\varepsilon_i)$, $i = 1, 2, 3, 4$, and $\beta_2, \beta_3, \beta_4$. (We use the symbol ϕ rather than ω here to denote the variances of the dependent variables. Since there are no independent variables, there should be no confusion.) The residual variances are $\text{var}(\zeta_{i+1}) = \phi_{i+1} - \beta_{i+1}^2 \phi_i$, $i = 1, 2, 3$. The covariance matrix of y_1, y_2, y_3 and y_4 is

$$\boldsymbol{\Sigma} = \begin{bmatrix} \phi_1 + \theta_1^2 & & & \\ \beta_2\phi_1 & \phi_2 + \theta_2^2 & & \\ \beta_2\beta_3\phi_1 & \beta_3\phi_2 & \phi_3 + \theta_3^2 & \\ \beta_2\beta_3\beta_4\phi_1 & \beta_3\beta_4\phi_2 & \beta_4\phi_3 & \phi_4 + \theta_4^2 \end{bmatrix}. \tag{40}$$

It is seen from (40) that the product $\beta_2\phi_1 = \sigma_{21}$ is identified, but β_2 and ϕ_1 are not separately identified. We can multiply β_2 by a constant and divide ϕ_1 by the same constant without changing the product; the change induced in ϕ_1 can be absorbed in θ_1^2 in such a way that σ_{11} remains unchanged. Hence $\theta_1^2 = \text{var}(\varepsilon_1)$ is not identified either. On the other hand, for f_2 and f_3 we have

$$\phi_2 = (\sigma_{32}\sigma_{21})/\sigma_{31}, \qquad \phi_3 = (\sigma_{43}\sigma_{32})/\sigma_{42},$$

so that ϕ_2 and ϕ_3, and hence also θ_2^2 and θ_3^2, are identified. With ϕ_2 and ϕ_3 in hand, β_3 and β_4 are determined by σ_{32} and σ_{43}. The central coefficient β_3 is overidentified since

$$\beta_3\phi_2 = (\sigma_{31}\sigma_{42})/\sigma_{41} = \sigma_{32}.$$

Since both ϕ_4 and θ_4^2 enter only into σ_{44}, only their sum is identified.

This analysis shows that for the 'inner' variables (y_2 and y_3), the parameters ϕ_2, ϕ_3, θ_2, θ_3 and β_3 are identified, whereas there is an indeterminacy associated with the 'outer' variables (y_1 and y_4). To eliminate these indeterminacies one of the parameters ϕ_1, θ_1 and β_2 must be specified and one of the parameters ϕ_4 and θ_4 must also be specified. Hence there are only nine independent parameters and the model has one degree of freedom. In the general case of $m \geq 4$ occasions there will be $3m-3$ free parameters and the degrees of freedom is $\frac{1}{2}m(m+1)-(3m-3)$.

The simplex model can be put into the LISREL format, with no $\boldsymbol{x}$. We write

$$\begin{pmatrix} y_1 \\ y_2 \\ y_3 \\ y_4 \end{pmatrix} = \begin{bmatrix} 1 & 0 & 0 & 0 \\ 0 & 1 & 0 & 0 \\ 0 & 0 & 1 & 0 \\ 0 & 0 & 0 & 1 \end{bmatrix} \begin{bmatrix} f_1 \\ f_2 \\ f_3 \\ f_4 \end{bmatrix} + \begin{pmatrix} 0 \\ \varepsilon_2 \\ \varepsilon_3 \\ 0 \end{pmatrix}, \tag{41}$$

$$\begin{bmatrix} 1 & 0 & 0 & 0 \\ -\beta_2 & 1 & 0 & 0 \\ 0 & -\beta_3 & 1 & 0 \\ 0 & 0 & -\beta_4 & 1 \end{bmatrix} \begin{bmatrix} f_1 \\ f_2 \\ f_3 \\ f_4 \end{bmatrix} = \begin{bmatrix} \zeta_1 \\ \zeta_2 \\ \zeta_3 \\ \zeta_4 \end{bmatrix}. \tag{42}$$

In (41) we have taken $\varepsilon_1 = \varepsilon_4 = 0$ to eliminate the indeterminacies and in (42) we have defined ζ_1 as f_1. Note that we have taken $\psi_i = \text{var}(\zeta_i)$ as free parameters, rather than the $\phi_i = \text{var}(f_i)$; there is obviously a one-to-one correspondence between them. The parameter matrices are specified as

$\boldsymbol{\Lambda}_y = \boldsymbol{I}$,

$\boldsymbol{B}$ is as in (42),

$$\boldsymbol{\Psi} = \text{diag}\,(\psi_1, \psi_2, \psi_3, \psi_4),$$

$$\boldsymbol{\Theta}_\varepsilon = \text{diag}\,(0, \sigma_{\varepsilon_2}, \sigma_{\varepsilon_3}, 0).$$

4.2. *Multi-wave, two-variable models*

Now suppose that we have two observed variables at each of four occasions as in Figure 8. This model is a direct generalization of the model in Figure 1b to four occasions.

With $\boldsymbol{x}' = (x_1, x_2, x_3, x_4)$, $\boldsymbol{y}' = (y_1, y_2, y_3, y_4)$, the measurement model is

$$\boldsymbol{x} = \boldsymbol{f} + \boldsymbol{\delta}, \tag{43}$$

$$\boldsymbol{y} = \boldsymbol{D}_\lambda \boldsymbol{f} + \boldsymbol{\varepsilon}, \tag{44}$$

where $\boldsymbol{D}_\lambda = \text{diag}\,(\lambda_1, \lambda_2, \lambda_3, \lambda_4)$. The covariance matrix of $\boldsymbol{z} = (\boldsymbol{x}', \boldsymbol{y}')'$ is

$$\boldsymbol{\Sigma} = \begin{pmatrix} \boldsymbol{\Sigma}_{xx} & \boldsymbol{\Sigma}_{xy} \\ \boldsymbol{\Sigma}_{yx} & \boldsymbol{\Sigma}_{yy} \end{pmatrix}, \tag{45}$$

with

$$\boldsymbol{\Sigma}_{xx} = \boldsymbol{\Phi} + \boldsymbol{\Theta}^*_\delta, \tag{46}$$

$$\boldsymbol{\Sigma}_{yx} = \boldsymbol{D}_\lambda \boldsymbol{\Phi}, \tag{47}$$

$$\boldsymbol{\Sigma}_{yy} = \boldsymbol{D}_\lambda \boldsymbol{\Phi} \boldsymbol{D}_\lambda + \boldsymbol{\Theta}^*_\varepsilon, \tag{48}$$

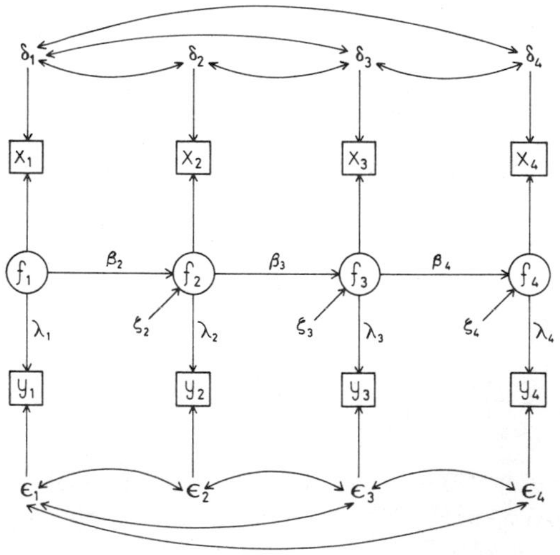

FIGURE 8. A four-wave, two-variable model with correlated errors.

where $\boldsymbol{\Phi}$, $\boldsymbol{\Theta}_\delta^*$ and $\boldsymbol{\Theta}_\varepsilon^*$ are the covariance matrices of f, $\boldsymbol{\delta}$ and $\boldsymbol{\varepsilon}$, respectively. It is seen that $\boldsymbol{D}_\lambda$ may be multiplied by a non-zero scalar, $\boldsymbol{\Phi}$ divided by the same scalar, and with $\boldsymbol{\Theta}_\delta^*$ and $\boldsymbol{\Theta}_\varepsilon^*$ properly adjusted, $\boldsymbol{\Sigma}$ will not change. Hence the model is not identified. One restriction is needed but there does not seem to be any natural way to choose it. We shall therefore consider three alternative models which are all identified. These models all represent different specification of the correlation structures for the errors in $\boldsymbol{\delta}$ and $\boldsymbol{\varepsilon}$ as follows:

Figure 9a: The errors are uncorrelated.
Figure 9b: The errors have one common factor.
Figure 9c: The errors have a simplex structure.

In all three models the covariance matrix $\boldsymbol{\Phi}$ is generated by a simplex (or first-order autoregressive) model,

$$f_{i+1} = \beta_{i+1} f_i + \zeta_{i+1}, \qquad i = 1, 2, 3.$$

This implies that

$$\boldsymbol{\Phi} = \begin{bmatrix} \phi_1 & & & \\ \beta_2\phi_1 & \phi_2 & & \\ \beta_2\beta_3\phi_1 & \beta_3\phi_2 & \phi_3 & \\ \beta_2\beta_3\beta_4\phi_1 & \beta_3\beta_4\phi_2 & \beta_4\phi_3 & \phi_4 \end{bmatrix}, \tag{49}$$

where, as before, $\phi_i = \text{var}(f_i)$, $i = 1, 2, 3, 4$. We now examine the three models in turn.

Model 9a

To show that the model in Figure 9a is identified we first show that $\boldsymbol{\Phi}$ is identified even if it is unrestricted. In Model 9a, $\boldsymbol{\Theta}_\delta^*$ and $\boldsymbol{\Theta}_\varepsilon^*$ are diagonal, and are denoted by $\boldsymbol{\Theta}_\delta^2$ and $\boldsymbol{\Theta}_\varepsilon^2$, respectively. We can use the off-diagonal part of (46) to solve for

$$\phi_{ij} = \sigma_{ij}^{(xx)}, \qquad i \neq j.$$

Then, from (47) we have, for a given i,

$$\sigma_{ij}^{(yx)} = \lambda_i \phi_{ij}, \qquad j \neq i,$$

which represents three equations in λ_i. Hence each λ_i is overidentified with two restrictions. With $\boldsymbol{D}_\lambda$ determined, we can now use the diagonal of (47) to determine the diagonal elements of $\boldsymbol{\Phi}$, i.e.,

$$\sigma_{ii}^{(yx)} = \lambda_i \phi_{ii}.$$

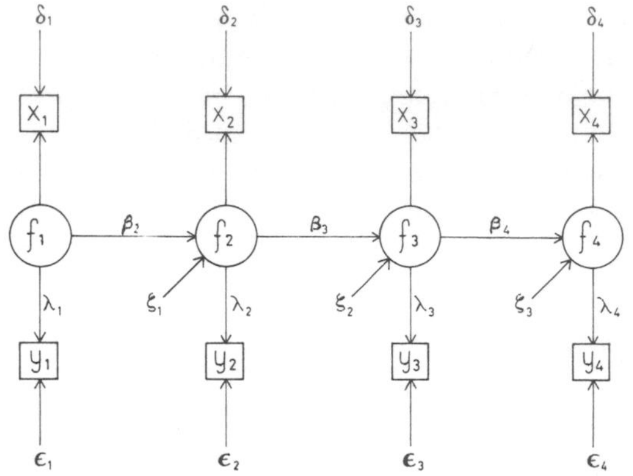

FIGURE 9a. A four-wave, two-variable model with uncorrelated errors.

The diagonal matrices $\boldsymbol{\Theta}_\delta$ and $\boldsymbol{\Theta}_\varepsilon$ are then determined from the diagonal parts of (46) and (48). The off-diagonal part of (48) then yields six overidentifying equations for $\boldsymbol{D}_\lambda$ and ϕ_{ij} $(i \neq j)$. This analysis shows that each diagonal element in $\boldsymbol{\Theta}_\delta$, $\boldsymbol{\Theta}_\varepsilon$, and $\boldsymbol{\Phi}$ is determined by one equation whereas the λ_i $(i = 1, 2, 3, 4)$ and the ϕ_{ij} $(i \neq j)$ are determined with fourteen overidentifying restrictions. Finally (49) is used to determine β_2, β_3 and β_4 from ϕ_{21}, ϕ_{32} and ϕ_{43}, respectively. Elements ϕ_{ij} with $i - j > 1$ yield three additional overidentifying restrictions. Hence the total number of overidentifying restrictions is 17, which checks with the number of degrees of freedom computed as number of distinct elements in $\boldsymbol{\Sigma}$ (36) minus the number of free parameters (4ϕ's, 3β's, 4λ's, 4 in $\boldsymbol{\Theta}_\delta$ and 4 in $\boldsymbol{\Theta}_\varepsilon$ for a total of 19). In the general case of $m \geq 2$ occasions the number of degrees of freedom will be $2m^2 - 4m + 1$.

Model 9a can be put into the LISREL format with no $\boldsymbol{x}$. The measurement model is

$$\begin{pmatrix} \boldsymbol{x} \\ \boldsymbol{y} \end{pmatrix} = \begin{pmatrix} \boldsymbol{I} \\ \boldsymbol{D}_\lambda \end{pmatrix} \boldsymbol{f} + \boldsymbol{e},$$

and the structural model is

$$\begin{bmatrix} 1 & 0 & 0 & 0 \\ -\beta_2 & 1 & 0 & 0 \\ 0 & -\beta_3 & 1 & 0 \\ 0 & 0 & -\beta_4 & 1 \end{bmatrix} \begin{pmatrix} f_1 \\ f_2 \\ f_3 \\ f_4 \end{pmatrix} = \begin{pmatrix} \zeta_1 \\ \zeta_2 \\ \zeta_3 \\ \zeta_4 \end{pmatrix}$$

As before, LISREL treats $\psi_i = \text{var}(\zeta_i)$ as primary parameters rather than $\phi_i = \text{var}(f_i)$. Estimates of ϕ_i are obtained as a by-product:

$$\phi_1 = \psi_1,$$
$$\phi_i = \psi_i + \beta_i^2 \phi_{i-1}, \qquad i = 2, 3, 4.$$

Model 9b

In Figure 9b it is assumed that correlations between the errors are accounted for by a common factor. The common factors ξ and η are *test-specific* in contrast to the factors f_1, f_2, f_3 and f_4, which are *occasion-specific* [Jöreskog (1970a)]. The test-specific factors ξ and η are assumed to be uncorrelated, uncorrelated with $\boldsymbol{f}$, $\boldsymbol{\delta}$ and $\boldsymbol{\varepsilon}$, and scaled to unit variance.

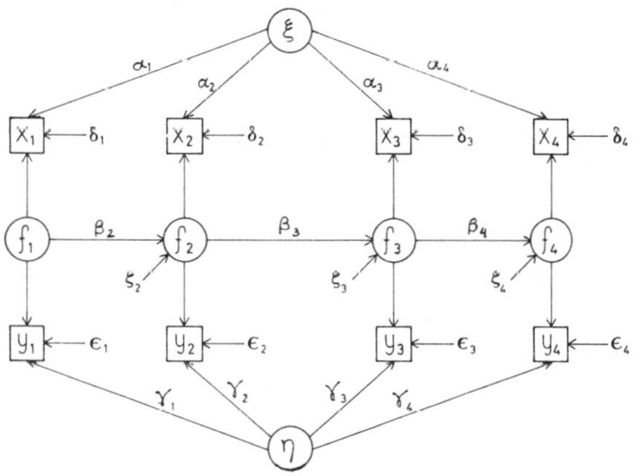

FIGURE 9b. A four-wave, two-variable model with test-specific factors.

The measurement model is

$$\boldsymbol{x} = \boldsymbol{f} + \boldsymbol{\alpha}\xi + \boldsymbol{\delta}, \tag{50}$$
$$\boldsymbol{y} = \boldsymbol{D}_\lambda \boldsymbol{f} + \boldsymbol{\gamma}\eta + \boldsymbol{\varepsilon}, \tag{51}$$

where $\boldsymbol{\alpha}$ and $\boldsymbol{\gamma}$ are factor loadings relating the observed variables $\boldsymbol{x}$ and $\boldsymbol{y}$ to the test-specific factors ξ and η, respectively. The blocks of the covariance matrix of the observed variables are

$$\boldsymbol{\Sigma}_{xx} = \boldsymbol{\Phi} + \boldsymbol{\alpha}\boldsymbol{\alpha}' + \boldsymbol{\Theta}_\delta^2, \tag{52}$$
$$\boldsymbol{\Sigma}_{yx} = \boldsymbol{D}_\lambda \boldsymbol{\Phi}, \tag{53}$$
$$\boldsymbol{\Sigma}_{yy} = \boldsymbol{D}_\lambda \boldsymbol{\Phi} \boldsymbol{D}_\lambda + \boldsymbol{\gamma}\boldsymbol{\gamma}' + \boldsymbol{\Theta}_\varepsilon^2. \tag{54}$$

The identification problem for Model 9b is slightly more complicated than for Model 9a. First we show that $\boldsymbol{D}_\lambda$ and $\boldsymbol{\Phi}$ can be determined from (53) alone. Since for $i \neq j$,

$$\sigma_{ij}^{(yx)} = \lambda_i \phi_{ij},$$

$$\sigma_{ji}^{(yx)} = \lambda_j \phi_{ji},$$

we have

$$\lambda_i / \lambda_j = \sigma_{ij}^{(yx)} / \sigma_{ji}^{(yx)}.$$

Hence the ratios λ_1/λ_2, λ_1/λ_3, λ_1/λ_4, λ_2/λ_3, λ_2/λ_4, λ_3/λ_4 are determined. This yields six equations in four unknowns which can be solved with two overidentifying restrictions on the λ's. With $\boldsymbol{D}_\lambda$ determined, $\boldsymbol{\Phi}$ is given by (53) as

$$\boldsymbol{\Phi} = \boldsymbol{\Sigma}_{yx} \boldsymbol{D}_\lambda^{-1}.$$

From the off-diagonal elements of $\boldsymbol{\Sigma}_{xx} - \boldsymbol{\Phi} = \boldsymbol{\Sigma}^*$, say, we have

$$\sigma_{ij}^* = \alpha_i \alpha_j, \qquad i \neq j,$$

which can be used to determine α_1 from

$$\alpha_1^2 = (\sigma_{21}^* \sigma_{31}^*)/\sigma_{32}^* = (\sigma_{21}^* \sigma_{41}^*)/\sigma_{42}^* = (\sigma_{31}^* \sigma_{41}^*)/\sigma_{43}^*,$$

and similarly for α_2, α_3 and α_4. This yields six equations in four unknowns which can be solved with two overidentifying restrictions on the α's. Next $\boldsymbol{\Theta}_\delta$ is determined from the diagonal elements of (52). Finally, $\boldsymbol{\gamma}$ and $\boldsymbol{\Theta}_\varepsilon$ are determined in a parallel way from (54). This analysis shows that if $\boldsymbol{\Phi}$ were free the model would be overidentified by six restrictions. The β's are determined by (49) as before, yielding three additional overidentifying restrictions. Hence, the total number of overidentifying restrictions is 9 which is the degrees of freedom of the model. In the general case with $m \geq 3$ occasions, the degrees of freedom will be $2m^2 - 6m + 1$.

Model 9a may be interpreted as the special case of Model 9b which arises when both $\boldsymbol{\alpha}$ and $\boldsymbol{\gamma}$ are zero. The hypothesis $\boldsymbol{\alpha} = \boldsymbol{0}$ and $\boldsymbol{\gamma} = \boldsymbol{0}$ may thus be tested with eight degrees of freedom.

Model 9b can be put into the LISREL format using the no-$\boldsymbol{x}$ option.

The parameter matrices are specified as

$$\boldsymbol{\Lambda}_y = \begin{bmatrix} 1 & 0 & 0 & 0 & \alpha_1 & 0 \\ 0 & 1 & 0 & 0 & \alpha_2 & 0 \\ 0 & 0 & 1 & 0 & \alpha_3 & 0 \\ 0 & 0 & 0 & 1 & \alpha_4 & 0 \\ \lambda_1 & 0 & 0 & 0 & 0 & \gamma_1 \\ 0 & \lambda_2 & 0 & 0 & 0 & \gamma_2 \\ 0 & 0 & \lambda_3 & 0 & 0 & \gamma_3 \\ 0 & 0 & 0 & \lambda_4 & 0 & \gamma_4 \end{bmatrix},$$

$$\boldsymbol{B} = \begin{bmatrix} 1 & 0 & 0 & 0 & 0 & 0 \\ -\beta_2 & 1 & 0 & 0 & 0 & 0 \\ 0 & -\beta_3 & 1 & 0 & 0 & 0 \\ 0 & 0 & -\beta_4 & 1 & 0 & 0 \\ 0 & 0 & 0 & 0 & 1 & 0 \\ 0 & 0 & 0 & 0 & 0 & 1 \end{bmatrix},$$

$$\boldsymbol{\Psi} = \text{diag}\,(\psi_1, \psi_2, \psi_3, \psi_4, 1, 1),$$

where $\psi_i = \text{var}\,(\zeta_i)$, $i = 1, 2, 3, 4$, $\boldsymbol{\Theta}_\delta$ and $\boldsymbol{\Theta}_\varepsilon$ are as before.

Model 9c

In Figure 9c it is assumed that the errors have a simplex structure of the

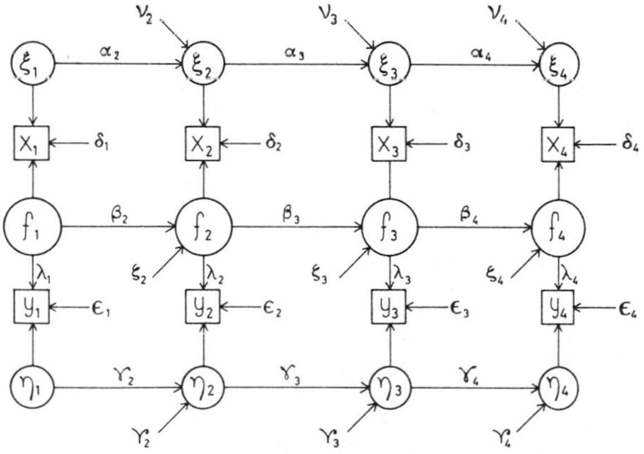

FIGURE 9c. A four-wave, two-variable model with simplex errors not identified.

type introduced in Section 4.1. The equations are

$$\boldsymbol{x} = \boldsymbol{f} + \boldsymbol{\xi} + \boldsymbol{\delta}, \tag{55}$$

$$\boldsymbol{y} = \boldsymbol{D}_\lambda \boldsymbol{f} + \boldsymbol{\eta} + \boldsymbol{\varepsilon}, \tag{56}$$

where

$$\xi_i = \alpha_i \xi_{i-1} + \nu_i,$$

$$\eta_i = \gamma_i \eta_{i-1} + \upsilon_i, \qquad i = 2, 3, 4. \tag{57}$$

The simplex variables $\boldsymbol{\xi}$ and $\boldsymbol{\eta}$ are assumed to be uncorrelated with $\boldsymbol{\delta}$, $\boldsymbol{\varepsilon}$ and f, and the residuals ν_i and υ_i are uncorrelated with each other and with ν_j, υ_j, ξ_j and η_j for $j < i$.

The covariance matrix of $\boldsymbol{\Sigma}$ is

$$\boldsymbol{\Sigma}_{xx} = \boldsymbol{\Phi} + \boldsymbol{\Sigma}_{\xi\xi} + \boldsymbol{\Theta}^2_\delta, \tag{58}$$

$$\boldsymbol{\Sigma}_{yx} = \boldsymbol{D}_\lambda \boldsymbol{\Phi}, \tag{59}$$

$$\boldsymbol{\Sigma}_{yy} = \boldsymbol{D}_\lambda \boldsymbol{\Phi} \boldsymbol{D}_\lambda + \boldsymbol{\Sigma}_{\eta\eta} + \boldsymbol{\Theta}^2_\varepsilon, \tag{60}$$

where $\boldsymbol{\Sigma}_{\xi\xi}$ and $\boldsymbol{\Sigma}_{\eta\eta}$ are the covariance matrices of $\boldsymbol{\xi}$ and $\boldsymbol{\eta}$.

Equation (59) determines $\boldsymbol{\Phi}$ and $\boldsymbol{D}_\lambda$ as before with two overidentifying restrictions. As in Section 4.1, the covariance matrix of $\boldsymbol{\xi} + \boldsymbol{\delta}$ is

$$\boldsymbol{\Sigma}_{\xi\xi} + \boldsymbol{\Theta}^2_\delta = \begin{bmatrix} \sigma^2_{\xi_1} + \theta^2_{\delta_1} & & & \\ \alpha_2 \sigma^2_{\xi_1} & \sigma^2_{\xi_2} + \theta^2_{\delta_2} & & \\ \alpha_2 \alpha_3 \sigma^2_{\xi_1} & \alpha_3 \sigma^2_{\xi_2} & \sigma^2_{\xi_3} + \theta^2_{\delta_3} & \\ \alpha_2 \alpha_3 \alpha_4 \sigma^2_{\xi_1} & \alpha_3 \alpha_4 \sigma^2_{\xi_2} & \alpha_4 \sigma^2_{\xi_3} & \sigma^2_{\xi_4} + \theta^2_{\delta_4} \end{bmatrix},$$

where $\sigma^2_{\xi_i} = \text{var}(\xi_i)$ and $\theta^2_{\delta_i} = \text{var}(\delta_i)$, $i = 1, 2, 3, 4$. Since $\boldsymbol{\Sigma}_{\xi\xi}$ and $\boldsymbol{\Theta}_\delta$ enter only into $\boldsymbol{\Sigma}_{xx}$ it is apparent that $\sigma^2_{\xi_1}$, α_2, σ_{δ_1}, $\theta^2_{\xi_4}$ and θ_{δ_4} are not identified. If one fixes $\sigma^2_{\xi_1}$ and $\sigma^2_{\xi_4}$ to unity, all the other parameters are identified with one overidentifying restriction. The situation is exactly the same as for the single simplex in Section 4.1. Similarly, fixing $\sigma^2_{\eta_1}$ and $\sigma^2_{\eta_4}$ to unity identifies all the parameters in $\boldsymbol{\Sigma}_{\eta\eta} + \boldsymbol{\Theta}^2_\varepsilon$ with one overidentifying restriction. The β's are determined from $\boldsymbol{\Phi}$ in (49) with three overidentifying restrictions as before. Altogether there will be seven overidentifying restrictions in Model 9c.

The indeterminacy in the two simplex models postulated for $\boldsymbol{\xi}$ and $\boldsymbol{\eta}$ means that all that one can do is to estimate the path from ξ_2 to x_1 and the path from ξ_3 to x_4. Similarly one can only estimate the paths from η_2 to y_1 and from η_3 to y_4. For this reason the model in Figure 9c is modified as in Figure 10, i.e., for $\boldsymbol{\xi}$ and $\boldsymbol{\eta}$ the end variables are eliminated. The model in

Figure 10 is an identified reparameterization of that in Figure 9c. However, with four occasions there are only two occasions between the first and the last occasion, so the simplex concept no longer applies. Still the model in Figure 10 may be interpreted as a generalization of Model 9b from one test-specific factor to two test-specific factors for each set of variables. Model 9b is a special case of the model in Figure 10 with $\text{var}(\nu) = \text{var}(\upsilon) = 0$. The hypothesis $\text{var}(\nu) = \text{var}(\upsilon) = 0$ may be tested with two degrees of freedom. As illustrated in Figure 11 the simplex concept for the errors would be meaningful when there are more than four occasions.

The model in Figure 10 can be put into the LISREL format. We do so in a way that facilitates generalization to more than four occasions. The measurement model is

$$\begin{pmatrix} \boldsymbol{x} \\ \boldsymbol{y} \end{pmatrix} = \begin{pmatrix} \boldsymbol{I} & \boldsymbol{P}_\alpha & \boldsymbol{0} \\ \boldsymbol{D}_\lambda & \boldsymbol{0} & \boldsymbol{P}_\gamma \end{pmatrix} \begin{pmatrix} \boldsymbol{f} \\ \boldsymbol{\xi} \\ \boldsymbol{\eta} \end{pmatrix} + \begin{pmatrix} \boldsymbol{\delta} \\ \boldsymbol{\varepsilon} \end{pmatrix},$$

where

$$\boldsymbol{P}_\alpha = \begin{bmatrix} \alpha_2 & 0 \\ 1 & 0 \\ 0 & 1 \\ 0 & \alpha_4 \end{bmatrix}, \qquad \boldsymbol{P}_\gamma = \begin{bmatrix} \gamma_2 & 0 \\ 1 & 0 \\ 0 & 1 \\ 0 & \gamma_4 \end{bmatrix},$$

$\boldsymbol{\xi} = (\xi_2, \xi_3)'$ and $\boldsymbol{\eta} = (\eta_2, \eta_3)'$. The structural model is

$$\begin{bmatrix} \boldsymbol{B}_\beta & \boldsymbol{0} & \boldsymbol{0} \\ \boldsymbol{0} & \boldsymbol{B}_\alpha & \boldsymbol{0} \\ \boldsymbol{0} & \boldsymbol{0} & \boldsymbol{B}_\gamma \end{bmatrix} \begin{pmatrix} \boldsymbol{f} \\ \boldsymbol{\xi} \\ \boldsymbol{\eta} \end{pmatrix} = \begin{pmatrix} \boldsymbol{\zeta} \\ \boldsymbol{\nu} \\ \boldsymbol{\upsilon} \end{pmatrix},$$

where

$$\boldsymbol{B}_\beta = \begin{bmatrix} 1 & 0 & 0 & 0 \\ -\beta_2 & 1 & 0 & 0 \\ 0 & -\beta_3 & 1 & 0 \\ 0 & 0 & -\beta_4 & 1 \end{bmatrix}, \quad \boldsymbol{B}_\alpha = \begin{pmatrix} 1 & 0 \\ -\alpha_3 & 1 \end{pmatrix}, \quad \boldsymbol{B}_\gamma = \begin{pmatrix} 1 & 0 \\ -\gamma_3 & 1 \end{pmatrix},$$

$\boldsymbol{\zeta} = (\zeta_1 \equiv f_1, \zeta_2, \zeta_3, \zeta_4)'$, $\boldsymbol{\nu} = (\nu_2 \equiv \xi_2, \nu_3)'$, and $\boldsymbol{\upsilon} = (\upsilon_2 \equiv \eta_2, \upsilon_3)'$. The LISREL program gives estimates of λ_1, λ_2, λ_3, λ_4, β_2, β_3, β_4, α_2, α_3, α_4, γ_2, γ_3, γ_4 and the variances of ζ_1, ζ_2, ζ_3, ζ_4, ν_2, ν_3, υ_2, υ_3. The estimates of the

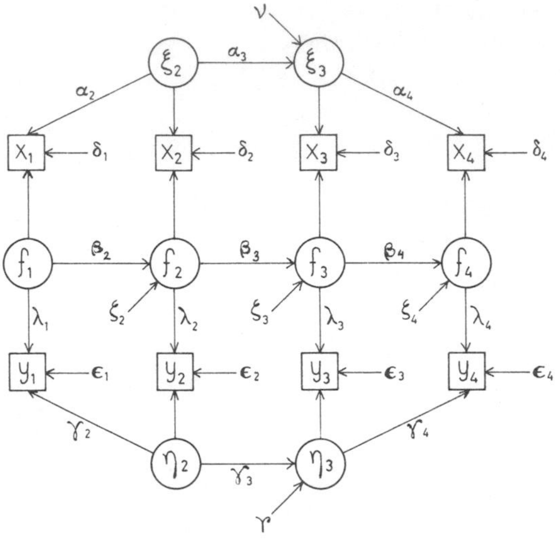

FIGURE 10. A four-wave, two-variable model with simplex errors identified.

variances of f_i, ξ_i and η_i may be recovered as

$$\text{var}(f_1) = \text{var}(\zeta_1),$$

$$\text{var}(f_i) = \beta_i^2 \,\text{var}(f_{i-1}) + \text{var}(\zeta_i), \qquad i = 2, 3, 4,$$

$$\text{var}(\xi_2) = \text{var}(\nu_2),$$

$$\text{var}(\xi_3) = \alpha_3^2 \,\text{var}(\xi_2) + \text{var}(\nu_3),$$

$$\text{var}(\eta_2) = \text{var}(\upsilon_2),$$

$$\text{var}(\eta_3) = \gamma_3^2 \,\text{var}(\eta_2) + \text{var}(\upsilon_3).$$

4.3. Mathematical aptitudes and achievements

To illustrate the theory we draw on a large growth study conducted at Educational Testing Service [Anderson and Maier (1963), Hilton (1969)]. A nationwide sample of fifth graders was tested in 1961 and then again in 1963, 1965 and 1967 as seventh, ninth and eleventh graders, respectively. The test scores included the verbal (SCATV) and quantitative (SCATQ) parts of SCAT (Scholastic Aptitude Test) and achievement tests in mathematics (MATH), science (SCI), social studies (SS), reading

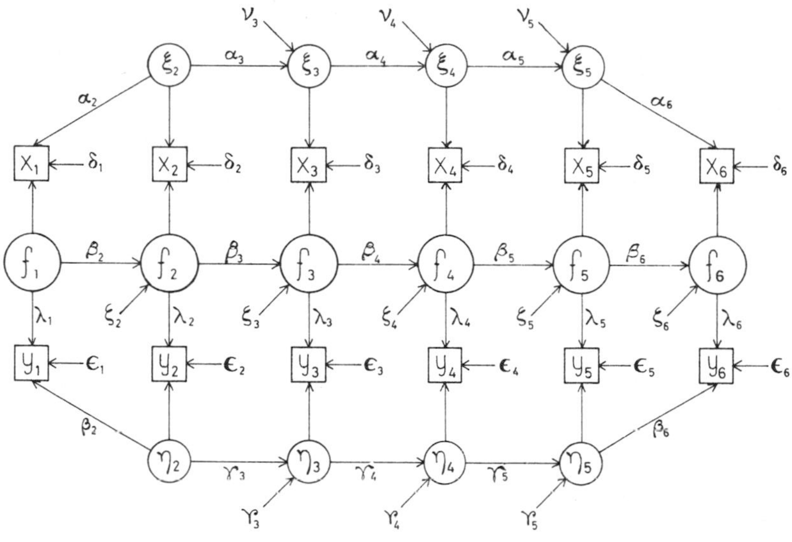

FIGURE 11. A six-wave, two-variable model with simplex errors.

(READ) and writing (WRIT). The analyses which we present are based on a subsample of 383 girls for which complete data were available for all the grades 5, 7, 9 and 11. In this section we analyze the data on MATH and SCATQ from all four occasions under Models 9a, 9b and 9c.

The maximum-likelihood estimates of the parameters are shown in Table 2. The χ^2-values clearly reject Model 9a in favor of Model 9b or 9c, but it is not so easy to discriminate between 9b and 9c. The difference in χ^2 between Model 9b and 9c is 9.85 with 2 degrees of freedom, which is significant at the 1% level. Hence it seems best to use Model 9c rather than 9b.

For interpretive purposes it may be preferable to work with transformations of the LISREL parameters. For Models 9b and 9c, Table 3 gives the estimates of the factor variances and of the squared correlations R_i^2 between f_i and f_{i-1}; Table 4 gives the covariance matrices of the errors $\boldsymbol{\varepsilon}^* = \boldsymbol{y} - \boldsymbol{D}_\lambda \boldsymbol{f}$ and $\boldsymbol{\delta}^* = \boldsymbol{x} - \boldsymbol{f}$ (i.e., the partial covariance matrices of $\boldsymbol{y}$ and $\boldsymbol{x}$ after elimination of $\boldsymbol{f}$). Table 5 gives the corresponding correlation matrices.

The squared correlations R_i^2 in Table 3 are quite high, indicating a very high stability of the quantitative factor over time. Table 4 shows that covariation among the errors is present for the SCATQ tests to a larger extent than for the MATH tests. Table 5 shows that the residual

TABLE 2

Parameter estimates for the models in Figure 9a–c with x = MATH and y = SCATQ, N = 383.[a]

Parameter	Model 9a	Model 9b	Model 9c
λ_1	0.88	0.85	0.83
λ_2	1.13	1.06	1.05
λ_3	1.23	1.14	1.14
λ_4	1.31	1.21	1.21
β_2	1.22	1.22	1.20
β_3	1.01	1.00	1.01
β_4	1.06	1.05	1.05
σ^2_{f1}	55.25	57.31	58.20
$\sigma^2_{\zeta 2}$	9.28	10.93	12.65
$\sigma^2_{\zeta 3}$	10.29	13.74	11.06
$\sigma^2_{\zeta 4}$	2.37	5.79	6.51
σ_{δ_1}	6.19	5.94	5.82
σ_{δ_2}	6.20	5.84	5.80
σ_{δ_3}	5.77	2.29	2.79
σ_{δ_4}	7.55	6.83	6.85
σ_{ϵ_1}	4.53	4.47	4.45
σ_{ϵ_2}	6.18	5.83	4.21
σ_{ϵ_3}	7.40	6.87	6.88
σ_{ϵ_4}	7.14	6.51	6.08
χ^2	72.49	23.18	13.33
d.f.	17	9	7

[a]Additional parameter estimates

Model 9b
$\alpha_1 = 0.97$, $\gamma_1 = 1.48$
$\alpha_2 = 0.26$, $\gamma_2 = 3.51$
$\alpha_3 = 4.59$, $\gamma_3 = 4.66$
$\alpha_4 = -0.92$, $\gamma_4 = 4.90$

Model 9c
$\alpha_2 = 5.89$, $\gamma_2 = 0.32$
$\alpha_3 = 16.06$, $\gamma_3 = 0.45$
$\alpha_4 = -0.20$, $\gamma_4 = 1.15$
$\sigma^2_{\xi 2} = 0.04$, $\sigma^2_{\nu} = 7.55$
$\sigma^2_{\eta 2} = 31.05$, $\sigma^2_{\upsilon} = 15.64$

TABLE 3

Factor variances and squared multiple correlations for Models 9b and 9c.

Model 9b		Model 9c	
σ^2_{fi}	R^2_i	σ^2_{fi}	R^2_i
57.31	–	58.20	–
96.23	0.886	96.46	0.869
109.97	0.875	109.46	0.899
127.03	0.954	127.19	0.949

TABLE 4
Covariance matrices of $\boldsymbol{\varepsilon}^*$ and $\boldsymbol{\delta}^*$ for Models 9b and 9c.

	Model 9b				Model 9c			
ε^*	22.17				22.98			
	5.19	46.31			9.94	48.77		
	6.90	16.36	68.91		4.47	13.97	69.26	
	7.25	17.20	22.83	66.39	5.14	16.07	25.22	65.97
δ^*	36.22				35.26			
	0.25	34.17			0.24	33.68		
	4.45	1.19	26.31		3.78	0.64	25.65	
	−0.89	−0.24	−4.22	47.49	−0.76	−0.13	−3.57	47.35

TABLE 5
Correlation matrices of $\boldsymbol{\varepsilon}^*$ and $\boldsymbol{\delta}^*$ for Models 9b and 9c.

	Model 9b				Model 9c			
ε^*	1.00				1.00			
	0.16	1.00			0.30	1.00		
	0.18	0.29	1.00		0.11	0.24	1.00	
	0.19	0.31	0.34	1.00	0.13	0.28	0.37	1.00
δ^*	1.00				1.00			
	0.01	1.00			0.01	1.00		
	0.14	0.04	1.00		0.13	0.02	1.00	
	−0.02	−0.01	−0.12	1.00	−0.02	−0.00	−0.10	1.00

correlations among the ε's are in general higher than those among the δ's; the latter are indeed very small. Hence the model accounts for the intercorrelations among the MATH tests much better than the intercorrelations among the SCATQ tests.

5. Multi-wave, multi-variable models

5.1. A general model for longitudinal data

In this section we develop a general model for longitudinal data which generalizes the models of Section 4 to the situation when several (more than two) variables are involved at several occasions and also allows for background variables as in Section 3.

We assume that p_t variables $y_t' = (y_{1t}, y_{2t}, \ldots, y_{p_t,t})$ are measured at occasion t ($t = 1, 2, \ldots, m$). At each occasion $\boldsymbol{y}_t$ has a factor structure with m_t correlated common factors $\boldsymbol{\eta}'_t = (\eta_{1t}, \eta_{2t}, \ldots, \eta_{m_t,t})$ so that

$$\boldsymbol{y}_t = \boldsymbol{\Lambda}_{yt}\boldsymbol{\eta}_t + \boldsymbol{\varepsilon}_t, \tag{61}$$

where $\boldsymbol{\varepsilon}_t$ is the vector of unique factors and $\boldsymbol{\Lambda}_{yt}$ is the factor-loading matrix of order $p_t \times m_t$.

In addition to the variables $\boldsymbol{y}_t$, we assume that q background variables $\boldsymbol{x}' = (x_1, x_2, \ldots, x_q)$ are measured, representing characteristics and conditions existing before the first occasion and assumed to influence all the $\boldsymbol{y}_t$ variables. The background variables are in general assumed to be fallible measures having a factor structure with common factors $\boldsymbol{\xi}' = (\xi_1, \xi_2, \ldots, \xi_n)$ so that

$$\boldsymbol{x} = \boldsymbol{\Lambda}_x\boldsymbol{\xi} + \boldsymbol{\delta}, \tag{62}$$

where $\boldsymbol{\delta}$ is the vector of unique factors and $\boldsymbol{\Lambda}_x$ is the factor-loading matrix of order $q \times n$.

The structural equations connecting the $\boldsymbol{\eta}$'s and $\boldsymbol{\xi}$'s are assumed to be

$$\boldsymbol{\eta}_1 = \boldsymbol{A}_1\boldsymbol{\xi} + \boldsymbol{\zeta}_1, \tag{63}$$

$$\boldsymbol{\eta}_t = \boldsymbol{A}_t\boldsymbol{\xi} + \boldsymbol{B}_t\boldsymbol{\eta}_{t-1} + \boldsymbol{\zeta}_t, \qquad t = 2, \ldots, m, \tag{64}$$

where $\boldsymbol{A}_t$ is a matrix of order $m_t \times n$, and $\boldsymbol{B}_t$ is a matrix of order $m_t \times m_{t-1}$. The residual vectors $\boldsymbol{\zeta}'_t = (\zeta_{1t}, \zeta_{2t}, \ldots, \zeta_{m_t,t})$ are assumed to be correlated within occasions but uncorrelated between occasions.

Model (61)–(64) can be put into the general LISREL format. To see this, we illustrate with $m = 4$ occasions and write (61) as

$$\begin{pmatrix} \boldsymbol{y}_1 \\ \boldsymbol{y}_2 \\ \boldsymbol{y}_3 \\ \boldsymbol{y}_4 \end{pmatrix} = \begin{pmatrix} \boldsymbol{\Lambda}_{y1} & \boldsymbol{0} & \boldsymbol{0} & \boldsymbol{0} \\ \boldsymbol{0} & \boldsymbol{\Lambda}_{y2} & \boldsymbol{0} & \boldsymbol{0} \\ \boldsymbol{0} & \boldsymbol{0} & \boldsymbol{\Lambda}_{y3} & \boldsymbol{0} \\ \boldsymbol{0} & \boldsymbol{0} & \boldsymbol{0} & \boldsymbol{\Lambda}_{y4} \end{pmatrix} \begin{pmatrix} \boldsymbol{\eta}_1 \\ \boldsymbol{\eta}_2 \\ \boldsymbol{\eta}_3 \\ \boldsymbol{\eta}_4 \end{pmatrix} + \begin{pmatrix} \boldsymbol{\varepsilon}_1 \\ \boldsymbol{\varepsilon}_2 \\ \boldsymbol{\varepsilon}_3 \\ \boldsymbol{\varepsilon}_4 \end{pmatrix}, \tag{65}$$

and (63) and (64) as

$$\begin{pmatrix} \boldsymbol{I} & \boldsymbol{0} & \boldsymbol{0} & \boldsymbol{0} \\ -\boldsymbol{B}_2 & \boldsymbol{I} & \boldsymbol{0} & \boldsymbol{0} \\ \boldsymbol{0} & -\boldsymbol{B}_3 & \boldsymbol{I} & \boldsymbol{0} \\ \boldsymbol{0} & \boldsymbol{0} & -\boldsymbol{B}_4 & \boldsymbol{I} \end{pmatrix} \begin{pmatrix} \boldsymbol{\eta}_1 \\ \boldsymbol{\eta}_2 \\ \boldsymbol{\eta}_3 \\ \boldsymbol{\eta}_4 \end{pmatrix} = \begin{pmatrix} \boldsymbol{A}_1 \\ \boldsymbol{A}_2 \\ \boldsymbol{A}_3 \\ \boldsymbol{A}_4 \end{pmatrix} \boldsymbol{\xi} + \begin{pmatrix} \boldsymbol{\zeta}_1 \\ \boldsymbol{\zeta}_2 \\ \boldsymbol{\zeta}_3 \\ \boldsymbol{\zeta}_4 \end{pmatrix}. \tag{66}$$

Particular cases may be useful in various situations. If each of the observed variables measures a different characteristic, trait, or construct, the assumptions that $\boldsymbol{y}$ and $\boldsymbol{x}$ have a common factor structure may not be reasonable. Each variable may represent its own factor, i.e., we have only a single indicator of each factor. In this case it is usually not possible to estimate the unique factor or the measurement error in the variables. This case is handled by specifying $m_t = p_t$, $\boldsymbol{\Lambda}_{yt} = \boldsymbol{I}$, $\boldsymbol{\varepsilon}_t = \boldsymbol{0}$ $(t = 1, 2, \ldots, m)$, $\boldsymbol{\Lambda}_x = \boldsymbol{I}$, and $\boldsymbol{\delta} = \boldsymbol{0}$. The structural equations (66) will then be defined in terms of the observed variables.

Another special case is when $\boldsymbol{A}_2 = \boldsymbol{A}_3 = \cdots = \boldsymbol{A}_m = \boldsymbol{0}$. Then the background variables $\boldsymbol{\xi}$ affect $\boldsymbol{\eta}_1$ directly and $\boldsymbol{\eta}_2, \boldsymbol{\eta}_3, \ldots, \boldsymbol{\eta}_m$ only indirectly via $\boldsymbol{\eta}_1$. Assumptions of this kind may be tested on the basis of available data.

A third special case is when measurements on background variables are not available. Then (62) and (63) are omitted and (64) is replaced by

$$\boldsymbol{\eta}_t = \boldsymbol{B}_t\boldsymbol{\eta}_{t-1} + \boldsymbol{\zeta}_t, \qquad t = 2, \ldots, m, \tag{67}$$

which is the direct generalization of the simplex model (39) to the multivariate case. This model is handled by using the no-$\boldsymbol{x}$ option of the LISREL program.

In some longitudinal studies it may be reasonable to assume that all the ε's are uncorrelated, corresponding to Model 9a of Section 4. However, as demonstrated previously, if the same measurements are used at each occasion, the corresponding ε's will tend to correlate, and in order to get consistent estimates of the $\boldsymbol{B}$'s, one must include such correlations in the model. If background variables are included in the model it is possible to estimate the intercorrelations among the ε's for the same variables directly as in Section 3.5. When background variables are not included, on the other hand, one must specify some structure for these intercorrelations, such as Model 9b or 9c of Section 4.2.

Each of the matrices $\boldsymbol{\Lambda}_{yt}$, $\boldsymbol{\Lambda}_x$, $\boldsymbol{A}_t$ and $\boldsymbol{B}_t$ may contain fixed, free, and constrained parameters. For the model to be identified it is necessary that there be an adequate number of fixed zeros in these matrices.

5.2. Building a model for aptitude and achievement measures

In longitudinal studies there are often many variables involved and a model of the kind introduced in the preceding subsection can get very complicated and involve many parameters, perhaps several hundreds. The estimation and testing of such a model with the LISREL program may present considerable difficulties even on today's high-speed computers. Furthermore, the model may only be tentative, or in any case, if it does not fit the data well (which is likely), one may want to modify the model during the process of analysis. In such a situation it is probably best to build up the model in steps, fitting various parts of the model separately, and only then putting the parts together into a whole model. We illustrate how this can be done using the ETS growth data from all four occasions. We should emphasize that there are a number of alternative ways in which the analysis can be done. Presumably these give essentially the same results but this needs to be investigated further.

We begin by finding the factor structure for each occasion. We postulate a two-factor solution for each occasion with two correlated factors V = Verbal Ability and Q = Quantitative Ability. The factor matrices Λ_{yt} $(t = 1, 2, 3, 4)$ are assumed to have the following pattern:

$$\begin{array}{r} \\ \text{MATH} \\ \text{SCI} \\ \text{SS} \\ \text{READ} \\ \text{SCATV} \\ \text{SCATQ} \end{array} \begin{array}{c} \begin{array}{cc} Q & V \end{array} \\ \begin{bmatrix} x & 0 \\ x & x \\ x & x \\ 0 & x \\ 0 & 1 \\ 1 & 0 \end{bmatrix} \end{array}, \tag{68}$$

where 0 is a loading of zero, 1 is a loading of one, and x is a non-zero loading to be estimated. The unit loadings merely fix the units of measurement in the factors V and Q. The number of factors and the factor pattern should be chosen such that the intercorrelations among the variables within occasions are accounted for sufficiently well. Fitting such a model to the data for each occasion separately gives the following goodness-of-fit-values

Grade 5: $\chi^2 = 13.20$ with 6 d.f.
Grade 7: $\chi^2 = 17.64$ with 6 d.f.
Grade 9: $\chi^2 = 2.62$ with 6 d.f.
Grade 11: $\chi^2 = 11.61$ with 6 d.f.

Although these do not represent very good fits, except for grade 9, we shall nevertheless retain these models as representing reasonable approximations.

In the next step we analyze all tests at all occasions simultaneously to see to what extent the occasion-specific factors V_t and Q_t at each occasion also account for intercorrelations between tests and between occasions. Hence we specify the following pattern as in (65):

$$
\begin{array}{llcccccccc}
 & & Q_5 & V_5 & Q_7 & V_7 & Q_9 & V_9 & Q_{11} & V_{11} \\
 & \text{Math} & x & 0 & & & & & & \\
 & \text{SCI} & x & x & & & & & & \\
 & \text{SS} & x & x & & & & & & \\
\text{Grade 5} & \text{READ} & 0 & x & & & & & & \\
 & \text{SCATV} & 0 & 1 & & & & & & \\
 & \text{SCATQ} & 1 & 0 & & & & & & \\
 & \text{MATH} & & & x & 0 & & & & \\
 & \text{SCI} & & & x & x & & & & \\
 & \text{SS} & & & x & x & & & & \\
\text{Grade 7} & \text{READ} & & & 0 & x & & & & \\
 & \text{SCATV} & & & 0 & 1 & & & & \\
 & \text{SCATQ} & & & 1 & 0 & & & & \\
 & \text{MATH} & & & & & x & 0 & & \\
 & \text{SCI} & & & & & x & x & & \\
 & \text{SS} & & & & & x & x & & \\
\text{Grade 9} & \text{READ} & & & & & 0 & x & & \\
 & \text{SCATV} & & & & & 0 & 1 & & \\
 & \text{SCATQ} & & & & & 1 & 0 & & \\
 & \text{MATH} & & & & & & & x & 0 \\
 & \text{SCI} & & & & & & & x & x \\
 & \text{SS} & & & & & & & x & x \\
\text{Grade 11} & \text{READ} & & & & & & & 0 & x \\
 & \text{SCATV} & & & & & & & 0 & 1 \\
 & \text{SCATQ} & & & & & & & 1 & 0
\end{array}
\qquad (69)
$$

All blank entries are zeros. At this step we let all eight factors be freely intercorrelated. This model gives $\chi^2 = 646.78$ with 216 degrees of freedom which represents a very poor fit.

Although the intercorrelations within occasions are accounted for reasonably well, the model does not account for intercorrelations of each test between occasions. This suggests that one should let the ε's be correlated for the same variable between occasions. Since background variables are not available here we shall attempt to account for these intercorrelations by introducing test-specific factors for each of the six tests (Model 9b). Denoting these test-specific factors by S_i $(i = 1, \ldots, 6)$, this amounts to adding six columns to the previous factor matrix so that the new factor-loading matrix becomes:

		Q_5	V_5	Q_7	V_7	Q_9	V_9	Q_{11}	V_{11}	S_1	S_2	S_3	S_4	S_5	S_6
Grade 5	MATH	x	0							x					
	SCI	x	x								x				
	SS	x	x									x			
	READ	0	x										x		
	SCATV	0	1											x	
	SCATQ	1	0												x
Grade 7	MATH			x	0					x					
	SCI			x	x						x				
	SS			x	x							x			
	READ			0	x								x		
	SCATV			0	1									x	
	SCATQ			1	0										x
Grade 9	MATH					x	0			x					
	SCI					x	x				x				
	SS					x	x					x			
	READ					0	x						x		
	SCATV					0	1							x	
	SCATQ					1	0								x
Grade 11	MATH							x	0	x					
	SCI							x	x		x				
	SS							x	x			x			
	READ							0	x				x		
	SCATV							0	1					x	
	SCATQ							1	0						x

The test-specific factors S_i are assumed to have unit-variance and to be mutually uncorrelated and uncorrelated with the occasion-specific factors Q_t and V_t. We still assume that the factors Q_t and V_t are freely intercorrelated. In addition to these factors there will be unique factors for each test and occasion. This model gives $\chi^2 = 313.29$ with 192 degrees of freedom, which represents a considerable improvement in fit. The estimate of the covariance matrix of the occasion-specific factors is

$$\begin{array}{l} Q_5 \\ V_5 \\ Q_7 \\ V_7 \\ Q_9 \\ V_9 \\ Q_{11} \\ V_{11} \end{array} \begin{bmatrix} 40.40 & & & & & & & \\ 53.34 & 90.34 & & & & & & \\ 60.38 & 79.78 & 104.79 & & & & & \\ 62.53 & 104.74 & 109.36 & 141.63 & & & & \\ 64.66 & 85.20 & 112.11 & 115.66 & 142.60 & & & \\ 58.74 & 99.08 & 103.16 & 134.71 & 122.35 & 139.37 & & \\ 72.27 & 96.69 & 130.43 & 132.50 & 154.46 & 133.28 & 183.35 & \\ 43.38 & 83.41 & 83.01 & 110.18 & 93.59 & 110.21 & 110.55 & 104.09 \end{bmatrix}$$

Finally we fit the above covariance matrix by the bivariate simplex model

$$\begin{aligned} Q_t &= \beta_{Qt} Q_{t-2} + q_t, \\ V_t &= \beta_{Vt} V_{t-2} + v_t, \end{aligned} \qquad t = 7, 9, 11,$$

where the residuals q_t and v_t are correlated contemporaneously but uncorrelated between occasions. This can be done by performing three separate bivariate regressions with the regression matrix constrained in each case to contain two zero coefficients. Tests of these zero constraints give

$$\begin{aligned} &\text{Grades 5–7:} \quad \chi^2 = 3.31 \quad \text{with 2 d.f.,} \\ &\text{Grades 7–9:} \quad \chi^2 = 9.11 \quad \text{with 2 d.f.,} \\ &\text{Grades 9–11:} \quad \chi^2 = 6.80 \quad \text{with 2 d.f.,} \end{aligned}$$

which represent reasonably good fits for these parts of the model. The full set of parameter estimates is given in Table 6.

Perhaps the most notable estimates are the large loadings 5.68, -5.72 and 5.16 of SCATQ9, MATH11 and SCATQ11 on the test-specific factors. These correspond to partial correlations, after Q_t and V_t are eliminated, of 0.39, -0.43 and 0.32, respectively. The most reasonable interpretation of these correlations is that they reflect changes in the tests SCATQ and MATH from the earlier to the later grades. In the earlier grades these tests consist mainly of arithmetic items whereas in the later

TABLE 6

Estimates for a four-wave, six-variable model with a bivariate simplex, $N = 383$.[a]

Grade	Test	Q_t	V_t	S_1	S_2	S_3	S_4	S_5	S_6	Unique variances
	MATH	1.24	0.0*	0.68						32.49
	SCI	0.07	0.97		1.99					23.57
	SS	0.12	0.90			0.78				20.96
5	READ	0.0*	1.31				4.69			46.10
	SCATV	0.0*	1.00*					2.77		28.37
	SCATQ	1.00*	0.0*						2.00	19.79
	MATH	0.97	0.0*	0.58						33.35
	SCI	0.01	0.62		2.37					23.11
	SS	−0.09	1.05			1.54				32.45
7	READ	0.0*	1.07				3.71			41.13
	SCATV	0.0*	1.00*					4.67		29.47
	SCATQ	1.00*	0.0*						4.02	37.44
	MATH	0.90	0.0*	1.11						19.86
	SCI	−0.02	0.84		3.31					25.00
	SS	0.11	0.88			2.02				41.04
9	READ	0.0*	0.86				4.03			30.81
	SCATV	0.0*	1.00*					2.74		57.18
	SCATQ	1.00*	0.00*						5.68	39.15
	MATH	0.85	0.0*	−5.72						12.08
	SCI	0.08	0.70		3.08					46.29
	SS	−0.02	1.16			2.36				27.74
11	READ	0.0*	1.18				3.72			38.11
	SCATV	0.0*	1.00*					2.68		73.47
	SCATQ	1.00*	0.0*						5.16	44.52

t	β_{Vt}	β_{Qt}	var (v_t)	var (q_t)	cov (v_t, q_t)	corr (v_t, q_t)
7	1.13	0.92	13.64	23.13	16.44	0.93
9	1.03	0.78	18.42	8.38	10.36	0.83
11	1.03	1.09	11.53	23.55	8.61	0.52

$x^2 = 313.29$ with 192 degrees of freedom

[a] Estimates obtained by stepwise analysis. Asterisks denote parameter values specified by hypothesis.

grades they are made up of items measuring logical reasoning and presented in verbal form. The fact that the loading of MATH11 is negative suggests that there is a relatively large verbal component in this test which is not accounted for by the model because of the constrained zero loading of MATH11 on V_{11}.

The relative contribution of each set of factors to the total test variance is relatively easily computed from Table 6 since the three sets of factors are uncorrelated. These variance components are given in Table 7. It is seen that the occasion-specific factors account for a large proportion of the total variance, and that the test-specific variances are relatively small even when compared with the unique variances.

6. Summary and conclusion

We have described a general model LISREL involving linear structural equations among a set of true variables which cannot be directly observed but which act as common factors of a set of observed variables. By allowing the parameters of the model to be fixed, free, or constrained,

TABLE 7

Relative variance contributions of the three sets of factors in Table 6.

Grade	Test	Occasion-specific	Test-specific	Unique
5	MATH	0.654	0.005	0.342
	SCI	0.770	0.033	0.197
	SS	0.799	0.006	0.196
	READ	0.696	0.098	0.206
	SCATV	0.715	0.061	0.224
	SCATQ	0.629	0.062	0.308
7	MATH	0.744	0.003	0.254
	SCI	0.665	0.066	0.270
	SS	0.798	0.014	0.188
	READ	0.747	0.064	0.190
	SCATV	0.734	0.113	0.153
	SCATQ	0.662	0.102	0.236
9	MATH	0.845	0.009	0.146
	SCI	0.727	0.083	0.190
	SS	0.749	0.023	0.229
	READ	0.687	0.108	0.205
	SCATV	0.683	0.037	0.280
	SCATQ	0.666	0.151	0.183
11	MATH	0.746	0.186	0.069
	SCI	0.535	0.079	0.386
	SS	0.801	0.033	0.166
	READ	0.736	0.070	0.194
	SCATV	0.563	0.039	0.398
	SCATQ	0.720	0.105	0.175

great flexibility is obtained in that the general model contains a wide range of specific models useful in the social and behavioral sciences. In this paper we focused attention on models for longitudinal or panel studies of the type often conducted in education, psychology and sociology. Using the LISREL program, each model may be estimated by the maximum-likelihood method based on the assumption that the observed variables have a multinormal distribution. When the estimates have been obtained the information matrix is computed and used to determine standard errors for the estimated parameters. The over-all fit of the model may be assessed by means of a χ^2-test.

We have discussed two-wave and multi-wave models with one, two, or more observed variables on each occasion and with or without background variables. For each model we have considered the identification problem and showed how the model may be put into the LISREL format for estimation purposes. Some of the models have been illustrated on educational and sociological data.

Some points for further research on LISREL are:

(i) How serious are departures from multinormality (a) for estimation and (b) for testing goodness-of-fit?
(ii) Can the assumption of multinormality of the observed variables be replaced by some other distributional assumption which is more valid for a wide range of problems?
(iii) How can LISREL be extended to handle data from several groups?
(iv) How can LISREL be extended to handle qualitative variables?

These topics will be studied in our future research.

References

Anderson, S.B. and M.H. Maier, 1963, 34,000 pupils and how they grew, Journal of Teacher Education 14, 212–216.

Anderson, T.W., 1959, Some stochastic process models for intelligence test scores, in: K.J. Arrow, S. Karlin and P. Suppes, eds., Mathematical methods in the social sciences (Stanford University Press, Stanford, CA).

Bohrnstedt, G.W., 1969, Observations on the measurement of change, in: E.F. Borgatta, ed., Sociological methodology (Jossey-Bass, San Francisco, CA).

Corballis, M.C., 1973, A factor model for analysing change, The British Journal of Mathematical and Statistical Psychology 26, 90–97.

Duncan, O.D., 1969, Some linear models for two-wave, two-variable panel analysis, Psychological Bulletin 72, 177–182.

Duncan, O.D., 1972, Unmeasured variables in linear models for panel analysis, in: H.L. Costner, ed., Sociological methodology (Jossey-Bass, San Francisco, CA).
Fisher, F.M., 1966, The identification problem in econometrics (McGraw-Hill, New York).
Fredriksen, C.R., 1974, Models for the analysis of alternative sources of growth in correlated stochastic variables, Psychometrika 39, 223–245.
Geraci, V.J., 1974, Simultaneous equation models with measurement error, Unpublished Ph.D. Dissertation (University of Wisconsin, Madison, WI).
Guttman, L., 1954, A new approach to factor analysis: The radex, in: P.F. Lazarsfeld, ed., Mathematical thinking in the social sciences (Columbia University Press, New York) 258–348.
Heise, D.R., 1969, Separating reliability and stability in test-retest correlation, American Sociological Review 34, 93–101.
Heise, D.R., 1970, Causal inference from panel data, in: E.F. Borgatta and G.W. Bohrnstedt, eds., Sociological methodology (Jossey-Bass, San Francisco, CA) 3–27.
Hilton, T.L., 1969, Growth study annotated bibliography, Progress Report 69–11 (Educational Testing Service, Princeton, NJ).
Jöreskog, K.G., 1970a, Factoring the multitest-multioccasion correlation matrix, in: C.E. Lunneborg, ed., Current problems and techniques in multivariate psychology, Proceedings of a conference honoring Professor Paul Horst (University of Washington, Seattle, WA) 68–100.
Jöreskog, K.G., 1970b, Estimation and testing of simplex models, The British Journal of Mathematical and Statistical Psychology 23, 121–145.
Jöreskog, K.G. and M. van Thillo, 1973, LISREL – A general computer program for estimating a linear structural equation system involving multiple indicators of unmeasured variables, Research Report 73-5 (Statistics Department, Uppsala University, Uppsala).
Kenny, D.A., 1973, Cross-lagged and synchronous common factors in panel data, in: A.S. Goldberger and O.D. Duncan, eds., Structural equation models in the social sciences (Seminar Press, New York) 153–165.
Lord, F. and M. Novick, 1968, Statistical theories of mental test scores (Addison-Wesley, Reading, MA).
Schmidt, W.H. and D.E. Wiley, 1974, Analytic problems in longitudinal data, Prepared for the Conference on Methodological Concerns in Evaluational Research (Chicago, IL).
Summers, G.F., R.L. Hough, J.T. Scott and C.L. Folse, 1969, Before industrialisation: A rural social system base study, Bulletin no. 736 (Illinois Agricultural Experiment Station, University of Illinois, Urbana, IL).
Werts, C.E., K.G. Jöreskog and R.L. Linn, 1971, Comment on "The estimation of measurement error in panel data", American Sociological Review 36, 110–113.
Werts, C.E., K.G. Jöreskog and R.L. Linn, 1972, A multitrait-multimethod model for studying growth, Educational and Psychological Measurement 32, 655–678.
Werts, C.E., K.G. Jöreskog and R.L. Linn, 1973, Identification and estimation in path analysis with unmeasured variables, American Journal of Sociology 78, 1469–1484.
Werts, C.E., K.G. Jöreskog and R.L. Linn, 1976, A simplex model for analyzing academic growth, Educational and Psychological Measurement 36.
Wheaton, B., B. Muthén, D. Alwin and G. Summers, 1977, Assessing reliability and stability in panel models with multiple indicators, in: D.R. Heise, ed., Sociological methodology, in press.
Wiley, D.E., 1973, The identification problem for structural equation models with unmeasured variables, in: A.S. Goldberger and O.D. Duncan, eds., Structural equation models in the social sciences (Seminar Press, New York) 69–83.
Wiley, D.E. and J.A. Wiley, 1970, The estimation of measurement error in panel data, American Sociological Review 35, 112–117.

Reprinted from: D. V. Aigner and A. S. Goldberg, eds., *Latent Variables in Socioeconomic Models* (Amsterdam: No. Holland Publishing Co., 1977), Chapter 16.

Chapter 6

Detection of Correlated Errors in Longitudinal Data

Dag Sörbom

A study of change in ability between two occasions may employ a number of tests believed to measure the ability in question. Either the same battery of tests is used on both occasions, or equivalent forms are used. For a variety of reasons, correlations may exist between certain errors remaining after eliminating variance due to true scores, and hence the classical factor analysis model is not applicable. A procedure for detecting correlations between errors is discussed. A search strategy is proposed since, even if the number of observed variables is small, the number of possible models is very large. A computer program is described, which produces maximum-likelihood estimates for the parameters in a factor analytic model in which the error variables may be correlated.

1. Introduction

Consider the following formal description of the classical factor analysis model (cf. Anderson, 1959).

The correlations of a number of observed variables are given. We are trying to explain these correlations by a number of variates called 'factors', fewer than the number of variables. Thus, when the influence of the factors on the variables is removed, the variables are supposed to be uncorrelated; that is,

$$\mathrm{cov}(\mathbf{x}|\mathbf{f}) = \boldsymbol{\psi}, \qquad (1)$$

where $\mathbf{x}$ is a random vector of observed variables, $\mathbf{f}$ is a random vector of factors and $\boldsymbol{\psi}$ is a diagonal matrix whose elements are variances of unique or residual variables assumed to be independent of $\mathbf{f}$.

In several recent papers there has been an interest in models for which this assumption, the assumption of local independence, does not hold (cf. Corballis, 1973; Costner & Schönberg, 1973; Kenny, 1973; Wiley & Hornik, 1973). In these papers, in contrast to the classical factor analysis model, there is no assumption that all error variates are uncorrelated. For example, when some traits are measured with the same tests at several occasions it may not be reasonable to assume that the measurement errors for the same tests at different occasions are uncorrelated (cf. Corballis, 1973); but, when similar tests are used to measure the same trait, a correlation between errors of measurement within occasions seems plausible. Conditions like these give rise to a great variety of different plausible models, and the search for 'the most plausible model' may be

Requests for reprints should be sent to Dr Dag Sörbom, Department of Statistics, University of Uppsala, P.O. Box 2300, S-750 02 Uppsala, Sweden.

hazardous. A strategy for such a search is proposed by Costner & Schönberg (1973). An alternative procedure is given in this paper.

The aim of the procedure is to guide a sequenced search for a model with an acceptable fit to data. At each stage the current model can be examined and perhaps reformulated to result in a model with more interpretable parameters. For example, an analysis might suggest a structure of the $\boldsymbol{\psi}$ matrix which is equivalent to a model with an extra factor added.

2. An Illustrative Example

The objective of the search procedure is to detect covariation between errors (measurement errors or residuals). In this section an example is given to illustrate how the procedure works.

Suppose we want to study the change in ability between two occasions for a group of individuals. Let ξ denote the random variable representing the ability at the first occasion for a random sample from the group, and η the ability for the same individuals at the second occasion. Suppose the structural equation

$$\eta = \alpha + \beta\xi + \zeta \tag{2}$$

describes the connexion between final and initial status. The residual ζ is supposed to be uncorrelated with ξ. Our main concern is the parameters α and β in (2). To estimate these we administer three tests at each occasion. Let $\mathbf{x}$ denote the measurements at the first occasion and $\mathbf{y}$ the measurements at the second. The following model for the measurements is assumed

$$\begin{aligned} x_1 &= \lambda_{11}\xi + \varepsilon_1, \\ x_2 &= \lambda_{21}\xi + \varepsilon_2, \\ x_3 &= \lambda_{31}\xi + \varepsilon_3 \end{aligned} \tag{3a}$$

and

$$\begin{aligned} y_1 &= \lambda_{12}\eta + \delta_1, \\ y_2 &= \lambda_{22}\eta + \delta_2, \\ y_3 &= \lambda_{32}\eta + \delta_3, \end{aligned} \tag{3b}$$

$\boldsymbol{\varepsilon}$ and $\boldsymbol{\delta}$ are vectors of errors, supposed to be uncorrelated with ξ and η. If we further assume that the variables in $\boldsymbol{\varepsilon}$ and $\boldsymbol{\delta}$ are uncorrelated and that

$$E(\boldsymbol{\varepsilon}) = E(\boldsymbol{\delta}) = \mathbf{0},$$

the model is a special case of that considered by Sörbom (1973).

Let us assume that x_1 and y_1 represent very similar tests administered at the two occasions. Then there may be covariation between x_1 and y_1 not only because of the covariation between the abilities at the two occasions, ξ and η, but also because of incidental features arising from the construction of the tests or because of memory effects. Thus it may not be true that $\text{cov}(\varepsilon_1, \delta_1) = 0$. Assume further that the tests x_2 and x_3 measure some traits, which are additional to but are uncorrelated with ξ. Thus covariation between x_2 and x_3 for a given value of ξ would be expected, and $\text{cov}(\varepsilon_2, \varepsilon_3) \neq 0$. If we assume that the errors for y_2 and y_3 are correlated too, the measurement model can be described as in Fig. 1.

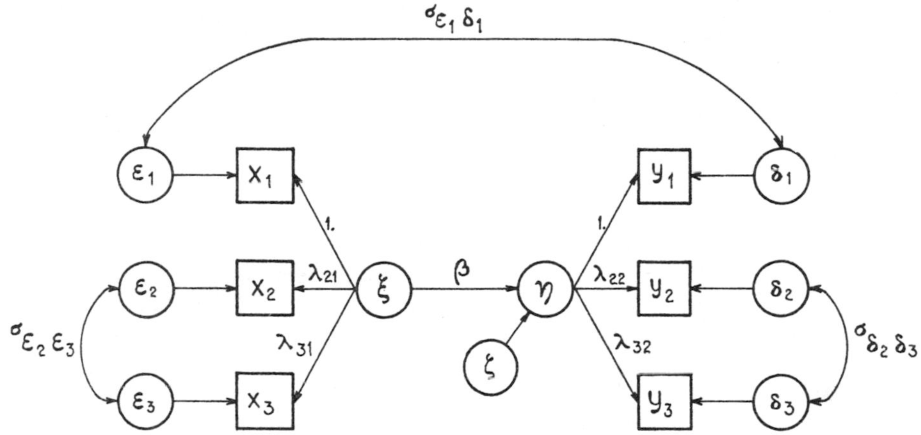

FIG. 1.—The true model. Circles denote unobservable variables and squares denote observable variables.

It should be noted that the parameters λ_{11}, λ_{21}, λ_{31} and σ_ξ^2 in (3*a*) cannot be identified simultaneously. We can, for example, multiply ξ by a constant and divide λ_{11}, λ_{21} and λ_{31} by the same constant without changing the measurements $\mathbf{x}$. Thus we have to fix at least one of these parameters. In the following λ_{11} is chosen to be equal to 1. Similarly λ_{12} is set equal to 1.

Let $\boldsymbol{\theta} = E\begin{pmatrix}\xi\\ \eta\end{pmatrix}$,

$$\boldsymbol{\Lambda} = \begin{bmatrix} 1 & 0 \\ \lambda_{21} & 0 \\ \lambda_{31} & 0 \\ 0 & 1 \\ 0 & \lambda_{22} \\ 0 & \lambda_{32} \end{bmatrix}, \quad \boldsymbol{\Phi} = E\left[\begin{pmatrix}\xi - \theta_1 \\ \eta - \theta_2\end{pmatrix}(\xi - \theta_1, \eta - \theta_2)\right] = \begin{bmatrix} \sigma_\xi^2 & \\ \sigma_{\xi\eta} & \sigma_\eta^2 \end{bmatrix},$$

$$\boldsymbol{\Psi} = E\left[\begin{pmatrix}\boldsymbol{\varepsilon} \\ \boldsymbol{\delta}\end{pmatrix}(\boldsymbol{\varepsilon}, \boldsymbol{\delta})\right] = \begin{bmatrix} \sigma_{\varepsilon_1}^2 & & & & & \\ 0 & \sigma_{\varepsilon_2}^2 & & & & \\ 0 & \sigma_{\varepsilon_2\varepsilon_3} & \sigma_{\varepsilon_3}^2 & & & \\ \sigma_{\varepsilon_1\delta_1} & 0 & 0 & \sigma_{\delta_1}^2 & & \\ 0 & 0 & 0 & 0 & \sigma_{\delta_2}^2 & \\ 0 & 0 & 0 & 0 & \sigma_{\delta_2\delta_3} & \sigma_{\delta_3}^2 \end{bmatrix}. \tag{4}$$

Consider the following parameter values for an artificial population:

$$\mathbf{\Lambda}_0 = \begin{bmatrix} 1{\cdot}0 & 0{\cdot}0 \\ 0{\cdot}8 & 0{\cdot}0 \\ 0{\cdot}5 & 0{\cdot}0 \\ 0{\cdot}0 & 1{\cdot}0 \\ 0{\cdot}0 & 0{\cdot}9 \\ 0{\cdot}0 & 0{\cdot}7 \end{bmatrix}, \quad \mathbf{\Phi}_0 = \begin{bmatrix} 16{\cdot}0 & \\ 12{\cdot}0 & 25{\cdot}0 \end{bmatrix}, \quad \sigma_{\varepsilon_0}{}^2 = \begin{pmatrix} 4{\cdot}0 \\ 2{\cdot}56 \\ 1{\cdot}0 \end{pmatrix}, \quad \sigma_{\delta_0}{}^2 = \begin{pmatrix} 6{\cdot}25 \\ 5{\cdot}0625 \\ 3{\cdot}0625 \end{pmatrix}$$

and

$$\mathbf{\theta}_0 = \begin{pmatrix} 10 \\ 20 \end{pmatrix}.$$

These parameters have been chosen so that the reliability of each test equals 0·8. This means, for example, that the variance of ε_2 is $\lambda_{21}{}^2\sigma_\xi{}^2(1{\cdot}0-0{\cdot}8)/0{\cdot}8 = 2{\cdot}56$. It is further assumed that the population correlation between ε_1 and δ_1, $\rho_{\varepsilon_1\delta_1}$, equals 0·5 and that $\rho_{\varepsilon_2\varepsilon_3}$ equals 0·3 and $\rho_{\delta_2\delta_3}$ equals 0·4.

The parameters of (2) are obtained by

$$\beta = \frac{\sigma_{\xi\eta}}{\sigma_\xi{}^2} = \frac{12}{16} = 0{\cdot}75,$$

$$\alpha = \theta_2 - \beta\theta_1 = 20-(0{\cdot}75)(10) = 12{\cdot}5,$$

and the mean values of the observed measurements are given by

$$E(\mathbf{x}) = \begin{bmatrix} \theta_1 \\ \lambda_{21}\theta_1 \\ \lambda_{31}\theta_1 \end{bmatrix} = \begin{bmatrix} 10 \\ 8 \\ 5 \end{bmatrix} \quad \text{and} \quad E(\mathbf{y}) = \begin{bmatrix} \theta_2 \\ \lambda_{22}\theta_2 \\ \lambda_{32}\theta_2 \end{bmatrix} = \begin{bmatrix} 20 \\ 18 \\ 14 \end{bmatrix}.$$

The population variance–covariance matrix for the observed measurements is given by

$$\mathbf{\Sigma}_0 = E\left[\begin{pmatrix} \mathbf{x}-E(\mathbf{x}) \\ \mathbf{y}-E(\mathbf{y}) \end{pmatrix} (\mathbf{x}-E(\mathbf{x}), \mathbf{y}-E(\mathbf{y}))\right]$$

$$= E\left[\mathbf{\Lambda}\begin{pmatrix} \xi \\ \eta \end{pmatrix} - \mathbf{\Lambda\theta} + \begin{pmatrix} \boldsymbol{\varepsilon} \\ \boldsymbol{\delta} \end{pmatrix}\right]\left[\mathbf{\Lambda}\begin{pmatrix} \xi \\ \eta \end{pmatrix} - \mathbf{\Lambda\theta} + \begin{pmatrix} \boldsymbol{\varepsilon} \\ \boldsymbol{\delta} \end{pmatrix}\right]' = \mathbf{\Lambda\Phi\Lambda}' + \mathbf{\Psi}$$

$$
= \begin{bmatrix}
\sigma_\xi^2+\sigma_{\epsilon_1}^2 & & & & & \\
\lambda_{21}\sigma_\xi^2 & \lambda_{21}^2\sigma_\xi^2+\sigma_{\epsilon_2}^2 & & & & \\
\lambda_{31}\sigma_\xi^2 & \lambda_{21}\lambda_{31}\sigma_\xi^2+\sigma_{\epsilon_2\epsilon_3} & \lambda_{31}^2\sigma_\xi^2+\sigma_{\epsilon_3}^2 & & & \\
\sigma_{\xi\eta}+\sigma_{\epsilon_1\delta_1} & \lambda_{21}\sigma_{\xi\eta} & \lambda_{31}\sigma_{\xi\eta} & \sigma_\eta^2+\sigma_{\delta_1}^2 & & \\
\lambda_{22}\sigma_{\xi\eta} & \lambda_{21}\lambda_{22}\sigma_{\xi\eta} & \lambda_{31}\lambda_{22}\sigma_{\xi\eta} & \lambda_{22}\sigma_\eta^2 & \lambda_{22}^2\sigma_\eta^2+\sigma_{\delta_2}^2 & \\
\lambda_{32}\sigma_{\xi\eta} & \lambda_{21}\lambda_{32}\sigma_{\xi\eta} & \lambda_{31}\lambda_{32}\sigma_{\xi\eta} & \lambda_{32}\sigma_\eta^2 & \lambda_{22}\lambda_{32}\sigma_\eta^2+\sigma_{\delta_2\delta_3} & \lambda_{32}^2\sigma_\eta^2+\sigma_{\delta_3}^2
\end{bmatrix}
$$

$$
= \begin{bmatrix}
20{\cdot}0000 & & & & & \\
12{\cdot}8000 & 12{\cdot}8000 & & & & \\
8{\cdot}0000 & 7{\cdot}3600 & 5{\cdot}0000 & & & \\
14{\cdot}5000 & 9{\cdot}6000 & 6{\cdot}0000 & 31{\cdot}2500 & & \\
10{\cdot}8000 & 8{\cdot}6400 & 5{\cdot}4000 & 22{\cdot}5000 & 25{\cdot}3125 & \\
8{\cdot}4000 & 6{\cdot}7200 & 4{\cdot}2000 & 17{\cdot}5000 & 17{\cdot}3250 & 15{\cdot}3125
\end{bmatrix}. \tag{5}
$$

To illustrate the procedure a sample variance–covariance matrix, $\mathbf{S}$, and a vector of sample means, $\bar{\mathbf{z}}' = (\bar{\mathbf{x}}, \bar{\mathbf{y}})'$, was obtained by generating 1000 normal variates according to a normal distribution with mean equal to $\boldsymbol{\Lambda}_0\boldsymbol{\theta}_0$ and variance–covariance matrix equal to $\boldsymbol{\Sigma}_0$.

Thus it is as though for six tests we have obtained independent measurements from a sample of 1000 individuals belonging to the population of individuals under study. Three of the tests were administered at one occasion and three were administered at a later occasion. At each occasion the tests are measuring approximately the same ability. We are mainly concerned with the development of this ability between the two occasions. Because of the influence of memory effects, imperfections in the construction of the tests and similar matters, the estimates of the parameters in (2) might be distorted. The procedure to be presented is concerned with how to detect these distorting effects, and how to take them into account.

Given the data from the above simulated study, it seems natural to start with a model in which the errors are uncorrelated; that is,

$$
\boldsymbol{\Sigma} = \boldsymbol{\Lambda\Phi\Lambda}' + \boldsymbol{\Psi}, \tag{6}
$$

where $\boldsymbol{\Psi}$ is a diagonal matrix and $\boldsymbol{\Lambda}$ is structured as in (4). Maximum-likelihood estimates of the parameters of the model may be obtained by minimizing (see Section 4)

$$f = \log|\mathbf{\Sigma}| + \mathrm{tr}(\mathbf{T\Sigma}^{-1}) - \log|\mathbf{S}| - p$$

and a χ^2 measure of goodness of fit may be obtained as N times the minimum value of f (see Sörbom, 1973). The minimization of f may be done by a computer program developed by Sörbom (1974*a*). The χ^2 goodness-of-fit statistic of this model is given in Table 1, row 1. Of course, as is seen, the fit is not acceptable.

Table 1.—*Goodness of Fit for the Different Models in the Simulated Example*

Model	χ^2	d.f.	Probability level	χ^2 for test of the hypothesis $\psi_{ij} = 0$ (d.f. = 1)
Uncorrelated errors	216·491	12	0·000	
ψ_{32} free	163·217	11	0·000	53·274
ψ_{32}, ψ_{41} free	35·227	10	0·000	127·990
ψ_{32}, ψ_{41}, ψ_{65} free	14·430	9	0·108	20·797
ψ_{32}, ψ_{41}, ψ_{65}, ψ_{62} free	13·924	8	0·084	0·506

To improve the fit one can relax some of the restrictions imposed on the elements of either $\mathbf{\Lambda}$ or $\mathbf{\Psi}$, the elements of $\mathbf{\Phi}$ being already unrestricted. However, those elements in $\mathbf{\Lambda}$ set equal to zero reflect the basic structure of the measurements; that is, the tests are measuring one ability at each occasion. Thus the only restrictions that one would be willing to relax are the zeros in the off-diagonal elements of $\mathbf{\Psi}$. The question is: which? As indicated by Costner & Schönberg (1973), an examination of the differences between the observed covariances and the covariances implied by the estimated model may be misleading, when we are dealing with maximum-likelihood estimates. If, on the other hand, we had obtained the estimates by minimizing $\mathrm{tr}(\mathbf{\Sigma}-\mathbf{S})^2$, these differences might have been useful.

The problem can be summarized as follows. We want to find those off-diagonal elements of $\mathbf{\Psi}$ which are different from zero. Our metric for the goodness of fit of a model is the function f which is in principle the negative of the log-likelihood function, if the observed variables have a multinormal distribution. Thus, since we want to find the restriction which is least probable, we should relax the zero-restriction for that element which gives the largest decrease in f. This leads us to find the indices i and j such that the absolute value of $\partial f/\partial\psi_{ij}$ is greatest. In Table 2 the derivatives $\partial f/\partial\boldsymbol{\psi}$ are given for the model (6). An inspection of this table suggests that ψ_{32} is not zero; that is, there is a covariation between ε_2 and ε_3. Thus, we release the restriction $\psi_{32} = 0$ and estimate the parameters of this new model. This can be done by another computer program developed by Sörbom (1974*b*). The fit of this model, as given in Table 1, is still not acceptable, but a test of $\psi_{32} = 0$ with one degree of freedom shows that the hypotheses $\psi_{32} = 0$ is not tenable. An inspection of the derivatives $\partial f/\partial\psi_{ij}$ for this model, as given in Table 3, suggests that ψ_{41} may not be equal to zero. Thus we relax the restriction $\psi_{41} = 0$ and estimate the

Table 2.—Derivatives and Residuals for the Model with Uncorrelated Errors

Derivatives ($\partial f/\partial \boldsymbol{\psi}$)

x_1	0·000					
x_2	0·013	0·000				
x_3	0·026	−0·057	0·000			
y_1	−0·045	0·027	0·031	0·000		
y_2	0·022	−0·012	−0·020	0·005	0·000	
y_3	0·027	−0·019	−0·008	0·011	−0·019	0·000

Residuals ($\hat{\boldsymbol{\Sigma}} - \mathbf{S} - [(\bar{\mathbf{x}}, \bar{\mathbf{y}})' - \hat{\boldsymbol{\Lambda}}\hat{\boldsymbol{\theta}}]\,[(\bar{\mathbf{x}}, \bar{\mathbf{y}})' - \hat{\boldsymbol{\Lambda}}\hat{\boldsymbol{\theta}}]'$)

x_1	3·764					
x_2	1·370	−0·413				
x_3	0·912	−0·258	−0·127			
y_1	−0·946	0·345	0·243	3·988		
y_2	0·777	−0·320	−0·157	1·338	−1·044	
y_3	0·868	−0·063	0·013	1·357	−0·698	−0·280

Table 3.—Derivatives and Residuals for the Model with ψ_{32} free

Derivatives

x_1	0·000					
x_2	−0·002	0·000				
x_3	0·003	0·000	0·000			
y_1	−0·046	0·024	0·025	0·000		
y_2	0·028	−0·014	−0·022	0·007	0·000	
y_3	0·037	−0·023	−0·012	0·013	−0·024	0·000

Residuals

x_1	0·433					
x_2	−0·015	−0·299				
x_3	0·043	−0·151	−0·072			
y_1	−1·062	0·062	0·064	3·824		
y_2	0·651	0·575	−0·318	1·308	−1·034	
y_3	0·772	−0·262	−0·112	1·334	−0·722	−0·276

parameters again. At this step we find by Table 4 that the element ψ_{65} may not be equal to zero. A new estimation gives the result reported in Table 5. Now the fit of the model is acceptable, and it is seen that the original structure of the measurements has been retrieved.

The above procedure may be compared with that obtained by considering the residuals, which are given in Table 2 to Table 4. It will be seen that at no step would the correct covariance have been detected by choosing the largest off-diagonal residual. Further, it may seem irrational to relax only one restriction at each step, but an inspection of the information matrix shows that the estimates of the parameters in $\boldsymbol{\psi}$ are correlated. Thus, a procedure which relaxes more than one restriction at a time will be quite complex.

In applications with real data the χ^2 values for the different models may not indicate when to stop the model-fitting. In particular, when the number of observations is large the χ^2 values may well indicate that any conceptually

Table 4.—Derivatives and Residuals for the Model with ψ_{32} and ψ_{41} free

Derivatives

x_1	0·000					
x_2	−0·002	0·000				
x_3	0·003	0·000	0·000			
y_1	0·000	0·001	−0·008	0·000		
y_2	0·001	0·000	−0·003	0·004	0·000	
y_3	0·004	−0·006	0·011	0·009	−0·016	0·000

Residuals

x_1	0·719					
x_2	0·107	−0·263				
x_3	0·118	−0·123	−0·051			
y_1	1·666	0·152	0·120	4·201		
y_2	0·728	−0·492	−0·267	1·471	−0·948	
y_3	0·482	−0·197	−0·072	1·460	−0·604	−0·228

Table 5.—Parameter Estimates for the Final Model and their Standard Errors (within parentheses)

Factor loadings (Λ)

x_1	1·000†	0·000†
x_2	0·809 (0·007)	0·000†
x_3	0·507 (0·004)	0·000†
y_1	0·000†	1·000†
y_2	0·000†	0·895 (0·004)
y_3	0·000†	0·695 (0·004)

Factor variance–covariance matrix (Φ)

ξ	15·914 (0·794)	
η	11·711 (0·804)	25·640 (1·264)

Free parameters of ψ

$\hat{\sigma}_{\varepsilon_1}{}^2 = 3{\cdot}793\ (0{\cdot}341)$	$\hat{\sigma}_{\varepsilon_2\varepsilon_3} = 0{\cdot}824\ (0{\cdot}129)$
$\hat{\sigma}_{\varepsilon_2}{}^2 = 2{\cdot}270\ (0{\cdot}220)$	$\hat{\sigma}_{\varepsilon_1\delta_1} = 2{\cdot}395\ (0{\cdot}259)$
$\hat{\sigma}_{\varepsilon_3}{}^2 = 0{\cdot}968\ (0{\cdot}089)$	$\hat{\sigma}_{\delta_2\delta_3} = 1{\cdot}361\ (0{\cdot}306)$
$\hat{\sigma}_{\delta_1}{}^2 = 6{\cdot}207\ (0{\cdot}536)$	
$\hat{\sigma}_{\delta_2}{}^2 = 4{\cdot}497\ (0{\cdot}434)$	
$\hat{\sigma}_{\delta_3}{}^2 = 2{\cdot}957\ (0{\cdot}271)$	

Factor means

$\hat{E}(\xi) = 10{\cdot}002\ (0{\cdot}139)$
$\hat{E}(\eta) = 20{\cdot}007\ (0{\cdot}178)$

† Fixed parameters.

plausible model is non-acceptable (cf. Jöreskog, 1971, p. 421). In such a case examination of the difference in χ^2 values between two consecutive models in the procedure may help decide when elaboration of the model should stop. After the fourth step of the example, the derivatives suggest that ψ_{62} might be different from zero (see Table 6). An estimation of this model results in a decrease in χ^2 by 0·506. Thus the hypothesis that $\psi_{62} = 0$ cannot be rejected. Since ψ_{62} is the parameter that gives the greatest decrease in χ^2, at least locally, it may be concluded that none of the other fixed ψ parameters is significantly different from zero.

Table 6.—Derivatives and Residuals for the Model with ψ_{32}, ψ_{41} and ψ_{65} free

Derivatives						
x_1	0·000					
x_2	−0·002	0·000				
x_3	0·003	0·000	0·000			
y_1	0·000	0·005	−0·003	0·000		
y_2	0·003	−0·003	−0·008	−0·001		
y_3	0·006	−0·010	0·006	0·001	0·000	0·000
Residuals						
x_1	0·808					
x_2	0·116	−0·361				
x_3	0·124	−0·186	−0·091			
y_1	1·186	−0·042	−0·002	1·213		
y_2	0·472	−0·704	−0·399	0·312	−0·469	
y_3	0·633	−0·362	−0·175	0·560	−0·124	0·090

3. An Application based on Longitudinal Growth Data

To exemplify the procedure described in the previous section, some data from the ETS longitudinal study of academic prediction and growth (cf. Anderson & Maier, 1963) are analysed. As part of that study, scores from the Sequential Test of Educational Progress (STEP) and School and College Ability Test (SCAT) were obtained. The data selected for the example concern 383 examinees for whom complete data are available, and three tests measuring verbal ability employed on two occasions. The following variables are identified:

x_1 = STEP Reading, Girls Academic, grade 9,

x_2 = STEP Writing, Girls Academic, grade 9,

x_3 = SCAT Verbal, Girls Academic, grade 9,

y_1 = STEP Reading, Girls Academic, grade 11,

y_2 = STEP Writing, Girls Academic, grade 11,

y_3 = SCAT Verbal, Girls Academic, grade 11.

In STEP Reading the testee is asked to read some written material and then to answer a number of multiple-choice items in connexion with each piece. Most of the items measure reading comprehension. In the items of STEP Writing the testee is usually asked to choose the correct evaluation of a given text with respect to grammatical or conceptual aspects. The SCAT Verbal test contains items of word completion and word synonyms. Thus, the information available about the measurements is rather complete, and it may seem meaningless to use the procedure in this case, since it is merely constructed to be used in exploratory studies. On the other hand, the real data may be used as a test of whether the procedure produces a reasonable model or not.

As in Section 2 the first model analysed was a model with uncorrelated errors. By means of the χ^2 value for the overall fit of this model, as given in Table 7,

Table 7.—Goodness of Fit for the Different Models in the Application Example

Model	χ^2	d.f.	Probability level	χ^2 for test of the hypothesis $\psi_{ij} = 0$ (d.f. = 1)
Uncorrelated errors	74·219	12	0·000	
ψ_{63} free	27·393	11	0·004	46·826
ψ_{63}, ψ_{52} free	22·906	10	0·011	4·487

Table 8.—Derivatives for the Model with Uncorrelated Errors†

x_1	0·000					
x_2	−0·285	0·000				
x_3	0·201	0·092	0·000			
y_1	−0·126	−0·062	0·389	0·000		
y_2	−0·084	−0·167	0·179	−0·100	0·000	
y_3	0·466	0·291	−0·903	−0·010	0·115	0·000

† All values multiplied by 100.

it is concluded that the data do not fulfil the assumptions imposed. An inspection of the derivatives, shown in Table 8, suggests that the element ψ_{63} is not equal to zero. This means that there might be a correlation between the errors for the test SCAT Verbal across occasions. This is reasonable since the test SCAT Verbal, as compared with the other two tests, measures 'vocabulary' rather than 'a verbal ability'. Thus, when the influence of the verbal ability has been removed from the test scores at the two occasions, there might be covariation left due to the variation in vocabularly among the testees.

If the constraint $\psi_{63} = 0$ is relaxed a considerable decrease in χ^2 is obtained as indicated in Table 7. An inspection of the derivatives for this model, shown in Table 9, suggests that the element ψ_{52} is not equal to zero. This interpretation

Table 9.—Derivatives for the Model with ψ_{63} free†

x_1	0·000					
x_2	−0·122	0·000				
x_3	0·130	−0·005	0·000			
y_1	0·039	0·005	0·049	0·000		
y_2	0·042	−0·132	−0·121	0·043	0·000	
y_3	0·037	0·081	0·000	−0·101	0·058	0·000

† All values multiplied by 100.

of the derivatives is not as clear cut as in the previous case, and a release of the constraint $\psi_{52} = 0$ does not produce a significant decrease in χ^2 at the 1 per cent level, χ^2 with 1 degree of freedom being equal to 4·487. Further, the element ψ_{52} is estimated to be equal to 8·981 with a standard error equal to 4·363. Thus, in addition, the hypothesis that the separate parameter ψ_{52} is equal to zero cannot be rejected.

The estimates of the parameters in the model with ψ_{63} treated as a free parameter are listed in Table 10. It is seen that the covariance between the

Table 10.—*Parameter Estimates for the Model with* ψ_{63} *treated as a Free Parameter, and Standard Errors of the Estimates (in parentheses)*

Factor loadings (Λ)

STEP Reading$_9$	1·000†	0·000†
STEP Writing$_9$	0·995 (0·002)	0·000†
SCAT Verbal$_9$	0·968 (0·001)	0·000†
STEP Reading$_{11}$	0·000†	1·000†
STEP Writing$_{11}$	0·000†	0·990 (0·002)
SCAT Verbal$_{11}$	0·000†	0·955 (0·001)

Factor variance–covariance matrix (Φ) and factor means (θ)

	Variances and covariance		Means
Verbal ability$_9$	125·394 (9·914)		291·418 (0·643)
Verbal ability$_{11}$	131·979 (10·362)	151·403 (12·026)	303·057 (0·718)

Free parameters of Ψ

$\sigma_{\varepsilon_1}{}^2$ = 32·753 (3·270)	$\sigma_{\varepsilon_3\delta_3}$ = 14·661 (2·521)
$\sigma_{\varepsilon_2}{}^2$ = 73·842 (6·045)	
$\sigma_{\varepsilon_3}{}^2$ = 25·302 (2·745)	
$\sigma_{\delta_1}{}^2$ = 45·962 (4·384)	
$\sigma_{\delta_2}{}^2$ = 65·962 (5·658)	
$\sigma_{\delta_3}{}^2$ = 33·344 (3·470)	

† Fixed parameters.

errors for the test SCAT Verbal is considerable. In fact, the corresponding correlation coefficient is estimated as 0·505, although this is an 'attenuated' estimate of the 'vocabulary correlation', since the variances of the errors also contain measurement error variances.

The obtained model may be considered as a model with three common factors, the two ability factors plus a SCAT Verbal factor common to the two occasions. However, a model of this kind cannot be made identifiable in any reasonable way when the tests are known to contain measurement errors. All we can do is to include a term for the covariance between the errors to reduce the effect of this covariance on estimates of the parameters in the structural eqn. (2).

4. Some Statistical Properties of the Procedure

In this section the procedure exemplified in the previous sections is shown to be a special case of the Lagrangian multipliers test as described by Aitchison &

Silvey (1958) and Silvey (1958). The derivation given here is rather heuristic. For a rigorous derivation the reader is referred to the papers cited. It should be noted that the procedure is rather an *ad hoc* rule than a statistical procedure. Byron (1972) proposes a similar procedure for econometric systems and he recommends us not to use it too literally.

At each step of the procedure the introduction of a free parameter should be accompanied by an examination of its empirical interpretation. Thus the aim of the procedure is merely to generate hints in searching for the 'truth'. In any case, as already demonstrated in Section 2, it should be better than a procedure based on the estimated residuals.

In the most general case, the model considered in Sörbom (1974*b*) deals with the following equation for the observable variables, $\mathbf{x}$:

$$\mathbf{x}^{(g)} = \boldsymbol{\mu} + \boldsymbol{\Lambda}^{(g)} \boldsymbol{\xi}^{(g)} + \boldsymbol{\varepsilon}^{(g)} \quad (g = 1, 2, \ldots, m). \tag{7}$$

The superscript, g, refers to a division of the observations into m exclusive groups (cf. Sörbom, 1974*c*). The term $\boldsymbol{\xi}^{(g)}$ is a random vector of common factors assumed to be distributed as $N(\boldsymbol{\theta}^{(g)}, \boldsymbol{\Phi}^{(g)})$, $\boldsymbol{\varepsilon}^{(g)}$ is a random vector of unique factors, assumed to be uncorrelated with $\boldsymbol{\xi}^{(g)}$ and distributed as $N(\mathbf{0}, \boldsymbol{\Psi}^{(g)})$, $\boldsymbol{\mu}$ is a constant vector representing a common origin for the observable variables over the groups. $\boldsymbol{\Lambda}^{(g)}$ is a matrix of factor loadings.

The variance–covariance matrix for the observed variables in (7) is given by

$$\boldsymbol{\Sigma}^{(g)} = \boldsymbol{\Lambda}^{(g)} \boldsymbol{\Phi}^{(g)} \boldsymbol{\Lambda}^{(g)\prime} + \boldsymbol{\Psi}^{(g)}, \tag{8a}$$

and the vector of expected values by

$$E(\mathbf{x}^{(g)}) = \boldsymbol{\mu} + \boldsymbol{\Lambda}^{(g)} \boldsymbol{\theta}^{(g)}. \tag{8b}$$

The maximum-likelihood estimators are defined as those values of $\boldsymbol{\Lambda}^{(g)}$, $\boldsymbol{\Phi}^{(g)}$, $\boldsymbol{\Psi}^{(g)}$, $\boldsymbol{\theta}^{(g)}$ and $\boldsymbol{\mu}$ which maximize the likelihood function or, equivalently, the arguments to the function

$$F = \sum_{g=1}^{m} N^{(g)} [\log|\boldsymbol{\Sigma}^{(g)}| + \operatorname{tr}(\mathbf{T}^{(g)} \boldsymbol{\Sigma}^{(g)-1}) - \log|\mathbf{S}^{(g)}| - p] \tag{9}$$

at the minimum (cf. Sörbom, 1974*c*). $\mathbf{T}^{(g)}$ in (9) is the matrix

$$\mathbf{T}^{(g)} = \mathbf{S}^{(g)} + (\bar{\mathbf{x}}^{(g)} - \boldsymbol{\Lambda}^{(g)} \boldsymbol{\theta}^{(g)})(\bar{\mathbf{x}}^{(g)} - \boldsymbol{\Lambda}^{(g)} \boldsymbol{\theta}^{(g)})',$$

where $\mathbf{S}^{(g)}$ is the sample variance–covariance matrix for the gth group and $\bar{\mathbf{x}}^{(g)}$ is the vector of sample means. p is the number of observed variables and $N^{(g)}$ is the number of observations in the gth group.

The function F in (9) is a function of the elements of the matrices $\boldsymbol{\Lambda}^{(g)}$, $\boldsymbol{\Phi}^{(g)}$, $\boldsymbol{\Psi}^{(g)}$ and the vectors $\boldsymbol{\mu}$ and $\boldsymbol{\theta}^{(g)}$ for $g = 1, 2, \ldots, m$. Define $\boldsymbol{\pi}$ as the vector containing all these elements, in the following way

$$\begin{aligned} \boldsymbol{\pi} = (&\lambda_{11}^{(1)}, \lambda_{12}^{(1)}, \ldots, \lambda_{pk}^{(1)}, \phi_{11}^{(1)}, \phi_{21}^{(1)}, \ldots, \phi_{kk}^{(1)}, \\ &\psi_{11}^{(1)}, \psi_{21}^{(1)}, \ldots, \psi_{pp}^{(1)}, \theta_{1}^{(1)}, \theta_{2}^{(1)}, \ldots, \theta_{k}^{(1)}, \\ &\lambda_{11}^{(2)}, \lambda_{12}^{(2)}, \ldots, \ldots, \ldots, \theta_{k}^{(m)}, \mu_1, \mu_2, \ldots, \mu_p). \end{aligned}$$

The computer program described in Sörbom (1974*b*) minimizes F in (9) as a function of $\boldsymbol{\pi}$ under certain constraints on the elements of $\boldsymbol{\pi}$. The constraints are of two kinds, namely (i) fixed elements and (ii) equalities among elements. Thus the minimization procedure may be stated as follows. Find the minimum of $F(\boldsymbol{\pi})$ subject to the constraint $\mathbf{H}\boldsymbol{\pi} = \mathbf{s}$. This means that if the jth element of $\boldsymbol{\pi}$ belongs to category (i), then there is a 1 in the ith row and jth column of $\mathbf{H}$, i is the index running over the total number of constraints. All other elements of the ith row of $\mathbf{H}$ are zeros and s_i is the constant to which π_j is fixed. A constraint of the category (ii) gives rise to a row in $\mathbf{H}$ consisting of 1's in the positions corresponding to the elements of $\boldsymbol{\pi}$ involved in the constraint, except for one equal to -1 and zeros elsewhere. In this case s_j is equal to zero. If the Lagrangian multipliers $\boldsymbol{\nu}$ are introduced, the estimates of the parameters of the model are found as the solution of

$$\left.\begin{aligned} \partial F(\boldsymbol{\pi})/\partial\boldsymbol{\pi} + \mathbf{H}'\boldsymbol{\nu} &= \mathbf{0}, \\ \mathbf{H}\boldsymbol{\pi} &= \mathbf{0}. \end{aligned}\right\} \tag{10}$$

As noted by Byron (1972), this implies that for a fixed element of $\boldsymbol{\pi}$ the derivative at the minimum is equal to minus the Lagrangian multiplier. Thus the procedure described in the previous sections is equivalent to selecting for relaxation that fixed element of $\boldsymbol{\pi}$ which has the greatest absolute value of its Lagrangian multiplier.

Byron (1972) proposes that we shall relax those fixed elements for which the ν_i's are significantly different from zero. To do this we have to find the standard errors for the vector $\boldsymbol{\nu}$. As shown by Silvey (1970), for example, this involves in principle the inversion of the information matrix for $\boldsymbol{\pi}$ and $\boldsymbol{\nu}$. This matrix is of order the total number of parameters in the model (7) minus the number of parameters constrained in order to make the model identified. Thus, the order of the matrix is so big that its storage may exhaust the memory of today's computers even for rather moderate-sized models. Further, if the null hypothesis, that is, the hypothesis that the data analysed are generated according to a specified model, is not accepted, then the estimates of $\boldsymbol{\nu}$ are biased, and Byron (1972, p. 755) concludes after a Monte Carlo study: 'it cannot be asserted that having rejected the null hypothesis, information can be gained from the rejected hypothesis about the true nature of the model which generated the sample'.

By the above, the *ad hoc* status of the procedure is apparent. It rather appeals to intuition in the sense that if a model is rejected we can find that fixed parameter which, when relaxed, gives a relatively large decrease in our measure of fit, the χ^2 value. It should be noted that there is no guarantee that we find the parameter that gives the greatest decrease, since, of course, we do not beforehand know the value of the parameter. A large change of one parameter with a small derivative could, at least theoretically, give rise to a greater decrease than a smaller change in another parameter with a large derivative. Thus the procedure shares the objection with the Lagrangian multiplier test, that as long as we do not know the population parameters we cannot be certain to find the misspecified parameters. The procedure only gives hints, and as long as these hints give interpretable results we should accept them.

The research reported in this paper has been supported by the Swedish Council for Social Science Research under the project, Statistical Methods for Analysis of Longitudinal Data, project director K. G. Jöreskog.

References

AITCHISON, J. & SILVEY, S. D. (1958). Maximum-likelihood estimation of parameters subject to restraints. *Ann. math. Statist.* **29**, 813–828.

ANDERSON, S. B. & MAIER, M. H. (1963). 34,000 pupils and how they grow. *J. Teach. Educ.* **14**, 212–216.

ANDERSON, T. W. (1959). Some scaling models and estimation procedures in the latent class model. In U. Grenander (ed.), *Probability and Statistics*. New York: Wiley.

BYRON, R. P. (1972). Testing for misspecification in econometric systems using full information. *Int. econ. Rev.* **13**, 745–756.

CORBALLIS, M. C. (1973). A factor model for analysing change. *Br. J. math. statist. Psychol.* **26**, 90–97.

COSTNER, H. L. & SCHÖNBERG, R. (1973). Diagnosing indicator ills in multiple indicator models. In A. S. Goldberger & O. D. Duncan (eds.), *Structural Equation Models in the Social Sciences*. New York: Seminar Press.

JÖRESKOG, K. G. (1971). Simultaneous factor analysis in several populations. *Psychometrika* **36**, 409–426.

KENNY, D. A. (1973). Cross-lagged and synchronous common factors in panel data. In A. S. Goldberger & O. D. Duncan (eds.), *Structural Equation Models in the Social Sciences*. New York: Seminar Press.

SILVEY, S. D. (1958). The Lagrangian multiplier test. *Ann. math. Statist.* **30**, 389–407.

SILVEY, S. D. (1970). *Statistical Inference*. Harmondsworth: Penguin Books.

SÖRBOM, D. (1973). A statistical model for the measurement of change. Research Report 73–6, Statistics Department, Uppsala University.

SÖRBOM, D. (1974*a*). FASPSM: a computer program for factor analysis in several populations with structured means. (In preparation.)

SÖRBOM, D. (1974*b*). FASCOE: a computer program for factor analysis in several populations with structured means and correlated errors. (In preparation.)

SÖRBOM, D. (1974*c*). A general method for studying differences in factor means and factor structure between groups. *Br. J. math. statist. Psychol.* **27**, 229–239.

WILEY, D. E. & HORNIK, R. (1973). Measurement error and the analysis of panel data. Report no. 5, Studies of Educative Processes, University of Chicago.

Reprinted from: *British Journal of Mathematical and Statistical Psychology*, 1975, Vol. 28, pp. 138-151.

Part III
New Models for Group Comparisons

Introduction to Part III

Different causal models may be in operation for different groups of populations. For example, anxiety about enrolling in a high school mathematics course may result from different sources for males and females and may have a stronger deterrent effect on females. The influence of home background on achievement may be reduced for participants in a compensatory education program while for similar nonparticipants it may maintain its high level of disadvantage. To identify such differences between groups it is important to analyze each group separately and to make comparisons across groups. Part III of this book describes the LISREL approach for performing simultaneous analyses across multiple groups.

Chapter 7 extends the factor analysis models, discussed in part I of this book, to the case of several populations. Statistical tests are described for a variety of sequential hypotheses including tests for an equal number of factors within each group and for invariance in factor structure or factor correlations. A method for scaling factors to make results more easily interpretable is also presented. A detailed example illustrates the general approach using data from nine psychological tests administered to seventh- and eighth-grade pupils in two different schools.

In chapter 8 Dag Sörbom presents a multiple group LISREL model that incorporates the factor means for each group. Two applications are presented to illustrate the approach using data from the Project Talent study in addition to the data on the nine psychological tests analyzed in chapter 7. The applications show how the factor means can be used to plot group profiles associated with the latent factors. Tests for the equality of factor means across groups are also discussed.

It is often the case in quasi (nonrandomized) experiments that the experimental group scores higher or lower than the control group in the outcome measure of performance even before the experiment. To determine the treatment effect in this confounded setting, one must be able to estimate the portion of the differences that should be attributed to the preexisting differences. Dag Sörbom tackles this difficult problem in chapter 9 as he extends the methods of the previous two chapters to yield methodology that offers substantial improvement over the traditional analysis of covariance for the assessment of treatment effects.

Chapter 7

Simultaneous Factor Analysis in Several Populations

Karl G. Jöreskog

This paper is concerned with the study of similarities and differences in factor structures between different groups. A common situation occurs when a battery of tests has been administered to samples of examinees from several populations.

A very general model is presented, in which any parameter in the factor analysis models (factor loadings, factor variances, factor covariances, and unique variances) for the different groups may be assigned an arbitrary value or constrained to be equal to some other parameter. Given such a specification, the model is estimated by the maximum likelihood method yielding a large sample χ^2 of goodness of fit. By computing several solutions under different specifications one can test various hypotheses.

The method is capable of dealing with any degree of invariance, from the one extreme, where nothing is invariant, to the other extreme, where everything is invariant. Neither the number of tests nor the number of common factors need to be the same for all groups, but to be at all interesting, it is assumed that there is a common core of tests in each battery that is the same or at least content-wise comparable.

1. *Introduction and Summary*

This paper is concerned with the study of similarities and differences in factor structures between different groups. A common situation is when a battery of tests has been administered to samples of examinees from several populations. Traditionally this type of problem has been solved by obtaining orthogonal unrotated solutions for each group separately, rotating these to similarity and examining various similarity indices.

Meredith [1964a], using Lawley's [1943–44] selection theorem, showed that, under certain conditions, when the various populations are derivable as subpopulations from a parent population under selection on some external variable, there is a factor pattern that is invariant over populations. Meredith [1964b] gives two methods for estimating the common factor pattern by least squares rotation of independent orthogonal solutions for each group into a common factor pattern. The common factor pattern may be rotated further, orthogonally or obliquely, to a more readily interpretable solution.

The method to be presented is in some respects more general and yields

This research was supported by grant NSF-GB-12959 from National Science Foundation. My thanks are due to Michael Browne for his comments on an earlier draft of this paper and to Marielle van Thillo who checked the mathematical derivations and wrote and debugged the computer program SIFASP.

efficient estimates in large samples. It is more general in several respects. Firstly, the method may be used regardless of whether the populations are derived by selection or not. The only requirement is that the populations be clearly defined and the samples independent. Secondly, the method is capable of dealing with any degree of invariance, from the one extreme, where nothing is invariant, to the other extreme, where everything is invariant. Thirdly, neither the number of tests nor the number of common factors need to be the same for all groups, but to be at all interesting it is assumed that there is a common core of tests in each battery that is the same or at least content-wise comparable.

A very general model is presented, in which any parameter in the factor analysis models (factor loading, factor variance, factor covariance, and unique variance) for the different populations may be assigned an arbitrary value or constrained to be equal to some other parameter. Given such a specification, the model is estimated by the maximum likelihood method assuming the observed variables to have a multinormal distribution in each population. This yields a large sample χ^2 test of the goodness of fit of the overall model. By computing several solutions under different specifications one can test various hypotheses. For example, one can test the hypothesis of an invariant factor pattern or the hypothesis of an invariant specified simple structure factor pattern.

2. A General Model

2.1 The Model

Consider a set of m populations $\Pi_1, \Pi_2, \cdots, \Pi_m$. These may be different nations, or culturally different groups, groups of individuals selected on the basis of some known or unknown selection variable, groups receiving different treatments, etc. In fact, they may be any set of exclusive groups of individuals that are clearly defined. It is assumed that a battery of tests has been administered to a sample of individuals from each population. The battery of tests need not be the same for each group, nor need the number of tests be the same. However, since we shall be concerned with characteristics of the tests that are invariant over populations, it is necessary that some of the tests in each battery are the same or at least content-wise equivalent.

Let p_g be the number of tests administered to group g and let x_g be a vector of order p_g, representing the measurements obtained in group g. We regard x_g as a random vector with mean vector μ_g and variance-covariance matrix Σ_g. It is assumed that a factor analysis model holds in each population so that x_g can be accounted for by k_g common factors f_g and p_g unique factors z_g, as

$$x_g = \mu_g + \Lambda_g f_g + z_g , \tag{1}$$

with $\mathcal{E}(f_g) = 0$ and $\mathcal{E}(z_g) = 0$ and Λ_g a factor pattern of order $p_g \times k_g$. The usual factor analytic assumptions then imply that

$$\Sigma_g = \Lambda_g \Phi_g \Lambda'_g + \psi_g^2 , \tag{2}$$

where Φ_g is the variance-covariance matrix of f_g and ψ_g^2 is the diagonal variance-covariance matrix of z_g.

In addition to assuming that a factor analytic model holds in each population the model may specify that certain parameters in Λ_g, Φ_g, ψ_g, $g = 1, 2, \cdots, m$ have assigned values and that some set of unknown elements in Λ_g, Φ_g and ψ_g are the same for all g. The most common situation is when the same battery has been administered to each group and when the whole factor pattern Λ_g is assumed to be invariant over groups. This case will be considered separately in section 3.

2.2 Identification of Parameters

Before an attempt is made to estimate a model of this kind, the identification problem must be examined. The identification problem depends on the specification of fixed, free and constrained parameters. Under a given specification, each Λ_g, Φ_g and ψ_g generates one and only one Σ_g but it is well known that different Λ_g and Φ_g can generate the same Σ_g. It should be noted that if Λ_g is replaced by $\Lambda_g T_g^{-1}$ and Φ_g by $T_g \Phi_g T'_g$, where T_g is an arbitrary non-singular matrix of order $k_g \times k_g$, then Σ_g is unchanged. Since T_g has k_g^2 independent elements, this suggests that k_g^2 independent conditions should be imposed on Λ_g and/or Φ_g to make these uniquely defined and hence that $\sum_{g=1}^{m} k_g^2$ independent conditions altogether should be imposed. However, when equality constraints over groups are taken into account, all the elements of all the transformation matrices are not independent of each other and therefore a lesser number of conditions need to be imposed. It is hard to give further specific rules in the general case. For the special case when the whole factor pattern is invariant over groups, however, a more precise consideration of the identification problem is given in section 3.2. In other cases one should verify that the only transformations T_1, T_2, $\cdots$, T_m that preserve the specification about fixed, free and constrained parameters are identity matrices.

2.3 Estimation and Testing of the Model

Let N_g be the number of individuals in the sample from the gth population and let $\bar{x}_g$ be the usual sample mean vector and S_g the usual sample variance-covariance matrix with $n_g = N_g - 1$ degrees of freedom. The only requirement for the sampling procedure is that it produces independent measurements for the different groups.

If we assume that x_g has a multinormal distribution it follows that S_g has a Wishart distribution based on Σ_g and n_g degrees of freedom. The logarithm of the likelihood for the gth sample is

$$\log L_g = -\tfrac{1}{2}n_g[\log|\Sigma_g| + \operatorname{tr}(S_g\Sigma_g^{-1})]. \tag{3}$$

Since the samples are independent, the log-likelihood for all the samples is

$$\log L = \sum_{g=1}^{m} \log L_g . \tag{4}$$

Maximum likelihood estimates of the unknown elements in Λ_g , Φ_g , ψ_g , $g = 1, 2, \cdots , m$, may be obtained by maximizing $\log L$. However, it is slightly more convenient to minimize

$$F = \tfrac{1}{2}\sum_{g=1}^{m} n_g[\log|\Sigma_g| + \operatorname{tr}(S_g\Sigma_g^{-1}) - \log|S_g| - p_g] \tag{5}$$

instead. At the minimum, F equals minus the logarithm of the likelihood ratio for testing the hypothesis implied by the model against the general alternative that each Σ_g is unconstrained. Therefore, twice the minimum value of F is approximately distributed, in large samples, as χ^2 with degrees of freedom equal to

$$d = \sum_{g=1}^{m} \tfrac{1}{2}p_g(p_g + 1) - t \tag{6}$$

where t is the total number of independent parameters estimated in the model.

2.4 Minimization Procedure

The function F will be minimized numerically with respect to the independent parameters using a modification of the method of Fletcher and Powell [1963]. The application of this method makes use of exact expressions for first-order derivatives and approximate expressions for second-order derivatives of F.

Let

$$\Omega_g = \Sigma_g^{-1}(\Sigma_g - S_g)\Sigma_g^{-1} , \qquad g = 1, 2, \cdots , m. \tag{7}$$

Then it follows from the corresponding results for a single population (see *e.g.*, Lawley & Maxwell, [1963], Chapter 6 or Jöreskog, [1969]) that for $g = 1, 2, \cdots , m$,

$$\partial F/\partial\Lambda_g = n_g\Omega_g\Lambda_g\Phi_g , \tag{8a}$$

$$\partial F/\partial\Phi_g = n_g\Lambda_g'\Omega_g\Lambda_g - \tfrac{1}{2}n_g \operatorname{diag}(\Lambda_g'\Omega_g\Lambda_g), \tag{8b}$$

$$\partial F/\partial\psi_g = n_g \operatorname{diag}(\Omega_g\psi_g). \tag{8c}$$

In (8b) the symmetry of Φ_g has been taken into account.

We shall also need expressions for $\mathcal{E}(\partial^2 F/\partial\theta_i\, \partial\theta_j)$, where θ_i and θ_j are any two parameters. If θ_i is an element of Λ_g, Φ_g or ψ_g and θ_j is an element of Λ_h, Φ_h or ψ_h, $g \neq h$, $\mathcal{E}(\partial^2 F/\partial\theta_i\, \partial\theta_j)$ is zero. Otherwise, if both θ_i and θ_j are elements of Λ_g, Φ_g or ψ_g, the required second-order derivatives can be expressed in terms of the elements of

(9a) $$\xi = \Sigma^{-1}\Lambda$$

(9b) $$\eta = \Sigma^{-1}\Lambda\Phi = \xi\Phi$$

(9c) $$\alpha = \Lambda'\Sigma^{-1}\Lambda = \Lambda'\xi$$

(9d) $$\beta = \Phi\Lambda'\Sigma^{-1}\Lambda = \Phi\alpha$$

(9e) $$\gamma = \Phi\Lambda'\Sigma^{-1}\Lambda\Phi = \beta\Phi,$$

as

(10a) $$\mathcal{E}(\partial^2 F/\partial\lambda_{ir}\, \partial\lambda_{js}) = n(\sigma^{ij}\gamma_{rs} + \eta_{is}\eta_{jr})$$

(10b) $$\mathcal{E}(\partial^2 F/\partial\lambda_{ir}\, \partial\phi_{st}) = (n/2)(2 - \delta_{st})(\xi_{is}\beta_{rt} + \xi_{it}\beta_{rs})$$

(10c) $$\mathcal{E}(\partial^2 F/\partial\lambda_{ir}\, \partial\psi_{jj}) = 2n\sigma^{ij}\eta_{jr}\psi_{jj}$$

(10d) $$\mathcal{E}(\partial^2 F/\partial\phi_{rs}\, \partial\phi_{tu}) = (n/4)(2 - \delta_{rs})(2 - \delta_{tu})(\alpha_{rt}\alpha_{su} + \alpha_{ru}\alpha_{st})$$

(10e) $$\mathcal{E}(\partial^2 F/\partial\phi_{rs}\, \partial\psi_{jj}) = n(2 - \delta_{rs})\xi_{jr}\xi_{js}\psi_{jj}$$

(10f) $$\mathcal{E}(\partial^2 F/\partial\psi_{ii}\, \partial\psi_{jj}) = 2n(\sigma^{ij})^2\psi_{ii}\psi_{jj} .$$

Here we have omitted the subscript g for simplicity of notation.

The function F is regarded as a function of the elements of Λ_g, Φ_g, ψ_g, $g = 1, 2, \cdots, m$, and is to be minimized with respect to these taking into account that some elements may be fixed and some may be constrained to be equal to others. Such a minimization problem may be solved as follows.

Let θ_g be a vector of all the elements in Λ_g, Φ_g and ψ_g arranged in a prescribed order. Since Φ_g is symmetric, only the elements in the lower half and the diagonal are counted. Then θ_g is of order $r_g = p_g k_g + \frac{1}{2}k_g(k_g + 1) + p_g$. Let $\theta' = (\theta_1', \theta_2', \cdots, \theta_m')$. Then θ consists of all the elements of all the parameter matrices and is of order $r = r_1 + r_2 + \cdots + r_m$. The function F may now be regarded as a function $f(\theta)$ of $\theta_1, \theta_2, \cdots, \theta_r$, which is continuous and has continuous derivatives $\partial F/\partial\theta_i$ and $\partial^2 F/\partial\theta_i\, \partial\theta_j$ of first and second order, except where any Σ_g is singular. The totality of these derivatives is represented by a gradient vector $\partial F/\partial\theta$ and a symmetric second order derivative matrix $\partial^2 F/\partial\theta\, \partial\theta'$.

The vector $\partial F/\partial\theta$ of order r is formed by arranging the elements of the derivative matrices (8a)–(8c) in the same order as the elements of Λ_g, Φ_g and ψ_g, $g = 1, 2, \cdots, m$, in θ. An as approximation to the $r \times r$ matrix

$\partial^2 F/\partial\theta\,\partial\theta'$ we use $\mathcal{E}(\partial^2 F/\partial\theta\,\partial\theta')$ which is of the form

$$(11)\quad \mathcal{E}(\partial^2 F/\partial\theta\,\partial\theta') = \begin{bmatrix} \mathcal{E}(\partial^2 F/\partial\theta_1\,\partial\theta_1') & 0 & \cdots & 0 \\ 0 & \mathcal{E}(\partial^2 F/\partial\theta_2\,\partial\theta_2') & \cdots & 0 \\ \vdots & \vdots & & \vdots \\ 0 & 0 & \cdots & \mathcal{E}(\partial^2 F/\partial\theta_m\,\partial\theta_m') \end{bmatrix}$$

where $\mathcal{E}(\partial^2 F/\partial\theta_g\,\partial\theta_g')$ is a symmetric matrix of order $r_g \times r_g$ formed by computing (10a)–(10f) and arranging these so that the order of rows and columns corresponds to the order of the parameters in θ_g .

Now let some $r - s$ of the θ's be fixed and denote the remaining θ's by $\pi_1, \pi_2, \cdots, \pi_s$, $s \leq r$. The function F is now regarded as a function $G(\pi)$ of $\pi_1, \pi_2, \cdots, \pi_s$. Derivatives $\partial G/\partial\pi$ and $\mathcal{E}(\partial^2 G/\partial\pi\,\partial\pi')$ are obtained from $\partial F/\partial\theta$ and $\mathcal{E}(\partial^2 F/\partial\theta\,\partial\theta')$ by omitting rows and columns corresponding to the fixed θ's. Among $\pi_1, \pi_2, \cdots, \pi_s$ let there be some t distinct and independent parameters denoted $\kappa_1, \kappa_2, \cdots, \kappa_t$, $t \leq s$, so that each π_i is equal to one and only one κ_j but possibly several π's equal the same κ. Let $K = (k_{ij})$ be a matrix of order $s \times t$ with elements $k_{ij} = 1$ if $\pi_i = \kappa_j$ and $k_{ij} = 0$ otherwise. The function G (or F) is now a function $H(\kappa)$ of the independent arguments $\kappa_1, \kappa_2, \cdots, \kappa_t$ and we have

$$(12)\qquad \partial H/\partial\kappa = K'(\partial G/\partial\pi)$$

$$(13)\qquad \mathcal{E}(\partial^2 H/\partial\kappa\,\partial\kappa') = K'\mathcal{E}(\partial^2 G/\partial\pi\,\partial\pi')K.$$

Thus, the first-order and expected second-order derivatives of H are simple sums of the corresponding derivatives of G.

For the minimization of $H(\kappa)$ we use a modification of the method of Fletcher and Powell [1963] for which a computer program has been written by Gruvaeus and Jöreskog [1970]. This method makes use of a symmetric matrix E of order $t \times t$, which is evaluated in each iteration. Initially E is any positive definite matrix approximating the inverse of $\partial^2 H/\partial\kappa\,\partial\kappa'$. In subsequent iterations E is improved, using information built up about the function so that ultimately E converges to an approximation of the inverse of $\partial^2 H/\partial\kappa\,\partial\kappa'$ at the minimum. If t is large, the number of iterations may be excessive but can be considerably decreased by the provision of a good starting point for κ and a good initial estimate of E.

In principle, a good initial estimate of E may be obtained by computing $\partial^2 H/\partial\kappa\,\partial\kappa'$ at the starting point and then inverting this matrix. However, in our problem, the second-order derivatives are rather complicated and time-consuming to compute. Instead, we therefore use estimates of the second-order derivatives provided by the information matrix

$$(14)\qquad \mathcal{E}(\partial^2 H/\partial\kappa\,\partial\kappa') = \mathcal{E}(\partial H|\partial\kappa\,\partial H/\partial\kappa').$$

In addition to being more easily evaluated, this matrix also yields other valuable information. The inverse of $\mathcal{E}(\partial^2 H/\partial\kappa\, \partial\kappa')$ evaluated at the minimum of H is an estimate of the variance covariance of the estimated parameters $\hat{\kappa}_1, \hat{\kappa}_2, \cdots, \hat{\kappa}_t$. This may be used to obtain standard errors of the estimated parameters.

The starting point κ may be chosen arbitrarily but the closer it is to the final solution the fewer iterations will be required to find the solution. The minimization method converges quadratically from an arbitrary starting point to a local minimum of the function. If several local minima exist there is no guarantee that the method will converge to the absolute minimum.

2.5 Computer Program

A computer program, SIFASP, that performs all the computations described in the previous sections has been written in FORTRAN IV and a write-up for this is available [van Thillo & Jöreskog, 1970]. This program reads an observed covariance matrix or a correlation matrix and a vector of standard deviations for each group, a set of pattern matrices specifying the fixed, free and constrained parameters and a set of matrices of start values for the minimization. It then minimizes the function F as described in the previous section to obtain the maximum likelihood solution for each group. These are then printed together with residuals, *i.e.*, differences between observed and reproduced variances and covariances, and a χ^2 measure of overall fit.

The computer program assumes that the number of variables and the number of common factors are the same for each group. This is no loss of generality, since it can always be achieved by the introduction of pseudovariables and pseudofactors in some groups as follows. Each pseudovariable has unit observed variance, zero observed covariances with every other variable, zero factor loadings on each factor including the pseudofactors and unit unique variance. Each pseudofactor has unit variance and zero covariance with every other factor and pseudofactor. It is readily verified that such pseudovariables and pseudofactors have no effect on the likelihood function whatsoever.

The observed variables may be rescaled initially as described in section 3.4. This is sometimes convenient when the observed variables have arbitrary units of measurements. In the special case of an invariant factor pattern, as described in the next section, the factors in the maximum likelihood solutions may be rescaled as shown in section 3.3.

The implementation of the minimization algorithm is simpler if all matrices are stored as singly subscripted arrays. This saves space, since only the lower halves of symmetric matrices need to be stored, and makes the program more efficient. The program makes use of a set of subroutines for matrix algebra on matrices stored as singly subscripted arrays. A further important advantage with this technique is the flexibility in the choice of m, p and k. Thus, in the same space as one can have $m = 4$, $p = 12$ and $k = 5$ one can also have $m = 2$, $p = 17$ and $k = 5$ or $m = 1$, $p = 24$ and $k = 12$.

The computer program works with one group ($m = 1$) as well as with more groups. When m is one the model is the same as that of Jöreskog [1969] but is is now possible to handle not only fixed parameters but also equality constraints between parameters. SIFASP can handle many types of factor analytic solutions.

3. A Model of Factorial Invariance

3.1 The Model

Perhaps the most common application of the method just described will be the case when the same tests have been administered in each population and when it is hypothesized that the factor pattern Λ is invariant over populations. Although this model is a special case of the general model described in the previous section, it deserves a separate discussion.

In this case $p_1 = p_2 = \cdots = p_m = p$ and $k_1 = k_2 = \cdots = k_m = k$ and the matrices Σ_g and ψ_g, $g = 1, 2, \cdots, m$ are all of the order $p \times p$ and the Φ_g, $g = 1, 2, \cdots, m$ are all of order $k \times k$. The common factor pattern Λ is of order $p \times k$. The regression of x_g on f_g is [c.f. (1)].

(15) $$x_g = \mu_g + \Lambda f_g + z_g$$

and the variance-covariance matrix Σ_g is

(16) $$\Sigma_g = \Lambda \Phi_g \Lambda' + \psi_g^2 .$$

In the special case of two populations, $m = 2$, a stricter form of invariance was considered by Lawley and Maxwell [1963, Chapter 8]. This requires not only the regression matrix Λ in (15) to be invariant but also the variances about the regression, *i.e.*, $\psi_1^2 = \psi_2^2$. This type of restriction can easily be incorporated using the general approach of the preceding section.

3.2 Identification of Parameters

Suppose that the Λ in (16) is replaced by $\Lambda^* = \Lambda T^{-1}$ and each Φ_g is replaced by $\Phi_g^* = T\Phi_g T'$, $g = 1, 2, \cdots, m$, where T is an arbitrary non-singular matrix of order $k \times k$. Then each Σ_g remains the same so that the function F in (5) is unaltered. Since the matrix T has k^2 independent elements, this means that at least k^2 independent conditions must be imposed on the parameters in Λ, Φ_1, Φ_2, $\cdots$, Φ_m to make these uniquely defined.

Within the framework of the general procedure of the previous section, the most convenient way of doing this is to let all the Φ_g be free and to fix one nonzero element and at least $k - 1$ zeros in each column of Λ. In an exploratory study one can fix exactly $k - 1$ zeros in almost arbitrary positions.

For example one may choose zero loadings where one thinks there should be "small" loadings in the factor pattern. The resulting solution may be rotated further, if desired, to facilitate better interpretation. In a confirmatory study, on the other hand, the positions of the fixed zeros, which often exceed $k - 1$ in each column, are given *a priori* by an hypothesis and the resulting solution cannot be rotated without destroying the fixed zeros.

3.3 Scaling of Factors

The fixed nonzero loading in each column of Λ can have any value. This is only used to fix a scale for each factor that is common to all groups. When the maximum likelihood solution has been obtained, the factors may be rescaled so that their average variance is unity. This rescaling is obtained as follows. Let

$$\hat{\Phi} = (1/n) \sum_{g=1}^{m} n_g \hat{\Phi}_g \,, \tag{17}$$

with $n = \sum_{g=1}^{m} n_g$, and

$$D = (\text{diag } \hat{\Phi})^{-1/2}. \tag{18}$$

Then the rescaled solution is

$$\hat{\Lambda}^* = \hat{\Lambda} D^{-1} \tag{19}$$

$$\hat{\Phi}_g^* = D \hat{\Phi}_g D, \qquad g = 1, 2, \cdots, m. \tag{20}$$

The matrix $\hat{\Lambda}^*$ has zeros wherever $\hat{\Lambda}$ has zeros but the fixed nonzeros in $\hat{\Lambda}$ have changed their values. The weighted average of the $\hat{\Phi}_g^*$ is a correlation matrix.

3.4 Scaling of Observed Variables

When the units of measurements in the different tests are arbitrary, it is usually convenient, though not necessary, to rescale the observed variables, before the factor analysis. Let

$$S = (1/n) \sum_{g=1}^{m} n_g S_g \,, \tag{21}$$

with $n = \sum_{g=1}^{m} n_g$ as before and let

$$D = (\text{diag } S)^{-1/2}. \tag{22}$$

Then the variance-covariance matrices for the rescaled variables are

$$S_g^* = D S_g D. \tag{23}$$

The weighted average of the S_g^* is a correlation matrix. The advantage of this rescaling is that, when combined with the rescaling of the factors of the previous section, the factor loadings are of the same order of magnitude as usual when correlation matrices are analyzed and when factors are standardized to unit variances. This makes it easier to choose start values for the minimization (see section 3.5) and interpret the results.

It should be pointed out that it is not permissible to standardize the variables in each group and to analyze the correlation matrices instead of the variance-covariance matrices. This violates the likelihood function (4) which is based on the distribution of the observed variances and covariances. Invariance of factor patterns is expected to hold only when the standardization of both tests and factors are relaxed.

3.5 Choice of Start Values

In a medium-sized study of say 4 groups, 20 variables and 5 factors, the number of free parameters to estimate may well exceed 200. To obtain the maximum likelihood estimates, a function of over 200 variables has to be minimized. This is not an easy task even on today's large computers. To reduce the computer time as much as possible it is necessary to choose good start values for the minimization. Although simpler and less costly procedures could be used, the most convenient way to get start values is to do some preliminary runs with the same computer program before the overall estimation is attempted.

1. Using $m = 1$ and the pooled correlation matrix $R = DSD$, where D is given by (22), obtain an oblique maximum likelihood solution with the fixed zeros in Λ and the diagonal elements of Φ equal to unity. Let the estimate of Λ so obtained be denoted $\hat{\Lambda}^{(0)}$.
2. For each group *separately*, using $m = 1$ and S_g^*, obtain an oblique maximum likelihood solution with the *whole* Λ fixed equal to $\hat{\Lambda}^{(0)}$ and with Φ_g and ψ_g free. Let the resulting estimates be denoted $\Phi_g^{(0)}$ and $\hat{\psi}_g^{(0)}$, $g = 1, 2, \cdots, m$.
3. Then $\hat{\Lambda}^{(0)}$, $\Phi_1^{(0)}$, $\Phi_2^{(0)}$, $\cdots$, $\hat{\Phi}_m^{(0)}$, $\hat{\psi}_1^{(0)}$, $\hat{\psi}_2^{(0)}$, $\cdots$, $\hat{\psi}_m^{(0)}$ provide good start values for the overall minimization with the largest element in each column of $\hat{\Lambda}^{(0)}$ fixed, in addition to the fixed zeros.

If the model also specifies that the ψ_g should be invariant over groups, one uses as start values $\hat{\psi}^{(0)}$ for the common ψ, the estimate obtained in step 1 and step 2 is done with ψ fixed at $\hat{\psi}^{(0)}$.

3.6 Testing of Hypotheses and Strategy of Analysis

Suppose H_0 represents one model under given specifications of fixed, free and constrained parameters, fitting the general framework of section 2.1. Then it is possible, in large samples, to test the model H_0 against any more general model H_1, by estimating each of them separately and comparing their χ^2 goodness of fit values. The difference in χ^2 is asymptotically a

χ^2 with degrees of freedom equal to the corresponding difference in degrees of freedom.

In an exploratory study there are various hypotheses that may be tested and it seems best to proceed stepwise in a certain order.

One begins by testing the *hypothesis of equality of covariance matrices, i.e.,*

$$(24) \qquad H_\Sigma : \Sigma_1 = \Sigma_2 = \cdots = \Sigma_m .$$

This may be tested by using the test statistic

$$(25) \qquad M = n \log |S| - \sum_{g=1}^{m} n_g \log |S_g|,$$

where S is given by (21). Under the hypothesis, M is distributed approximately as χ^2 with $d_\Sigma = \frac{1}{2}(m - 1)p(p + 1)$ degrees of freedom. As shown by Box [1949], the approximation to the χ^2 distribution is improved if M is multiplied by a certain constant. When p or m is larger than 4, Box suggests a transformation to an F distribution.

It should be noted that the test statistic (25) may be obtained in SIFASP by specifying $k_g = p$, $\Lambda_g = I$, $\psi_g = 0$, $g = 1, 2, \cdots, m$ and $\Phi_1 = \Phi_2 = \cdots = \Phi_m$. The maximum likelihood estimate of the common Φ will then be the pooled S as defined in (21). If the tests are scaled originally as described in section 3.4, this S is a correlation matrix R.

If the hypothesis is found to be tenable every characteristic common to all groups can be obtained from the pooled covariance matrix S or the correlation matrix R and there is no need to analyze each group separately or simultaneously.

If, on the other hand, the hypothesis of equality of covariance matrices is untenable, one may want to investigate similarities and difference in factor structures. For this purpose, a sequence of hypotheses, such that each hypothesis is a special case of the preceding, will now be considered. The first hypothesis is *the hypothesis of equality of number of common factors, i.e.,*

$$(26) \qquad H_k : k_1 = k_2 = \cdots = k_m = \text{a specified number } k.$$

This may be tested by doing an unrestricted factor analysis [Jöreskog, 1969] on each S_g (or S_g^* or the corresponding correlation matrix) separately, using the same number of common factors for each group. The analyses may be done by Jöreskog's [1967] method UMLFA but can also be done with the computer program SIFASP. In SIFASP one uses $m = 1$ and fixes k^2 elements in Λ_g and/or Φ_g ; for example, to obtain an orthogonal solution one can choose $\Phi_g = I$ and $\frac{1}{2}k(k - 1)$ zeros in Λ_g and to obtain an oblique solution one can choose diag $\Phi_g = I$ and $k(k - 1)$ zeros in Λ_g. Each analysis gives a χ^2 with $\frac{1}{2}[(p - k)^2 - (p + k)]$ degrees of freedom. Since these χ^2's are independent they may be added for each group to obtain a χ^2, χ_k^2 say, with $d_k = \frac{1}{2}m[(p - k)^2 - (p + k)]$ degrees of freedom, which may be used to test the overall hypothesis.

If the hypothesis of a common number of factors is found tenable, one may proceed to test *the hypothesis of an invariant factor pattern, i.e.,*

$$H_\Lambda : \Lambda_1 = \Lambda_2 = \cdots = \Lambda_m \,. \tag{27}$$

The common factor pattern Λ may either be completely unspecified or be specified to have zeros in certain positions. If Λ is unspecified, one fixes $k - 1$ zeros and one nonzero value in each column almost arbitrarily. If Λ is specified to have zeros in certain positions, one fixes an arbitrary nonzero element in each column in addition. There will then be k^2 fixed elements in Λ in the unspecified case and $q \geq k^2$ in the specified case. To obtain a χ^2 for this hypothesis, one estimates $\Lambda, \Phi_1, \Phi_2, \cdots, \Phi_m, \psi_1, \psi_2, \cdots, \psi_m$ from $S_1, S_2, \cdots, S_m$ simultaneously, yielding a minimum value of the function F. Twice this minimum value is a χ^2, χ^2_Λ say, with degrees of freedom

$$\tfrac{1}{2}mp(p + 1) - pk + q - \tfrac{1}{2}mk(k + 1) - mp,$$

where $q = k^2$ in the unspecified case. To test the hypothesis H_Λ, given that H_k holds, one uses $\chi^2_{\Lambda \cdot k} = \chi^2_\Lambda - \chi^2_k$ with $d_{\Lambda \cdot k} = d_\Lambda - d_k$ degrees of freedom.

If this hypothesis is found tenable one may proceed to test the hypothesis

$$H_{\Lambda\psi} : \Lambda_1 = \Lambda_2 = \cdots = \Lambda_m \,; \qquad \psi_1 = \psi_2 = \cdots = \psi_m \,. \tag{28}$$

To do so one has to estimate $\Lambda, \Phi_1, \Phi_2, \cdots, \Phi_m, \psi$ under $H_{\Lambda\psi}$. This again gives a minimum value of F which when multiplied by two gives $\chi^2_{\Lambda\psi}$ with

$$d_{\Lambda\psi} = \tfrac{1}{2}mp(p + 1) - pk + q - \tfrac{1}{2}mk(k + 1) - p$$

degrees of freedom. To test $H_{\Lambda\psi}$ against H_Λ one uses $\chi^2_{\psi \cdot \Lambda} = \chi^2_{\Lambda\psi} - \chi^2_\Lambda$ with $d_{\psi \cdot \Lambda} = d_{\Lambda\psi} - d_\Lambda$ degrees of freedom.

If the hypothesis $H_{\Lambda\psi}$ is found tenable one may want to test the hypothesis

$$H_{\Lambda\Phi\psi} : \Lambda_1 = \Lambda_2 = \cdots = \Lambda_m \,; \qquad \Phi_1 = \Phi_2 = \cdots = \Phi_m \,; \qquad \psi_1 = \psi_2 = \cdots = \psi_m \,. \tag{29}$$

This hypothesis is included in H_Σ but is stronger than H_Σ since H_Σ includes also the cases when the common Σ is not of the form

$$\Sigma = \Lambda\Phi\Lambda' + \psi^2. \tag{30}$$

This hypothesis $H_{\Lambda\Phi\psi}$ can be tested directly on the basis of the pooled S in (21). The test of $H_{\Lambda\Phi\psi}$ against H_Σ uses a χ^2 with

$$d_{\Lambda\Phi\psi \cdot \Sigma} = \tfrac{1}{2}p(p + 1) - pk + q - \tfrac{1}{2}k(k + 1) - p$$

degrees of freedom.

Various other types of hypotheses may also be tested. For example, one may assume that some factors are orthogonal and some are oblique (see Jöreskog [1969]).

It should be emphasized that significance levels are unknown when a sequence of tests like these are carried out and that even if a χ^2 is large, there may still be reasons to consider the model. After all, the basic model with its assumptions of linearity and normality is only regarded as an approximation to reality. The true population covariance matrix will not in general be exactly of the form specified by the hypothesis, but there will be discrepancies between the true population covariance matrix and the formal model postulated. These discrepancies will not get smaller when the sample size increases but will tend to give large χ^2 values. Therefore, a model may well be accepted even though χ^2 is large. Whether to accept or reject a model cannot be decided on a purely statistical basis. This is largely a matter of the experimenter's interpretations of the data, based on substantive theoretical and conceptual considerations. Ultimately the criteria for goodness of the model depends on the usefulness of it and the results it produces.

3.7 A Numerical Illustration

To illustrate the methods previously discussed we use the same data as Meredith [1964b] used to illustrate his rotational procedure. The data consist of nine tests selected from a battery of 26 psychological tests described by Holzinger and Swineford [1939]. The tests were administered to 7th and 8th grade children in two schools, the Pasteur and the Grant-White Schools in the Chicago area. The nine tests were selected so that each of the three factors—*space, verbal* and *memory*—would be represented by three tests. The nine tests used, with their original code numbers in parentheses, were: Visual Perception (1), Cubes (2), Paper Form Board (3), General Information (5), Sentence Completion (7), Word Classification (8), Figure Recognition (16), Object Number (17) and Number-Figure (18). On the basis of a speeded addition test, Meredith divided each of the samples from the two schools into two approximately equal groups by splitting at the median score within each school. This yielded four groups that will be used for this illustration. The correlation matrices taken from Meredith's Table 2 are shown in Table 1a with unscaled and scaled standard deviations in Table 1b. The sample sizes are: Group 1: Pasteur Low $N_1 = 77$, Group 2: Pasteur High $N_2 = 79$, Group 3: Grant-White Low $N_3 = 74$ and Group 4: Grant-White High $N_4 = 71$. Because of the way the two groups within schools were selected, it is doubtful that the assumption of multinormality is valid. This departure from multinormality will have no effect on the estimates but may be serious for the χ^2 values. In particular, the χ^2 test of H_Σ is known to be sensitive to departures from multinormality. For this reason and also because the sample sizes are relatively small, the χ^2 values that will be

TABLE 1a-b

Intercorrelation Matrices (a)

	Group 1 Above Main Diagonal Group 2 Below Main Diagonal								
	1	2	3	4	5	6	7	8	9
Visual Perception	—	.32	.48	.28	.26	.40	.42	.12	.23
Cubes	.24	—	.33	.01	.01	.26	.32	.05	−.04
Paper Form Board	.23	.22	—	.06	.01	.10	.22	.03	.01
General Information	.32	.05	.23	—	.75	.60	.15	−0.8	−0.5
Sentence Completion	.35	.23	.18	.68	—	.63	.07	.06	.10
Word Classification	.36	.10	.11	.59	.66	—	.36	.19	.24
Figure Recognition	.22	.01	−0.7	.09	.11	.12	—	.29	.19
Object-Number	−.02	−0.1	−.13	.05	.08	.03	.19	—	.38
Number-Figure	.09	−.14	−.06	.16	.02	.12	.15	.29	—

	Group 3 Above Main Diagonal Group 4 Below Main Diagonal								
	1	2	3	4	5	6	7	8	9
Visual Perception	—	.34	.41	.38	.40	.42	.35	.16	.35
Cubes	.32	—	.21	.32	.16	.13	.27	.01	.27
Paper Form Board	.34	.18	—	.31	.24	.35	.30	.09	.09
General Information	.31	.24	.31	—	.69	.55	.17	.31	.34
Sentence Completion	.22	.16	.29	.62	—	.65	.20	.30	.27
Word Classification	.27	.20	.32	.57	.61	—	.31	.34	.27
Figure Recognition	.48	.31	.32	.18	.20	.29	—	.31	.38
Object-Number	.20	.01	.15	.06	.19	.15	.36	—	.38
Number-Figure	.42	.28	.40	.11	.07	.18	.35	.44	—

Standard Deviations (b)

	Unscaled				Scaled			
	1	2	3	4	1	2	3	4
Visual Perception	7.4	6.7	6.6	7.2	1.06	0.96	0.95	1.03
Cubes	5.6	4.0	4.8	4.0	1.20	0.86	1.03	0.86
Paper Form Board	2.9	2.8	2.6	3.0	1.02	0.99	0.92	1.06
General Information	11.8	11.0	11.3	11.5	1.03	0.96	0.99	1.01
Sentence Completion	5.2	5.2	4.7	4.5	1.08	1.06	0.96	0.91
Word Classification	5.2	5.3	5.0	5.5	0.99	1.01	0.95	1.05
Figure Recognition	8.8	7.6	6.1	7.4	1.17	1.01	0.81	0.98
Object-Number	4.7	5.2	3.9	4.9	1.00	1.10	0.83	1.04
Number-Figure	4.6	4.4	3.9	4.7	1.04	1.00	0.88	1.07

reported should be interpreted very cautiously. It should be emphasized that these data have been chosen merely to illustrate the procedures of this paper. Another application, with more substantive interest and with larger and widely varying sample sizes, is given by McGaw and Jöreskog [1971].

We begin by testing the hypothesis H_Σ that $\Sigma_1 = \Sigma_2 = \Sigma_3 = \Sigma_4$. This gives the test statistic $M = 146.95$ with 135 degrees of freedom. Transformation to Box's F-statistic gives $F_{135,\infty} = 1.03$. In view of the remark just made, this value is inconclusive. However, for the purpose of illustrating a simultaneous analysis of all four scaled dispersion matrices, we shall follow the procedure of section 3.6 and test various hypotheses of interest. The results are summarized in Table 2.

The first hypothesis, $H_{k=3}$, is that three factors adequately reproduce the correlation in each population. This gives $\chi^2 = 47.73$ with 48 degrees of freedom. This χ^2 is the sum of four χ^2's one from each population and each with 12 degrees of freedom. These are $\chi_1^2 = 15.33$, $\chi_2^2 = 10.44$, $\chi_3^2 = 14.40$ and $\chi_4^2 = 7.56$. Thus we cannot reject the hypothesis that the number of factors is three for each population. We therefore proceed by investigating whether there is an invariant factor pattern or not.

The next hypothesis, H_{Λ_u}, is that there is an invariant unspecified (unrestricted) factor pattern Λ_u. To test this hypothesis we fix one nonzero element and two zero elements in each column of Λ_u and leave Φ_1, Φ_2, Φ_3, Φ_4 and ψ_1, ψ_2, ψ_3, ψ_4 completely unconstrained. A convenient way to choose the fixed elements in Λ_u is to use a reference variables solution as, for example,

$$\Lambda_u = \begin{bmatrix} 1 & 0 & 0 \\ x & x & x \\ x & x & x \\ 0 & 1 & 0 \\ x & x & x \\ x & x & x \\ 0 & 0 & 1 \\ x & x & x \\ x & x & x \end{bmatrix}. \tag{31}$$

Here the zeros and ones stand for fixed values and x's for parameters to be estimated. Tests 1, 4 and 7 have been chosen to be pure in their respective factors. The test of H_{Λ_u} gives $\chi^2 = 90.57$ with 102 degrees of freedom, which is not significant. Thus, we cannot reject the hypothesis that there is an invariant factor pattern with three factors. To strengthen the model we

TABLE 2
Summary of Analyses

Hypothesis	χ^2	No. par.	d.f.	P
H_{Σ}	146.95	45	135	0.23
$H_{k=3}$	47.73	132	48	0.47
H_{Λ_u}	90.57	78	102	0.78
H_{Λ}	131.24	66	114	0.13
$H_{\Lambda\psi}$	172.14	39	141	0.04
$H_{\Lambda\Phi\Psi}$	212.80	21	159	0.00
$H_{\Lambda_u\Phi\psi}$	36.20	33	147	1.00

now make use of our knowledge about the tests and hypothesize that the invariant factor pattern has a specific form, namely

(32)
$$\Lambda = \begin{bmatrix} 1 & 0 & 0 \\ x & 0 & 0 \\ x & 0 & 0 \\ 0 & 1 & 0 \\ 0 & x & 0 \\ 0 & x & 0 \\ 0 & 0 & 1 \\ 0 & 0 & x \\ 0 & 0 & x \end{bmatrix},$$

i.e., we assume that Λ has a nonoverlapping group structure, where the first three tests are loaded on the first factor only, the next three tests on the second factor only and the last three tests on the third factor only. As before, we put no constraints on the Φ's and ψ's. A test of this hypothesis gives $\chi^2 = 131.24$ with 114 degrees of freedom. This has a probability level of about 0.13. Thus we cannot reject the hypothesis that the invariant factor pattern is of the specified form. An examination of the ψ's, in relation to their standard errors in the solution under H_{Λ} , revealed that many of these were not sufficiently different to be considered different. This suggests that one should also examine the hypothesis $H_{\Lambda\psi}$, that a stricter form of invariance holds, namely where also the ψ's are the same for all populations. A test of $H_{\Lambda\psi}$ gives $\chi^2 = 172.14$ with 141 degrees of freedom. This is just significant at the 5% level. The maximum likelihood solution under $H_{\Lambda\psi}$ is shown in Table 3. Finally, to complete the sequence of hypotheses we consider the hypothesis $H_{\Lambda\Phi\psi}$, that the whole factor structure is invariant, with the same factor pattern Λ as before. This gives $\chi^2 = 212.90$ with 159 degrees of freedom which is highly significant. This would seem to contradict the test of H_Σ. This is not so, however, since $H_{\Lambda\Phi\psi}$ is a much more restrictive hypothesis than H_Σ. The hypothesis $H_{\Lambda\Phi\psi}$ requires that the common Σ has a factor structure with three common factors with a factor pattern of the restricted type (32), but there may be many other possible representations of the common Σ. In fact, if Σ is represented by the unrestricted factor pattern Λ_u in (31) instead, one obtains $\chi^2 = 36.20$ with 147 degrees of freedom, so that $H_{\Lambda_u\Phi\psi}$ cannot be rejected although $H_{\Lambda\Phi\psi}$ was rejected.

Altogether these results suggest two alternative descriptions of the data. One is that the whole factor structure is invariant over populations with a three-factor solution of a fairly complex form. The other is to represent the tests in each population by three factors of a particularly simple form, but these factors have different variance-covariance matrices in the different populations. Additional studies with larger sample sizes are needed to discriminate statistically between the two models. Perhaps, the second alternative has the most intuitive appeal. Inspecting the factor variances

TABLE 3

Maximum Likelihood Solution under $H_{\Lambda\psi}$

(Asterisks Denote Parameter Values Specified by Hyptohesis)

	$\hat{\Lambda}$ S	V	M	$\hat{\psi}$
Visual Perception	.72	0*	0*	.69
Cubes	.43	0*	0*	.90
Paper Form Board	.51	0*	0*	.86
General Information	0*	.80	0*	.60
Sentence Completion	0*	.85	0*	.53
Word Classification	0*	.75	0*	.67
Figure Recognition	0*	0*	.58	.81
Object-Number	0*	0*	.48	.88
Number-Figure	0*	0*	.57	.83

$$\hat{\Phi}_1 = \begin{array}{c} S \\ V \\ M \end{array}\overset{\begin{array}{ccc} S & V & M \end{array}}{\begin{bmatrix} 1.02 & & \\ 0.53 & 0.91 & \\ 1.03 & 0.36 & 1.30 \end{bmatrix}} \qquad \hat{\Phi}_2 = \begin{array}{c} S \\ V \\ M \end{array}\overset{\begin{array}{ccc} S & V & M \end{array}}{\begin{bmatrix} 0.89 & & \\ 0.62 & 0.93 & \\ 0.59 & 0.50 & 0.58 \end{bmatrix}}$$

$$\hat{\Phi}_3 = \begin{array}{c} S \\ V \\ M \end{array}\overset{\begin{array}{ccc} S & V & M \end{array}}{\begin{bmatrix} 0.72 & & \\ 0.52 & 1.06 & \\ 0.08 & 0.20 & 0.90 \end{bmatrix}} \qquad \hat{\Phi}_4 = \begin{array}{c} S \\ V \\ M \end{array}\overset{\begin{array}{ccc} S & V & M \end{array}}{\begin{bmatrix} 1.38 & & \\ 0.42 & 1.12 & \\ 0.71 & 0.27 & 1.25 \end{bmatrix}}$$

in Table 4, it is seen that for the Pasteur school they tend to be higher for the Low group than for the High group, whereas for the Grant-White school generally the opposite holds. Also for the two High groups the variances are generally lower for the Pasteur school than for the Grant-White school. Note also the low covariance of 0.08 between S and M for the Low Grant-White group and the corresponding high covariance 1.03 for the Low Pasteur group. This seems to indicate that the Low Pasteur group cannot fully discriminate between the spatial and the memory tasks whereas the Low Grant-White group can do so clearly.

REFERENCES

Box, G. E. P. A general distribution theory for a class of likelihood criteria. *Biometrika*, 1949, **36**, 317–346.

Fletcher, R. & Powell, M. J. D. A rapidly convergent descent method for minimization. *The Computer Journal*, 1963, **6**, 163–168.

Gruvaeus, G. T. & Jöreskog, K. G. A computer program for minimizing a function of several variables. Research Bulletin 70-14. Princeton, N. J.: Educational Testing Service, 1970.

Holzinger, K. J. & Swineford, F. A study in factor analysis: The stability of a bi-factor solution. *Supplementary Educational Monograph No.* **48**. Chicago: University of Chicago, 1939.

Jöreskog, K. G. Some contributions to maximum likelihood factor analysis. *Psychometrika*, 1967, **32**, 443–482.

Jöreskog, K. G. A general approach to confirmatory maximum likelihood factor analysis. *Psychometrika*, 1969, **34**, 183–220.

Lawley, D. N. A note on Karl Pearson's selection formulae. *Proceedings of the Royal Society of Edinburgh*, Section A, 1943–44, **62**, 28–30.

Lawley, D. N. & Maxwell, A. E. *Factor analysis as a statistical method.* London: Butterworths, 1963.

McGaw, B. & Jöreskog, K. G. Factorial invariance of ability measures in groups differing in intelligence and socioeconomic status. *British Journal of Mathematical and Statistical Psychology*, 1971, **24**, *in press.*

Meredith, W. Rotation to achieve factorial invariance. *Psychometrika*, 1964, **19**, 187–206. (a)

Meredith, W. Notes on factorial invariance. *Psychometrika*, 1964, **29**, 177–185. (b)

van Thillo, M. & Jöreskog, K. G. A general computer program for simultaneous factor analysis in several populations. Research Bulletin 70–62. Princeton, N. J.: Educational Testing Service, 1970.

Reprinted from: *Psychometrika*, December 1971, Vol. 36, No. 4.

Chapter 8

A General Method for Studying Differences in Factor Means and Factor Structure Between Groups

Dag Sörbom

A statistical model is developed for the study of similarities and differences in factor structure between several groups. The model assumes that the observed variables satisfy a factor analysis model in each group. A method of data analysis is presented which, in contrast to earlier work, makes use of information in the observed means as well as the observed variances and covariances to estimate the parameters in each group, i.e. factor means, factor loadings, factor variances and covariances and unique variances. Usually the units of measurement in the observed variables have no intrinsic meaning and therefore it is only meaningful to compare the relative magnitudes of the parameters for the different groups. The method estimates the parameters for all groups simultaneously and can take into account *a priori* information about factorial invariance of various degrees.

1. Introduction

The model described by Jöreskog (1971) is concerned with the 'study of similarities and differences in factor structure between different groups'. The aim of this paper is to discuss a model which, in addition, allows one to study the means of the factors for the groups. Usually these means are estimated in a manner similar to the estimation of factor scores (Lawley & Maxwell, 1971), i.e. the means are obtained from the estimated parameters of the model as if these were equal to the population paramcters. For instance, consider the model in classical factor analysis

$$\mathbf{x} = \boldsymbol{\Lambda}\boldsymbol{\xi} + \boldsymbol{\varepsilon}.$$

If we denote $E(\boldsymbol{\xi})$ by $\boldsymbol{\theta}$ it follows that $E(\mathbf{x}) = \boldsymbol{\Lambda}\boldsymbol{\theta}$. An estimate of the factor means may be obtained by minimizing (cf. Lawley & Maxwell, 1971; McGaw & Jöreskog, 1971):

$$(\bar{\mathbf{x}} - \hat{\boldsymbol{\Lambda}}\boldsymbol{\theta})'\hat{\boldsymbol{\Sigma}}^{-1}(\bar{\mathbf{x}} - \hat{\boldsymbol{\Lambda}}\boldsymbol{\theta}),$$

where $\bar{\mathbf{x}}$ is the vector of sample means of the manifest variables, $\hat{\boldsymbol{\Lambda}}$ is the matrix of estimated factor loadings, and $\hat{\boldsymbol{\Sigma}}$ is the population variance–covariance matrix estimated from the factor analysis model. Although this procedure gives the maximum-likelihood (ML) estimator of $\boldsymbol{\theta}$ given $\boldsymbol{\Sigma}$, it is not the ML estimator of the parameter vector $\boldsymbol{\theta}$ in the model, since, as will be shown in Section 3, information of $\boldsymbol{\Lambda}$ is contained also in $\bar{\mathbf{x}}$.

In a recent paper, Please (1973) gives a model very similar to the one considered in this paper and develops a procedure for obtaining ML estimators. However, his estimation procedure cannot take into account *a priori* information about the measurements, such as zero factor loadings in specified positions, which is a feature of some importance for the application of the model (see, e.g., Jöreskog, 1971; Sörbom, 1973*b*). In addition, the procedure described in this paper is more general than that of Please, which will be shown to be a special case.

2. The Model

For m groups consider the following model for an observable vector $\mathbf{x}_g$ of order p for the gth group,

$$\mathbf{x}_g = \boldsymbol{\mu} + \boldsymbol{\Lambda}_g \boldsymbol{\xi}_g + \boldsymbol{\varepsilon}_g \quad (g = 1, 2, \ldots, m), \tag{1}$$

where $\boldsymbol{\mu}$ is a constant vector of order p representing the origin of the measurements, $\boldsymbol{\xi}_g$ is a random vector of order k representing the common factors, $\boldsymbol{\Lambda}_g$ is a $p \times k$ matrix of factor loadings and $\boldsymbol{\varepsilon}_g$ is a random vector of order p of unique factors.

By the usual factor analytic assumptions we obtain the variance–covariance matrix of the measurements for the gth group as

$$\boldsymbol{\Sigma}_g = E[(\mathbf{x}_g - \boldsymbol{\mu} - \boldsymbol{\Lambda}_g \boldsymbol{\theta}_g)(\mathbf{x}_g - \boldsymbol{\mu} - \boldsymbol{\Lambda}_g \boldsymbol{\theta}_g)'] = \boldsymbol{\Lambda}_g \boldsymbol{\Phi}_g \boldsymbol{\Lambda}_g' + \boldsymbol{\Psi}_g^2, \tag{2}$$

where $\boldsymbol{\theta}_g$ and $\boldsymbol{\Phi}_g$ are the vector of means and the variance–covariance matrix of $\boldsymbol{\xi}_g$ respectively, and $\boldsymbol{\Psi}_g^2$ is a diagonal matrix consisting of the variances of the unique factors. The expected values of the measurements are given by

$$E(\mathbf{x}_g) = \boldsymbol{\mu} + \boldsymbol{\Lambda}_g \boldsymbol{\theta}_g. \tag{3}$$

To make the interpretation of the parameters in the model easier we may distinguish between

1. *Parameters differentiating the measurements.* The vector $\boldsymbol{\mu}$ indicates the level of the measurements, i.e. given $\boldsymbol{\xi}_g = \mathbf{0}$, the expected values of the measurements equal $\boldsymbol{\mu}$. The ith row of $\boldsymbol{\Lambda}_g$ gives the regression of x_i on the common factors and the ith diagonal element of $\boldsymbol{\Psi}_g^2$ is the corresponding residual variance.

2. *Parameters describing the factor space.* The matrix $\boldsymbol{\Phi}_g$ is the variance–covariance matrix of the common factors and $\boldsymbol{\theta}_g$ is their mean vector.

3. Estimation

Model (1) is not identified. We can premultiply $\boldsymbol{\Phi}_g$ in (1) by a non-singular matrix of order $k \times k$, $\mathbf{A}_g$, and postmultiply $\boldsymbol{\Lambda}_g$ by $\mathbf{A}_g^{-1}$. This does not change the observable measurements $\mathbf{x}_g$ and hence *a fortiori* their mean vector $E(\mathbf{x}_g)$ and their variance–covariance matrix $\boldsymbol{\Sigma}_g$. Thus we have to introduce at least k^2 restrictions on $\boldsymbol{\Lambda}_g$, $\boldsymbol{\Phi}_g$ and $\boldsymbol{\theta}_g$ $(g = 1, 2, \ldots, m)$, in such a way that the only $\mathbf{A}_g$ matrices that maintain the restrictions imposed are identity matrices. This can be done in several ways but, as indicated by Jöreskog (1971), 'it is hard to give further specific rules in the general case'.

The main purpose of the model just described is to study differences in the factor spaces of the measurements arising from different groups of observations. Suppose, for example, that we want to study differences in abilities for groups of individuals belonging to different socio-economic classes. To do this we administer the same battery of tests to all individuals. This means that the connexion between the factor space and the space of the measurements is the same for all groups. Thus we should have

$$\mathbf{\Lambda}_1 = \mathbf{\Lambda}_2 = \ldots = \mathbf{\Lambda}_m = \mathbf{\Lambda}. \tag{4}$$

Furthermore, each measurement is often constructed in such a way that it measures only a subclass of the abilities to be studied. This means that we have *a priori* information of the kind that $\lambda_{ij} = 0$, i.e. the jth ability does not affect the ith measurement. In Section 6 two examples are given, which show how *a priori* information of this kind may be used to make Model (1) identifiable.

To obtain ML estimates of the parameters, a computer program (Sörbom, 1973*a*) has been developed, which allows one to utilize information of the kind just discussed. Because the procedure is similar to that of Jöreskog (1971), only a brief account of it is given here.

If it can be assumed that the measurements obtained in the different groups are independent, the ML estimates can be calculated by minimizing the function

$$F = \sum_{g=1}^{m} (N_g/2) f_g, \tag{5}$$

where N_g is the number of observations in the gth group and f_g is minus the log likelihood function for the measurements in the gth group, i.e.

$$f_g = \log|\mathbf{\Sigma}_g| + \mathrm{tr}(\mathbf{\Sigma}_g^{-1}\mathbf{T}_g),$$

where

$$\mathbf{T}_g = (1/N_g) \sum_{\alpha=1}^{N_g} (\mathbf{x}_g^{(\alpha)} - \boldsymbol{\mu} - \mathbf{\Lambda}_g \boldsymbol{\theta}_g)(\mathbf{x}_g^{(\alpha)} - \boldsymbol{\mu} - \mathbf{\Lambda}_g \boldsymbol{\theta}_g)',$$

and

$\mathbf{x}_g^{(\alpha)}$ = the observed vector of measurements for the αth observation in the gth group.

The minimization procedure is a modification of the Fletcher & Powell (1963) method as described by Gruvaeus & Jöreskog (1970). This method makes use of the first derivatives of F to find the minimum by an iterative process. These first derivatives are almost the same as those given by Jöreskog (1971). The main differences are the derivatives for $\boldsymbol{\mu}$, $\mathbf{\Lambda}_g$ and $\boldsymbol{\theta}_g$, which are given by

$$\partial F/\partial \boldsymbol{\mu} = \sum_{g=1}^{m} N_g \mathbf{\Sigma}_g^{-1}[\mathbf{I} - \mathbf{\Lambda}_g(\mathbf{\Lambda}_g'\mathbf{\Sigma}_g^{-1}\mathbf{\Lambda}_g)^{-1}\mathbf{\Lambda}_g']\mathbf{\Sigma}_g^{-1}(\bar{\mathbf{x}}_g - \boldsymbol{\mu}), \tag{6}$$

where $\bar{\mathbf{x}}_g$ is the vector of sample means for the gth group,

$$\partial F/\partial \mathbf{\Lambda}_g = N_g[\mathbf{\Sigma}_g^{-1}(\mathbf{\Sigma}_g - \mathbf{T}_g)\mathbf{\Sigma}_g^{-1}\mathbf{\Lambda}_g\mathbf{\Phi}_g - \mathbf{\Sigma}_g^{-1}(\bar{\mathbf{x}}_g - \boldsymbol{\mu} - \mathbf{\Lambda}_g\boldsymbol{\theta}_g)\boldsymbol{\theta}_g'], \tag{7}$$

and

$$\partial F/\partial \boldsymbol{\theta}_g = N_g \mathbf{\Lambda}_g'\mathbf{\Sigma}_g^{-1}(\boldsymbol{\mu} + \mathbf{\Lambda}_g\boldsymbol{\theta}_g - \bar{\mathbf{x}}_g). \tag{8}$$

Eqn. (6) can be considerably simplified by the use of the identity

$$\mathbf{\Sigma}_g^{-1} = \mathbf{\Psi}_g^{-2} - \mathbf{\Psi}_g^{-2}\mathbf{\Lambda}_g(\mathbf{\Lambda}_g'\mathbf{\Psi}_g^{-2}\mathbf{\Lambda}_g + \mathbf{\Phi}_g^{-1})^{-1}\mathbf{\Lambda}_g'\mathbf{\Psi}_g^{-2}$$

and the definition

$$\mathbf{A}_g = \mathbf{\Psi}_g^{-1}\mathbf{\Lambda}_g,$$

to give

$$\partial F/\partial \boldsymbol{\mu} = \sum_{g=1}^{m} N_g \mathbf{\Psi}_g^{-1}(\mathbf{I} - \mathbf{A}_g(\mathbf{A}_g'\mathbf{A}_g)^{-1}\mathbf{A}_g')\mathbf{\Psi}_g^{-1}.$$

The estimates of $\boldsymbol{\theta}_g$ obtained by solving $\partial F/\partial \boldsymbol{\theta}_g = \mathbf{0}$ are the same as will be obtained by minimizing

$$N_g(\bar{\mathbf{x}}_g - \boldsymbol{\mu} - \mathbf{\Lambda}_g\boldsymbol{\theta}_g)'\mathbf{\Sigma}_g^{-1}(\bar{\mathbf{x}}_g - \boldsymbol{\mu} - \mathbf{\Lambda}_g\boldsymbol{\theta}_g).$$

On the other hand, the estimates of $\mathbf{\Lambda}_g$ obtained by solving $\partial F/\partial \mathbf{\Lambda}_g = \mathbf{0}$ are not the same as those obtained in classical factor analysis, where

$$\partial F/\partial \mathbf{\Lambda}_g = N_g \mathbf{\Sigma}_g^{-1}(\mathbf{\Sigma}_g - \mathbf{S}_g)\mathbf{\Sigma}_g^{-1}\mathbf{\Lambda}_g\mathbf{\Phi}_g. \tag{9}$$

The matrix $\mathbf{S}_g$ in (9) is the sample variance–covariance matrix. This means that the two-stage procedure mentioned in Section 1 does not produce ML estimates in the sense of Model (1).

When minimizing F, *a priori* information of the kind mentioned above can be handled by the following partition of the parameters in the model: (1) Free parameters, viz. independent parameters which are allowed to vary without any restrictions; (2) fixed parameters, viz. parameters whose values are known *a priori*; (3) constrained parameters, viz. parameters each of which is constrained to equal a free parameter. Thus measurements obtained in accordance with (4) may be assumed to be structured by a partition of the parameters in the following manner:

1. *Free parameters*: $\boldsymbol{\mu}$, $\mathbf{\Phi}_g$, $\mathbf{\Psi}_g$, $\boldsymbol{\theta}_g$ $(g = 1, 2, \ldots, m)$. Also, if it is assumed that the jth ability is involved in the ith test, λ_{ij} is free.

2. *Fixed parameters*: If it is assumed that the lth ability is not involved in the kth test, λ_{kl} is zero.

3. *Constrained parameters*: As the same tests have been administered to all groups we let the non-fixed parameters in $\mathbf{\Lambda}$ for the first group be free and let the corresponding parameters in $\mathbf{\Lambda}_2, \mathbf{\Lambda}_3, \ldots, \mathbf{\Lambda}_m$ equal these free parameters.

This method of specifying the parameters makes it possible to generate a wide variety of models. It should be noted that it is not necessary that the number of factors is the same in all populations. If, for example, the ith population has an extra factor, we can introduce an artificial factor for the other populations by specifying zeros in a column of $\mathbf{\Lambda}_g$, a row and column in $\mathbf{\Phi}_g$ and a zero in $\boldsymbol{\theta}_g$, for $g = 1, 2, \ldots, i-1, i+1, \ldots, m$. This procedure makes it possible to treat the 'basic factor model' studied by Please (1973).

4. Test of Hypothesis

The value of F at the minimum, $\hat{F}$, may be used in large samples to test hypotheses in the likelihood-ratio sense.

If we let

$$G = \hat{F} - \sum_{g=1}^{m} (N_g/2)(\log|\mathbf{S}_g| + p),$$

it can be shown that $2G$ is distributed asymptotically as χ^2 with degrees of freedom equal to

$$\text{d.f.} = p + p(p+1)/2 - q,$$

where q is the number of free parameters in the model estimated. This test is a test of the hypothesis that the measurements are structured in a manner defined by the model and the restrictions of fixed, free and constrained parameters as described in Section 3, as against the hypothesis that the measurements are not structured, i.e. that the mean vector $E(\mathbf{x}_g)$ and the variance–covariance matrix $\mathbf{\Sigma}_g$ are unconstrained in all groups. The value of G may also be used to test more specific hypotheses about the parameters. Suppose that G_1 is the value obtained for one model and that G_2 is the value obtained for a more restricted model. Then

$$\chi^2 = 2(G_2 - G_1)$$

is distributed as χ^2 with degrees of freedom equal to the difference in degrees of freedom for the two models. It should be noted, however, that if the same data are used to estimate the two models, the test is not a test of hypothesis in the common statistical sense. The χ^2 test should only be regarded as a tool to generate hypotheses in exploratory studies. These hypotheses must later be tested in a confirmatory study based on new data. Two very simple examples may elucidate this. Suppose an analysis has given a value of, say, $\lambda_{ij} = 1{\cdot}3$ with $G = G_1$. We do the analysis once again, but this time with λ_{ij} fixed to be equal to 1·3. We obtain this second time a value of G, G_2, which of course is equal to G_1. Thus we obtain $\chi^2 = 0$ with one degree of freedom, but this should not lead us to propose that it is confirmed that $\lambda_{ij} = 1{\cdot}3$. Similarly, if we obtained two λ values, say, $\lambda_{ij} = 1{\cdot}3$ and $\lambda_{kl} = 1{\cdot}4$, we could do the analysis again with the restriction that $\lambda_{kl} = \lambda_{ij}$. Again, a low value of χ^2 does not indicate a confirmation of the hypothesis that λ_{kl} equals λ_{ij}. Rather, it gives us a hint that in the metric used, the differences between the λ's is not very large. Only a confirmatory study of a new independent set of data can give support to the hypothesis.

5. Identification of $\boldsymbol{\theta}_g$ in a Special Case

In the special case

$$\mathbf{\Lambda}_1 = \mathbf{\Lambda}_2 = \ldots \mathbf{\Lambda}_m = \mathbf{\Lambda},$$

an indeterminacy of the parameters $\boldsymbol{\mu}$ and $\boldsymbol{\theta}_g$ appears. The observable means of the groups, $E(\mathbf{x}_g)$ $(g = 1, 2, \ldots, m)$, are structured as

$$E(\mathbf{x}_g) = \boldsymbol{\mu} + \mathbf{\Lambda}\boldsymbol{\theta}_g.$$

Thus we can add a k-dimensional vector, $\mathbf{a}$, to the $\boldsymbol{\theta}$'s in the groups and subtract $\mathbf{\Lambda a}$ from $\boldsymbol{\mu}$ to obtain:

$$E(\mathbf{x}_g) = \boldsymbol{\mu} - \mathbf{\Lambda a} + \mathbf{\Lambda}(\boldsymbol{\theta}_g + \mathbf{a}) \equiv \boldsymbol{\mu} + \mathbf{\Lambda}\boldsymbol{\theta}_g.$$

Thus the estimation of $\boldsymbol{\mu}$ and the $\boldsymbol{\theta}$ vectors can be obtained only if we introduce at least k constraints. In the estimation procedure this can be done by letting, for example, $\boldsymbol{\theta}_1 = \mathbf{0}$. Afterwards the estimates, $\hat{\boldsymbol{\theta}}_g$ and $\hat{\boldsymbol{\mu}}$, could be rescaled to obtain new estimates, $\boldsymbol{\theta}_g{}^*$ and $\boldsymbol{\mu}^*$, such that

$$\sum_{g=1}^{m} N_g \boldsymbol{\theta}_g{}^* = \mathbf{0},$$

that is,

$$\boldsymbol{\theta}_g{}^* = \hat{\boldsymbol{\theta}}_g - (1/N)\sum_{g=1}^{m} N_g \hat{\boldsymbol{\theta}}_g, \quad N = \sum_{g=1}^{m} N_g, \quad \boldsymbol{\mu}^* = \hat{\boldsymbol{\mu}} + \boldsymbol{\Lambda}(1/N)\sum_{g=1}^{m} N_g \hat{\boldsymbol{\theta}}_g.$$

6. Two Illustrations

Meredith (1964) and Jöreskog (1971) used data from Holzinger & Swineford (1939) to illustrate differences in factor spaces among four groups of individuals. The data were obtained from the same tests that had been administrated to all groups, and nine tests defining a three-dimensional factor space were selected. The groups consisted of 7th- and 9th-grade children from two schools in Chicago, the Pasteur and the Grant-White schools. The children from each school were divided into two groups according to whether they scored above or below the median on a speeded addition test. Thus the following four groups were analysed: (1) Pasteur, Low ($N_1 = 77$), (2) Pasteur, High ($N_2 = 79$), (3) Grant-White, Low ($N_3 = 74$) and (4) Grant-White, High ($N_4 = 71$).

By testing several hypotheses and by 'appealing to intuition' when faced with the interpretation, Jöreskog (1971) suggested that (1) the factors were the same in all groups, i.e. $\boldsymbol{\Lambda}_1 = \boldsymbol{\Lambda}_2 = \ldots = \boldsymbol{\Lambda}_m$; (2) the matrix of factor loadings was of a particularly simple form; (3) the specific variances (the $\boldsymbol{\Psi}_g{}^2$s) were the same in all groups. Thus it was proposed that differences in the sample variance–covariance matrices and the sample mean vectors among the groups could be explained solely by differences in the factor spaces. In this case the factor spaces were spanned by the three abilities designated space, verbal and memory.

It should be noted that since the elements in the $\boldsymbol{\Phi}_g$ matrices were free parameters, we have to fix at least one non-zero element in each column of $\boldsymbol{\Lambda}$. Otherwise it would have been possible to multiply, say, the ith column of $\boldsymbol{\Lambda}$ by a constant, c, and then multiply the ith row and column of $\boldsymbol{\Phi}_g$ by $1/c$ and the ith element of $\boldsymbol{\theta}_g$ by $1/c$: this would not change $E(\mathbf{x}_g)$ or $\boldsymbol{\Sigma}_g$. Afterwards the factors could be rescaled by postmultiplying $\boldsymbol{\Lambda}$ by a diagonal matrix, $\mathbf{D}^{\frac{1}{2}}$, premultiplying $\boldsymbol{\theta}_g$ by $\mathbf{D}^{-\frac{1}{2}}$, and by pre- and postmultiplying the $\boldsymbol{\Phi}_g$ matrices by $\mathbf{D}^{-\frac{1}{2}}$ in such a way that the diagonal elements of

$$\left(1 \Big/ \sum_{g=1}^{m} N_g\right) \sum_{g=1}^{m} N_g \mathbf{D}^{-\frac{1}{2}} \boldsymbol{\Phi}_g \mathbf{D}^{-\frac{1}{2}}$$

equal unity. This has been done for the results reported in Table 1. The parameters that were fixed during the analysis are marked by an asterisk.

The estimates of the factor loadings, $\boldsymbol{\Phi}$ matrices and specific variances are almost identical with those reported by Jöreskog (1971). It should be noted that in Table 3 of Jöreskog's paper the numbering of the groups is given in reverse order. Thus the $\mathbf{\Phi}_4$ matrix is in fact the $\mathbf{\Phi}$ matrix for the Pasteur Low group, and not, as stated by Jöreskog in Section 3.7, the $\mathbf{\Phi}$ matrix for the Grant-White High group. The $\mathbf{\Phi}_3$ matrix belongs to the Pasteur High group and so on. To further illustrate the differences in ability among the groups, the factor means are plotted

Table 1.—*Holzinger–Swineford Data*

Test	Factor loadings: Space	Verbal	Memory	Specific variances
Visual perception	0·720*	0·0*	0·0*	0·484
Cubes	0·425	0·0*	0·0*	0·844
Paper form board	0·516	0·0*	0·0*	0·736
General information	0·0*	0·821*	0·0*	0·352
Sentence completion	0·0*	0·799	0·0*	0·327
Word classification	0·0*	0·781	0·0*	0·432
Figure recognition	0·0*	0·0*	0·495*	0·769
Object–number	0·0*	0·0*	0·551	0·769
Number–figure	0·0*	0·0*	0·580	0·669

Estimated factor variance–covariance matrices

School	Low level			High level		
Pasteur	1·374			0·721		
	0·441	1·110		0·515	1·037	
	0·627	0·254	1·189	0·050	0·203	0·917
Grant-White	0·887			1·023		
	0·631	0·920		0·542	0·923	
	0·567	0·525	0·575	1·002	0·349	1·331

Estimated factor means and standard errors (within parentheses)

	Space	Verbal	Memory
Pasteur Low	−0·047 (0·145)	−0·636 (0·134)	−0·132 (0·134)
Pasteur High	−0·005 (0·117)	−0·215 (0·129)	0·247 (0·125)
Grant-White Low	0·038 (0·129)	0·257 (0·127)	−0·345 (0·114)
Grant-White High	0·016 (0·140)	0·661 (0·131)	0·229 (0·151)

in Fig. 1. It is seen that the profiles are similar within schools, with those scoring high on the addition test at a higher level. Further, the space ability does not differentiate the groups. For the verbal ability there is a difference also between schools with pupils from the Grant-White school on the average scoring higher. This reflects the fact that the Pasteur school 'enrols children of factory workers, a large percentage of whom were foreign-born and the Grant-White school enrols children in a middle-class suburban area' (Meredith, 1964). With regard to the memory factor, both groups of the Pasteur school seem to be superior to the corresponding groups of the Grant-White school, although the difference between the high groups is small.

Table 1 also gives estimates of standard errors for the estimated factor means. The formulae for these have not yet been derived, but in Sörbom (1973*b*) a method of obtaining them for the case $\boldsymbol{\mu} = \mathbf{0}$ is given. Thus, if we subtract the estimate of $\boldsymbol{\mu}$ from the sample means, approximate estimates of the standard errors for the estimated parameters in Model (1) can be obtained.

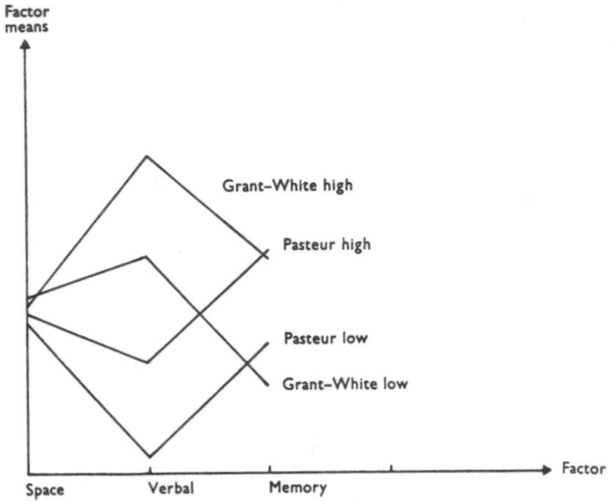

FIG. 1.—Holzinger–Swineford data: factor means. (The constant 0·7 has been added to the entries of Table 1.)

As a second illustration, data from the Project Talent study, earlier analysed by McGaw & Jöreskog (1971), are used. These consist of 12 tests administered to 11,743 high school subjects. The subjects were divided into four groups according to high and low intelligence (IQ) and high and low socio-economic status (SES), thus constituting the following four groups: (1) low IQ–low SES ($N_1 = 4491$), (2) low IQ–high SES ($N_2 = 1336$), (3) high IQ–low SES ($N_3 = 939$) and (4) high IQ–high SES ($N_4 = 4977$).

Table 2.—Project Talent Data

Test	Factor loadings			
	I General knowledge	II Verbal mechanics	III Spatial	IV Speed of perception
Vocabulary	0·659	0·275	0·0*	0·0*
Information I	0·812*	0·312	0·0*	0·0*
Information II	0·626	0·412	0·0*	0·0*
Spelling	0·0*	0·476	0·0*	0·0*
Punctuation	0·0*	0·706*	0·0*	0·0*
English usage	0·0*	0·550	0·0*	0·0*
Mechanical reasoning	0·224	0·0*	0·604	0·0*
Visualization I	0·0*	0·0*	0·452	0·190
Visualization II	0·0*	0·0*	0·695*	0·0*
Table reading	0·0*	0·0*	0·0*	0·639
Clerical checking	0·0*	0·087	0·0*	0·451
Object inspection	0·0*	0·0*	0·064	0·494*

Table 2—contd

Estimated specific variances

	Low IQ		High IQ	
	Low SES	High SES	Low SES	High SES
Vocabulary	0·453	0·534	0·352	0·327
Information I	0·068	0·153	0·000	0·105
Information II	0·382	0·563	0·439	0·446
Spelling	0·656	0·659	0·759	0·783
Punctuation	0·464	0·650	0·508	0·519
English usage	0·855	1·134	0·362	0·456
Mechanical reasoning	0·559	0·786	0·440	0·458
Visualization I	0·789	0·901	0·589	0·590
Visualization II	0·502	0·625	0·469	0·530
Table reading	1·073	1·334	0·113	0·142
Clerical checking	1·102	1·364	0·389	0·372
Object inspection	0·878	1·252	0·411	0·426

Estimated factor variance–covariance matrices

		Low SES				High SES			
		I	II	III	IV	I	II	III	IV
Low IQ	I	0·907				1·507			
	II	0·220	1·063			0·261	1·032		
	III	0·106	0·283	0·922		0·273	0·144	1·071	
	IV	0·110	0·469	0·485	1·481	−0·025	0·105	0·278	1·876
High IQ	I	1·025				0·943			
	II	0·386	0·995			0·265	0·936		
	III	0·034	0·078	1·125		0·364	0·193	1·028	
	IV	0·096	0·266	0·201	0·455	−0·018	0·099	0·077	0·434

Estimated factor means and standard errors (within parentheses)

Group	I General knowledge	II Verbal mechanics	III Spatial	IV Speed of perception
1 Low IQ–low SES	−1·63 (0·027)	−1·80 (0·020)	−1·20 (0·020)	−0·53 (0·024)
2 Low IQ–high SES	−1·12 (0·038)	−1·73 (0·038)	−1·00 (0·039)	−0·59 (0·043)
3 High IQ–low SES	0·76 (0·033)	1·27 (0·042)	0·93 (0·043)	0·42 (0·024)
4 High IQ–high SES	1·63 (0·015)	1·85 (0·019)	1·17 (0·023)	0·56 (0·010)

McGaw & Jöreskog proposed a structure of the factor loadings, invariant over the groups, as indicated in Table 2, where an asterisk denotes the factor loadings held fixed during the estimation procedure. The $\mathbf{\Phi}$ and the $\mathbf{\Psi}$ matrices of the groups were free, as were the vector of factor means. The estimates reported in Table 2 are very similar to those obtained by McGaw & Jöreskog, with the exception of the estimates for the factor means. The means given by McGaw & Jöreskog are plotted in Fig. 3, and the estimates from the present analysis are plotted in Fig. 2. It can be seen that in the latter case the profiles for the two SES groups within IQ classes are very similar, with those in the high SES groups in general on a higher level, especially for the general knowledge factor.

An inspection of the estimated standard errors for the factor means shows that all differences, except that between low IQ–low SES and low IQ–high SES for Speed of Perception, are significant.

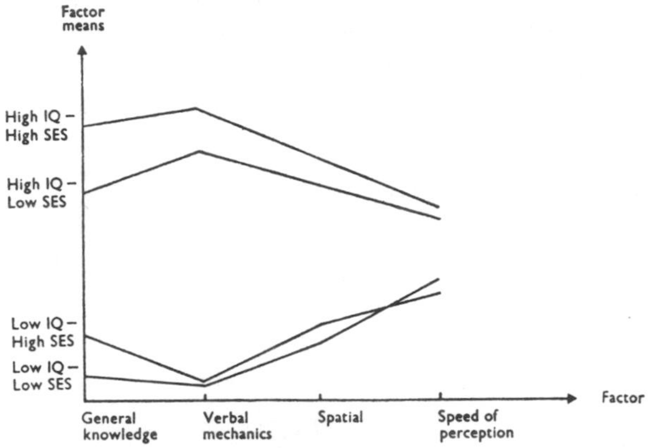

FIG. 2.—Talent data: factor means. (The constant 2·0 has been added to the entries of Table 2.)

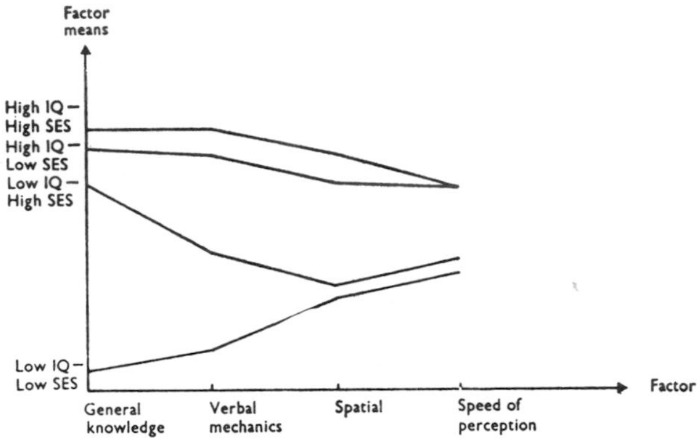

FIG. 3.—Talent data: factor means obtained by McGaw & Jöreskog (1971). (The constant 2·0 has been added to McGaw & Jöreskog's Table 7.)

This research has been supported by the Swedish Council for Social Science Research under project 'Statistical methods for analysis of longitudinal data', project director K. G. Jöreskog.

REFERENCES

FLETCHER, R. & POWELL, M. J. D. (1963). A rapidly convergent descent method for minimization. *Comp. J.* **6**, 163–168.

GRUVAEUS, G. T. & JÖRESKOG, K. G. (1970). A computer program for minimizing a function of several variables. *Res. Bull.* 70–14. Princeton, N.J.: Educational Testing Service.

HOLZINGER, K. & SWINEFORD, F. (1939). A study in factor analysis: the stability of a bi-factor solution. *Suppl. Educ. Monogr.* no. 48. Chicago: University Chicago Press.

Jöreskog, K. G. (1971). Simultaneous factor analysis in several populations. *Psychometrika* **36**, 409–426.

Lawley, D. N. & Maxwell, A. E. (1971). *Factor Analysis as a Statistical Method*, 2nd ed. London: Butterworth.

McGaw, B. & Jöreskog, K. G. (1971). Factorial invariance of ability measures in groups differing in intelligence and socioeconomic status. *Br. J. math. statist. Psychol.* **24**, 154–168.

Meredith, W. (1964). Rotation to achieve factorial invariance. *Psychometrika* **29**, 187–206.

Please, N. W. (1973). Comparison of factor loadings in different populations. *Br. J. math. statist. Psychol.* **26**, 61–89.

Sörbom, D. (1973*a*). FASPM, a computer program for factor analysis in several populations with structured means. (In preparation.)

Sörbom, D. (1973*b*). A statistical model for the measurement of change. *Res. Rep.* 73–6. Department of Statistics, University of Uppsala.

Reprinted from: *British Journal of Mathematical and Statistical Psychology*, 1974, Vol. 27, pp. 229-239.

Chapter 9

An Alternative to the Methodology for Analysis of Covariance

Dag Sörbom

A general statistical model for simultaneous analysis of data from several groups is described. The model is primarily designed to be used for the analysis of covariance. The model can handle any number of covariates and criterion variables, and any number of treatment groups. Treatment effects may be assessed when the treatment groups are not randomized. In addition, the model allows for measurement errors in the criterion variables as well as in the covariates. A wide variety of hypotheses concerning the parameters of the model can be tested by means of a large sample likelihood ratio test. In particular, the usual assumptions of ANCOVA may be tested.

Key words: confirmatory factor analysis, simultaneous factor analysis, measurement errors.

Introduction

In the usual methodology for analysis of covariance (ANCOVA) there are several more or less concealed assumptions inherent, which have been revealed and discussed extensively in the past [see e.g. Lord, 1960, 1963, 1967; Elashoff, 1969; Cronbach et al., Note 2]. In this paper we discuss a methodology to analyze data in which the different assumptions are more explicitly stated and testable, and the resulting model for the data is to a greater extent dictated by properties of the data than is possible in the classical ANCOVA analysis. In the kind of studies where the ANCOVA technique has been used, the variables that are observed are usually not exact measures of the variables one is really interested in. Therefore, the present model is designed to handle fallible measures. Especially, the influence of errors in the covariates are examined and discussed in the paper.

A great deal of the discussion of ANCOVA is concerned with the questions whether or not the subjects are randomly assigned to treatment groups and whether or not the treatments are randomized among the groups (see Elashoff, 1969 and further references therein). For the methodology discussed in this paper these questions are of less importance. Rather, we take the following view, which seems to be applicable in most ANCOVA studies: Assume we have data from a number of groups G. These G groups are supposed to be representative samples from G populations and at the end of the analysis we want to make conclusions regarding differences among these populations. For each population we formulate a model and if the models for the groups are consistent with data we are able to make inferences about the differences in terms of the estimated values for the parameters of these models.

The main objective of the paper is to discuss a statistical model, its identification and an estimation method which makes it possible to estimate the regression of the true criterion score on the true covariate score for a number of treatment groups. Using this

---

Research reported in this paper has been partly supported by the Swedish Council for Social Science Research under project "Statistical methods for analysis of longitudinal data", project director Karl G. Jöreskog, and partly by the Bank of Sweden Tercentenary Foundation under project "Structural Equation Models in the Social Sciences", project director Karl G. Jöreskog.

Requests for reprints should be sent to Dag Sörbom, Department of Statistics, University of Uppsala, P.O. Box 513, S-751 20 Uppsala, SWEDEN.

model, it is possible to test the assumption of equal slopes in these regressions among the groups, as well as assumptions of equal error variances, equal true score variances, whether the measurements are parallel or tau-equivalent [Lord & Novick, 1968] and so on.

The use of the model is illustrated by analysis of two small sets of data. Also, a strategy for modifying the model when it does not fit the data sufficiently well is exemplified.

Analysis of Covariance, a Brief Introduction

This section gives a very brief introduction to the general aspects of ANCOVA and introduces the notation to be used in the sections to follow. For simplicity the case of only one criterion variable and only one covariate is treated.

The main reason for the use of ANCOVA is that one is interested in studying the effect of a number of treatments. To this end, a number of cases, $N^{(g)}$, are considered in the gth treatment group. The main focus is on the expected values of a criterion variable, $\eta^{(g)}$. We assume that there exists a covariate, $\xi^{(g)}$, a variable which to some extent accounts for preexisting differences among the groups. Thus, it is desirable to eliminate the effect of the covariate, which, in addition, will result in a more powerful analysis [see e.g. Bock, 1975, p.355ff]. This is done by considering the linear regressions

$$\eta^{(g)} = \alpha^{(g)} + \gamma^{(g)}\,\xi^{(g)} + \zeta^{(g)} \tag{1}$$

for each group $g = 1, 2, \cdots, G$, where $\zeta^{(g)}$ denotes the error of the regression. The main focus is now on the intercept, $\alpha^{(g)}$. However, often the covariate $\xi^{(g)}$ cannot be measured without error, and it is a well-known fact [see e.g. Lord, 1960] that the errors can cause serious bias in the estimates of $\alpha^{(g)}$. Suppose that we have two variables, $x_1^{(g)}$ and $x_2^{(g)}$, that give us information about $\xi^{(g)}$, and that these measure $\xi^{(g)}$ in the following sense:

$$\begin{aligned} x_1^{(g)} &= \mu_1 + \lambda_1\,\xi^{(g)} + \epsilon_1^{(g)} \\ x_2^{(g)} &= \mu_2 + \lambda_2\,\xi^{(g)} + \epsilon_2^{(g)} \end{aligned} \tag{2}$$

where $\epsilon_1^{(g)}$ and $\epsilon_2^{(g)}$ represent measurement errors in $x_1^{(g)}$ and $x_2^{(g)}$, respectively. This will be referred to as the measurement model [Keesling & Wiley, Note 5]. In (2), μ_i is an arbitrary location parameter describing the level of the observable variate $x_i^{(g)}$, $i = 1, 2$. The inclusion of these parameters in the model implies that the other parameters are unaffected by adding or subtracting a constant to the observed variables. This feature of the model is important in several instances, since it is often not possible to determine a natural origin for the observed measures. Each λ_i is a parameter describing the scale of the variable $x_i^{(g)}$ as compared with the scale of other variables measuring the same $\xi^{(g)}$. It is seen from (2) that the variance of $\xi^{(g)}$ and the two λ-parameters are not identified, since we can multiply $\xi^{(g)}$ by a constant and divide the λ's by the same constant, without changing the observed measurements at all. Therefore, in the following, λ_1 is fixed equal to 1. The parameters μ_i and λ_i have no superscript, (g), because in general identical measurement operations are conducted in the treatment groups in order to make it at all possible to draw inferences about differences between the groups. However, by using the methodology discussed in this paper, it is possible to test this assumption, and if it is not tenable, to relax the invariance of the measurement models over groups in some way or other. This causes no trouble as far as the model formulation is concerned, but it may result in difficulties in interpreting the possible differences among groups obtained from the analysis. The error variables, $\epsilon_i^{(g)}$, on the other hand, have a superscript since it is often reasonable to assume that the measured variables have different reliabilities in different groups, and if this fact is not taken into account, these reliability differences may cause false conclusions [Campbell, 1963]. Of course, if desired, one can specify that these errors are invariant over groups and test this assumption.

If it is assumed that $\mathcal{E}(\epsilon_i^{(g)}) = 0$ for all groups, it follows from (2) that

$$\mathcal{E}(x_i^{(g)}) = \mu_i + \lambda_i \theta^{(g)}, \tag{3}$$

where

$$\theta^{(g)} = \mathcal{E}(\xi^{(g)}).$$

From (3) it can be seen that all μ- and θ-parameters cannot be identified simultaneously. We can add a constant a_i to μ_i and compensate for this by subtracting a_i/λ_i from $\theta^{(g)}$ for $g = 1, 2, \cdots G$. The observable variables have not been changed by this operation. Thus, in the following development we have fixed $\theta^{(1)}$ to be equal to zero. After estimates have been obtained we are free to do any rescaling, e.g. such that $\Sigma_{g=1}^{G} N^{(g)}\theta^{(g)} = 0$, where $N^{(g)}$ is the sample size of group g. For simplicity we can assume that the criterion variable, $\eta^{(g)}$, has been measured without error by a variable $y^{(g)}$. As noted by Lord [1960] this does not introduce any severe restrictions into the model, since an error in the y-variable can be regarded as being absorbed by the error in the equation variable, $\zeta^{(g)}$, in (1). Thus, if it is assumed that $\mathcal{E}(\zeta^{(g)}) = 0$ for all groups, it follows from (1) that

$$\mathcal{E}(y^{(g)}) = \mathcal{E}(\mu_y + \eta^{(g)}) = \mu_y + \alpha^{(g)} + \gamma^{(g)}\theta^{(g)}. \tag{4}$$

In (4) we can add a constant to μ_y and compensate for this by subtracting the same constant from $\alpha^{(g)}$ for all groups. This does not alter the observable variables $y^{(g)}$, so for the same reason as for the $x^{(g)}$- variables we have to fix at least one $\alpha^{(g)}$. Hereafter, $\alpha^{(1)}$ is set equal to zero.

If it is assumed that the error variables $\epsilon_1^{(g)}$ and $\epsilon_2^{(g)}$ are uncorrelated and each is uncorrelated with $\xi^{(g)}$, the variance-covariance matrix for the observable variables, $x_1^{(g)}$, $x_2^{(g)}$, and $y^{(g)}$, is given by

$$\mathbf{\Sigma}^{(g)} = \begin{bmatrix} \sigma^2_{\xi^{(g)}} + \sigma^2_{\epsilon_1^{(g)}} & \lambda_2\sigma^2_{\xi^{(g)}} & \gamma^{(g)}\sigma^2_{\xi^{(g)}} \\ \lambda_2\sigma^2_{\xi^{(g)}} & \lambda_2^2\sigma^2_{\xi^{(g)}} + \sigma^2_{\epsilon_2^{(g)}} & \lambda_2\gamma^{(g)}\sigma^2_{\xi^{(g)}} \\ \gamma^{(g)}\sigma^2_{\xi^{(g)}} & \lambda_2\gamma^{(g)}\sigma^2_{\xi^{(g)}} & \gamma^{(g)^2}\sigma^2_{\xi^{(g)}} + \sigma^2_{\zeta^{(g)}} \end{bmatrix} \tag{5}$$

We have three observable variables, which means that for each group there are 9 observable moments, 6 variances and covariances and 3 means, as long as we restrict our interest to first and second order moments. In the model there are $7G + 2$ parameters: $3G$ parameters from the regression functions, $\sigma^2_{\xi^{(g)}}$, $\gamma^{(g)}$, and $\sigma^2{}_{\zeta}$; $2G$ parameters for the error variances; $2(G - 1)$ parameters for $\theta^{(g)}$ and $\alpha^{(g)}$, since $\theta^{(1)}$ and $\alpha^{(1)}$ are restricted to be equal to zero; 3 μ-parameters; and 1 λ-parameter. Thus there are at least $9G - 7G - 2 = 2(G - 1)$ overidentifying restrictions on the model. This means that the estimation of the parameters cannot be done unequivocally by identifying $\mathbf{\Sigma}^{(g)}$ with the sample variance-covariance matrix and the population means with the sample means. The estimation problem is treated for the general model in a following section.

The model is identified, which can be shown as follows. Let the elements of $\mathbf{\Sigma}^{(g)}$ be denoted by $\sigma_{ij}^{(g)}$, $i, j = 1, 2, 3$; $\gamma^{(g)}$ can be obtained as $\sigma_{32}^{(g)}/\sigma_{21}^{(g)}$ and $\sigma^2_{\xi^{(g)}}$ as $\sigma_{31}^{(g)}/\gamma^{(g)} = \sigma_{31}^{(g)}\sigma_{21}^{(g)}/\sigma_{32}^{(g)}$. For λ_2 there are G possible ways: $\lambda_2 = \sigma_{32}^{(g)}/\sigma_{31}^{(g)}$, $g = 1, 2, \cdots, G$. This means that there are $G - 1$ overidentifying restrictions on λ_2. The parameters $\sigma_{\epsilon_1}^{2(g)}$, $\sigma_{\epsilon_2}^{2(g)}$ and $\sigma_{\zeta}^{2(g)}$ can be obtained from $\sigma_{11}^{(g)}$, $\sigma_{22}^{(g)}$ and $\sigma_{33}^{(g)}$ respectively. The means μ_1, μ_2, and μ_3 can be obtained from $\mathcal{E}(x_1^{(1)})$, $\mathcal{E}(x_2^{(1)})$ and $\mathcal{E}(y^{(1)})$ respectively and $\theta^{(g)}$ and $\alpha^{(g)}$ from $\mathcal{E}(x_1^{(g)})$ and $\mathcal{E}(y^{(g)})$ respectively for $g = 2, 3, \cdots, G$. Now all parameters have been identified and still we have not used $\mathcal{E}(x_2^{(g)})$, $g = 2, 3, \cdots, G$. Thus, in total there are $2(G - 1)$ overidentifying restrictions in the model. There are several possible ways of calculating estimates of the parameters using the observed variances, covariances, and means. In the next section the maximum likelihood method is suggested. With this method all sample information is used, and if the distributional assumptions are met, the estimates are guaranteed to be the most efficient in large samples.

The direct use of the observed variances and covariances is hazardous [cf. e.g. Lord, 1960; Cochran, 1968; Jöreskog & Sörbom, Note 4]. Suppose we are using y and x_1. Then γ is estimated by

$$\tilde{\gamma} = \frac{\text{cov}(x_1, y)}{\text{var}(x_1)} ,$$

but

$$\mathcal{E}(\tilde{\gamma}) = \gamma \, \rho_{x_1x_1} ,$$

where

$$\rho_{x_1x_1} = \frac{\sigma_\xi^2}{\sigma_\xi^2 + \sigma_{\epsilon_1}^2} ,$$

so that $\tilde{\gamma}$ is biased downwards. This in turn implies that the estimated treatment effect, $\tilde{\alpha}$, is biased, since

$$\tilde{\alpha} = \bar{y} - \tilde{\gamma}\bar{x}.$$

As noted by several authors [see e.g. Smith, 1957; Lord, 1960; Porter, Note 7; Bergman, Note 1] this bias can have detrimental effects on the conclusions made from an analysis of covariance. Depending on the actual values of $\bar{x}$ in the treatment groups, the bias can lead to a rejection of the hypothesis of no treatment effect when there is no such effect. Also, it can happen that the analysis fails to detect an existing effect because of this bias.

If, for example, we are studying two groups with the same criterion-covariate slope, γ, the effect of the treatment in Group 2 as compared with the effect of the treatment in Group 1 would have been computed as

$$\tilde{\alpha}^{(2)} - \tilde{\alpha}^{(1)} = \bar{y}^{(2)} - \bar{y}^{(1)} - \tilde{\gamma}(\bar{x}^{(2)} - \bar{x}^{(1)}).$$

Thus, by (3) and (4)

$$\begin{aligned}\mathcal{E}(\tilde{\alpha}^{(2)} - \tilde{\alpha}^{(1)}) &= \alpha^{(2)} - \alpha^{(1)} + \gamma(\theta^{(2)} - \theta^{(1)}) - \gamma\rho_{xx}(\theta^{(2)} - \theta^{(1)}) \\ &= \alpha^{(2)} - \alpha^{(1)} + \gamma(\theta^{(2)} - \theta^{(1)})(1 - \rho_{xx}).\end{aligned}$$

This means that as long as x is not measured without error, i.e. $\rho_{xx} \neq 1$, we obtain a bias in the estimated effect, and this bias can be positive or negative depending on the difference in level of the covariate among the groups. An obvious way to eliminate the bias is to include an estimate of the reliability of x, r_{xx}, in the analysis. That is, instead of $\tilde{\gamma}$ we use $\tilde{\gamma}/r_{xx}$ [Cochran, 1968]. If r_{xx} is obtained from a previous study, and thus is independent of the observed scores on x and y, the bias equals $\gamma(\theta^{(2)} - \theta^{(1)})(1 - \rho_{xx}/r_{xx})$, and the bias is removed whenever r_{xx} is an exact estimate of ρ_{xx}. However, it is quite seldom that estimates of ρ_{xx} are available, especially when the reliabilities vary among the treatment groups [Campbell, 1963].

The General Model

In this section we consider a generalization of the model in the previous section. When each group is viewed separately, the model is very similar to that in Keesling and

Wiley [Note 5] and to the LISREL model [Jöreskog & Sörbom, 1976a, 1976b, 1977]. LISREL does not make use of the means of the observed variables. The present model makes it possible to undertake a simultaneous analysis of several groups and to take into account the means of the unobserved variables. The last feature is of importance if we are interested in the effects of the treatments. Information about the parameters of the model is contained in the sample variance-covariance matrix as well as in the means of the observed variables, and these two sources of information are stochastically independent. However, the maximum likelihood estimates of the parameters of the model are dependent, e.g. the estimates of the $\mathbf{\Lambda}$-matrices and the factor means are correlated. This implies that in order to obtain estimates which are fully efficient one must find them in one step, which cannot be done by the estimation procedure suggested by Keesling and Wiley [Note 5].

Let $\mathbf{y}^{(g)}$ denote a vector of p criterion variables for the gth treatment group, and $\mathbf{x}^{(g)}$ a vector of covariates. These covariates may be considered as a set of variables containing information of any kind on preexisting differences among the groups. It is supposed that the $\mathbf{y}^{(g)}$- and $\mathbf{x}^{(g)}$-variables are measuring the unobservable $\boldsymbol{\eta}^{(g)}$- and $\boldsymbol{\xi}^{(g)}$-variables according to the following measurement model:

$$\mathbf{y}^{(g)} = \boldsymbol{\mu}_y + \mathbf{\Lambda}_y \boldsymbol{\eta}^{(g)} + \boldsymbol{\varepsilon}_y^{(g)} \tag{6}$$

$$\mathbf{x}^{(g)} = \boldsymbol{\mu}_x + \mathbf{\Lambda}_x \boldsymbol{\xi}^{(g)} + \boldsymbol{\varepsilon}_x^{(g)}.$$

Our main interest is focused on the parameters of the structural equations for the groups, that is

$$\boldsymbol{\eta}^{(g)} = \boldsymbol{\alpha}^{(g)} + \mathbf{\Gamma}^{(g)} \boldsymbol{\xi}^{(g)} + \boldsymbol{\zeta}^{(g)}. \tag{7}$$

Differences of the $\boldsymbol{\alpha}$-vectors are associated with the usual ANCOVA treatment effects.

Suppose we have $N^{(g)}$ observations of the q-dimensional vector $\mathbf{x}^{(g)}$ and of the p-dimensional vector $\mathbf{y}^{(g)}$ for $g = 1, 2, \cdots, G$, where G is the number of groups. If it is assumed that the means of the error terms $\boldsymbol{\varepsilon}_y^{(g)}$ and $\boldsymbol{\varepsilon}_x^{(g)}$ in (6) are equal to zero it follows that

$$\mathcal{E}(\mathbf{y}^{(g)}) = \boldsymbol{\mu}_y + \mathbf{\Lambda}_y \boldsymbol{\theta}_\eta^{(g)} \tag{8}$$

$$\mathcal{E}(\mathbf{x}^{(g)}) = \boldsymbol{\mu}_x + \mathbf{\Lambda}_x \boldsymbol{\theta}_\xi^{(g)},$$

where $\boldsymbol{\theta}_\eta^{(g)}$ and $\boldsymbol{\theta}_\xi^{(g)}$ denote the expectations of $\boldsymbol{\eta}^{(g)}$ and $\boldsymbol{\xi}^{(g)}$ respectively. If there are no restrictions on $\boldsymbol{\alpha}^{(g)}$ in (7), this vector is obtained as

$$\boldsymbol{\alpha}^{(g)} = \boldsymbol{\theta}_\eta^{(g)} - \mathbf{\Gamma}^{(g)} \boldsymbol{\theta}_\xi^{(g)}. \tag{9}$$

All the $\boldsymbol{\theta}_\eta^{(g)}$ and $\boldsymbol{\theta}_\xi^{(g)}$ vectors cannot be identified, since analogous to the case considered in the previous section, we can add a vector $\mathbf{a}_\eta$ to $\boldsymbol{\theta}_\eta^{(g)}$, and a vector $\mathbf{a}_\xi$ to $\boldsymbol{\theta}_\xi^{(g)}$ in (8) for $g = 1, 2, \cdots, G$ and compensate for this by subtracting $\mathbf{\Lambda}_y \mathbf{a}_\eta$ and $\mathbf{\Lambda}_x \mathbf{a}_\xi$ from $\boldsymbol{\mu}_y$ and $\boldsymbol{\mu}_x$ respectively. In the following, these indeterminacies are circumvented by letting $\boldsymbol{\theta}_\eta^{(1)} = 0$ and $\boldsymbol{\theta}_\eta^{(1)} = 0$. Afterwards, when estimates of the parameters are obtained, we are free to make any translation of the above type. For example, it can be done in such a way that

$$\sum_{g=1}^{G} N^{(g)} \boldsymbol{\alpha}^{(g)} = 0 \tag{10}$$

$$\sum_{g=1}^{G} N^{(g)} \boldsymbol{\theta}_\xi^{(g)} = 0.$$

To simplify the notation, let

$$\mathbf{z}^{(g)} = \begin{pmatrix} \mathbf{y}^{(g)} \\ \mathbf{x}^{(g)} \end{pmatrix}, \; \mathbf{\Lambda} = \begin{pmatrix} \mathbf{\Lambda}_y & 0 \\ 0 & \mathbf{\Lambda}_x \end{pmatrix}, \; \boldsymbol{\omega}^{(g)} = \begin{pmatrix} \boldsymbol{\eta}^{(g)} \\ \boldsymbol{\xi}^{(g)} \end{pmatrix}, \; \boldsymbol{\varepsilon}^{(g)} = \begin{pmatrix} \boldsymbol{\varepsilon}_y^{(g)} \\ \boldsymbol{\varepsilon}_x^{(g)} \end{pmatrix},$$

$$\boldsymbol{\mu} = \begin{pmatrix} \boldsymbol{\mu}_y \\ \boldsymbol{\mu}_x \end{pmatrix}, \; \boldsymbol{\theta}^{(g)} = \begin{pmatrix} \boldsymbol{\theta}_\eta^{(g)} \\ \boldsymbol{\theta}_\xi^{(g)} \end{pmatrix}.$$

Then (6) can be written as

(11) $$\mathbf{z}^{(g)} = \boldsymbol{\mu} + \mathbf{\Lambda}\boldsymbol{\omega}^{(g)} + \boldsymbol{\varepsilon}^{(g)},$$

and (8) as

(12) $$\mathcal{E}(\mathbf{z}^{(g)}) = \boldsymbol{\mu} + \mathbf{\Lambda}\boldsymbol{\theta}^{(g)}.$$

If it is assumed that $\boldsymbol{\varepsilon}^{(g)}$ is uncorrelated with $\boldsymbol{\omega}^{(g)}$ and has expectation 0, then it follows that the variance-covariance matrix for $\mathbf{z}^{(g)}$ equals

(13)
$$\begin{aligned}
\mathbf{\Sigma}^{(g)} &= \mathcal{E}[\mathbf{z}^{(g)} - \mathcal{E}(\mathbf{z}^{(g)})][\mathbf{z}^{(g)} - \mathcal{E}(\mathbf{z}^{(g)})]' = \\
&= \mathcal{E}[\mathbf{z}^{(g)} - \boldsymbol{\mu} - \mathbf{\Lambda}\boldsymbol{\theta}^{(g)}][\mathbf{z}^{(g)} - \boldsymbol{\mu} - \mathbf{\Lambda}\boldsymbol{\theta}^{(g)}]' = \\
&= \mathcal{E}[\mathbf{\Lambda}(\boldsymbol{\omega}^{(g)} - \boldsymbol{\theta}^{(g)}) + \boldsymbol{\varepsilon}^{(g)}][\mathbf{\Lambda}(\boldsymbol{\omega}^{(g)} - \boldsymbol{\theta}^{(g)}) + \boldsymbol{\varepsilon}^{(g)}]' = \\
&= \mathbf{\Lambda}\mathbf{\Phi}^{(g)}\mathbf{\Lambda}' + \mathbf{\Psi}^{*(g)},
\end{aligned}$$

where $\mathbf{\Phi}^{(g)}$ and $\mathbf{\Psi}^{*(g)}$ are the variance-covariance matrices for $\boldsymbol{\omega}^{(g)}$ and $\boldsymbol{\varepsilon}^{(g)}$ respectively. In each group, the model for $\mathbf{z}^{(g)}$ is similar to a restricted factor analysis model [Lawley & Maxwell, 1971]. However, it should be noted that there is no assumption of diagonality for the $\mathbf{\Psi}^{*(g)}$ matrices. This means that the model allows for covariances among the $\boldsymbol{\varepsilon}^{(g)}$-variables in (11), and this feature of the model can be of importance for some special data designs.

If it is further assumed that $\boldsymbol{\zeta}^{(g)}$ in (7) is uncorrelated with $\boldsymbol{\varepsilon}^{(g)}$ and $\boldsymbol{\xi}^{(g)}$ and has expectation 0, it follows that $\mathbf{\Phi}^{(g)}$ has the structure

(14) $$\mathbf{\Phi}^{(g)} = \begin{bmatrix} \mathbf{\Gamma}^{(g)}\mathbf{\Phi}_{\xi\xi}^{(g)}\mathbf{\Gamma}'^{(g)} + \mathbf{\Theta}^{(g)} & \\ \mathbf{\Phi}_{\xi\xi}^{(g)}\mathbf{\Gamma}'^{(g)} & \mathbf{\Phi}_{\xi\xi} \end{bmatrix} = \begin{bmatrix} \mathbf{\Phi}_{\eta\eta}^{(g)} & \\ \mathbf{\Phi}_{\xi\eta}^{(g)} & \mathbf{\Phi}_{\eta\eta}^{(g)} \end{bmatrix},$$

where $\mathbf{\Theta}^{(g)}$ is the variance-covariance matrix for the errors $\boldsymbol{\zeta}^{(g)}$ in (7).

Estimation of the Model

As noted above, the model is similar to a factor analysis model for several groups, where the means of the factors are taken into account. This model has been treated in Sörbom [1974], and the only difference here is that the variance-covariance matrix for the factors has the structure (14). The estimation of the parameters of the model can be treated in the same fashion as in Sörbom [1974], where the Fletcher and Powell [1963] procedure as modified by Gruvaeus and Jöreskog [1970] was used.

Let $\mathbf{z}_i^{(g)}$ denote the ith observation in the gth group. Assuming that $\mathbf{z}^{(g)}$ has a multinormal distribution and that the observations are obtained independently, it follows that minus the natural logarithm of the likelihood function for the gth group is given by

(15) $$f^{(g)} = \frac{N^{(g)}}{2}\left[\log\left|\mathbf{\Sigma}^{(g)}\right| + \mathrm{tr}\left(\mathbf{\Sigma}^{-1(g)}\mathbf{T}^{(g)}\right)\right],$$

where $|.|$ denotes the determinant and tr(.) the trace of a matrix. $\mathbf{T}^{(g)}$ in (15) is the matrix

(16) $$\mathbf{T}^{(g)} = \frac{1}{N^{(g)}}\sum_{i=1}^{N^{(g)}}(\mathbf{z}_i^{(g)} - \boldsymbol{\mu} - \mathbf{\Lambda}\boldsymbol{\theta}^{(g)})(\mathbf{z}_i^{(g)} - \boldsymbol{\mu} - \mathbf{\Lambda}\boldsymbol{\theta}^{(g)})'.$$

The maximum likelihood (ML) estimates of the parameters of the model are defined as those values of the parameters that make the function

$$F = \sum_{g=1}^{G} f^{(g)} \tag{17}$$

attain its minimum. With the Fletcher and Powell procedure, the minimum of (17) is obtained by an iterative algorithm which make use of the first derivatives. These derivatives are given in Jöreskog [1971a] and Sörbom [1974] except for the parameters in $\mathbf{\Phi}_{\xi\xi}^{(g)}$, $\mathbf{\Gamma}^{(g)}$ and $\mathbf{\Theta}^{(g)}$ in (14). It can be shown in a similar manner as in Jöreskog [1973] that these are given by

$$\frac{1}{N^{(g)}}\frac{\delta F}{\delta \mathbf{\Phi}_{\xi\xi}} = \mathbf{\Gamma}'^{(g)}\mathbf{\Lambda}_y'\mathbf{\Omega}_{yy}^{(g)}\mathbf{\Lambda}_y\mathbf{\Gamma}^{(g)} + \mathbf{\Lambda}_x'\mathbf{\Omega}_{xy}^{(g)}\mathbf{\Lambda}_y\mathbf{\Gamma}^{(g)} + \mathbf{\Gamma}'^{(g)}\mathbf{\Lambda}_y'\mathbf{\Omega}_{yx}^{(g)}\mathbf{\Lambda}_x + \mathbf{\Lambda}_x'\mathbf{\Omega}_{xx}^{(g)}\mathbf{\Lambda}_x$$

$$\frac{1}{N^{(g)}}\frac{\delta F}{\delta \mathbf{\Gamma}^{(g)}} = \mathbf{\Lambda}_y'(\mathbf{\Omega}_{yy}^{(g)}\mathbf{\Lambda}_y\mathbf{\Gamma}^{(g)} + \mathbf{\Omega}_{yx}^{(g)}\mathbf{\Lambda}_x)\mathbf{\Phi}_{\xi\xi}^{(g)} \tag{18}$$

$$\frac{1}{N^{(g)}}\frac{\delta F}{\delta \mathbf{\Theta}^{(g)}} = \mathbf{\Lambda}_y'\mathbf{\Omega}_{yy}^{(g)}\mathbf{\Lambda}_y$$

where $\mathbf{\Omega}^{(g)}$ denotes the matrix

$$\mathbf{\Omega}^{(g)} = \begin{pmatrix} \mathbf{\Omega}_{xx}^{(g)} & \mathbf{\Omega}_{xy}^{(g)} \\ \mathbf{\Omega}_{yx}^{(g)} & \mathbf{\Omega}_{yy}^{(g)} \end{pmatrix} = \mathbf{\Sigma}^{-1(g)}\left(\mathbf{\Sigma}^{(g)} - \mathbf{T}^{(g)}\right)\mathbf{\Sigma}^{-1(g)}.$$

As in Sörbom [1974] the parameters of the model are divided into three categories

(i) fixed parameters, i.e. parameters specified to have a given value;
(ii) free parameters, i.e. parameters the values of which are unknown and are to be estimated from data;
(iii) constrained parameters, i.e. parameters specified to be equal to one or more other parameters.

By this division of the parameters it is possible to specify a wide variety of different models and to take into account prior information about the data.

The minimum value of F in (17) can be used in hypothesis testing. For example, the equality of the slopes in the regressions of the criterion variable on the true covariate, that is the $\gamma^{(g)}$ in (1), can be tested in the following manner: first estimate the model with no restrictions on the $\gamma^{(g)}$ parameters. This gives a value of F equal to F_0, say. Then we can estimate the model with the restriction $\gamma^{(1)} = \gamma^{(2)} = \cdots = \gamma^{(G)}$, which gives rise to another value of F, say F_1. This value is greater than F_0 and we can examine the difference $F_1 - F_0$ to see whether the hypothesis of equal γs is acceptable. In fact, the difference is the likelihood ratio test statistic, and for large samples $F_1 - F_0$ is approximately distributed as χ^2 with $G - 1$ degress of freedom. In a similar fashion we can make a test for equalities of the αs among groups, i.e. to test the hypothesis of equal treatment effects.

An Example

To illustrate the use of the model and a procedure for model modification, a small subset of data from a study conducted by Olsson [Note 6] is analyzed in this section. The example is chosen mainly for the purpose of clarifying the use of the model and to illustrate how the parameters of the model can be interpreted in a real situation. In fact, the data consist of the smallest possible set for an analysis of the kind proposed, and, no doubt, the degrees of freedom for the models analyzed are too small to make any broader

TABLE 1.

Sample Variance-covariance Matrices and Sample Means for Olsson's Data.

Control Group (N=105)

Test	S				Means
$X_S^{(C)}$	37.626				18.381
$X_O^{(C)}$	24.933	34.680			20.229
$Y_S^{(C)}$	26.639	24.236	32.013		20.400
$Y_O^{(C)}$	23.649	27.760	23.565	33.443	21.343

Experimental Group (N=108)

Test	S				Means
$X_S^{(E)}$	50.084				20.556
$X_O^{(E)}$	42.373	49.872			21.241
$Y_S^{(E)}$	40.760	36.094	51.237		25.667
$Y_O^{(E)}$	37.343	40.396	39.890	53.641	25.870

conclusions. Rather, the analyses reported should be regarded as pure illustrations, which for the sake of clarity have been chosen to be as simple as possible.

The main data in the Olsson [Note 6] study consist of eight tests from the DBA-test battery [Härnqvist, 1962]. These tests were administered to about 400 11-year old pupils on two occasions, approximately one month apart. During this time period, four groups of about 100 pupils each were given different degrees of training on two of the verbal tests, Synonyms (S) and Opposites (O). The children were in principle randomly assigned to the treatment groups. In this example, data from two of the groups, Experiment group 1 (E) and Control group (C), are used. The members of the E-group were given tests similar to the S- and O-tests three times between the two administrations. These training tests contained all the items in the original test plus a number of similar items. The pupils were given the correct answers of the items after each training session. For group C, there was no such training.

The sample variance-covariance matrices and the sample means for the two groups are given in Table 1.

The tests S and O are both assumed to measure verbal ability, so we assume that the common part of these tests is a factor which we can call verbal ability (VA). Intuitively, the model shown in the path diagram in Figure 1 should be adopted first. In Figure 1, observed variables are enclosed in squares and unobserved variables in circles. An arrow between variables indicates a possible direct causal influence. Thus, for example, the score of the S-test for an individual is supposed to be composed of the individual's verbal ability and an error term, ϵ. The error should here be interpreted in a rather broad sense. It is that part of the test score that remains after the influence of the verbal ability has been removed. Thus, apart from what is usually meant by measurement errors, the ϵ-variable also contains influences from other abilities and traits of the individual which are not involved in other tests used to measure the same ability.

In accordance with the general model in (6) we can explicitly write the model for the test scores in the gth group as

$$
\begin{aligned}
x_S^{(g)} &= \mu_1 + \xi^{(g)} + \epsilon_1^{(g)} \\
x_O^{(g)} &= \mu_2 + \lambda_2\xi^{(g)} + \epsilon_2^{(g)} \\
y_S^{(g)} &= \mu_3 + \eta^{(g)} + \epsilon_3^{(g)} \\
y_O^{(g)} &= \mu_4 + \lambda_4\eta^{(g)} + \epsilon_4^{(g)}
\end{aligned}
\tag{19}
$$

where $x_S^{(g)}$ and $y_S^{(g)}$ denote test scores for the Synonyms test on the first and second occasion, respectively, and $x_O^{(g)}$ and $y_O^{(g)}$ denote the corresponding scores for the Opposites test. The values $\xi^{(g)}$ and $\eta^{(g)}$ are the verbal abilities at the first and second occasion, respectively. The superscript, g, has the value E for the Experiment group and C for the Control group. The structural equation to be studied is given by

$$\eta^{(g)} = \alpha^{(g)} + \gamma^{(g)}\xi^{(g)} + \zeta^{(g)}. \tag{20}$$

When the model in Figure 1 is estimated by the method outlined in the previous section, the obtained overall chi-square measure of fit is equal to 35.1 with 6 degrees of freedom. The differences between the observed variance-covariance matrices and the estimated Σ-matrices in (13), and the differences between the observed means and the estimated expected values in (12) are rather large. Thus, it seems that the model cannot be used to describe the data sufficiently well. The model has to be modified in some way. In Sörbom [1975], a procedure is described which uses the magnitudes of the derivatives of the fixed parameters of the function F in (17) to indicate in what respect the model has been incorrectly specified. A large value of the derivative for a fixed parameter means that

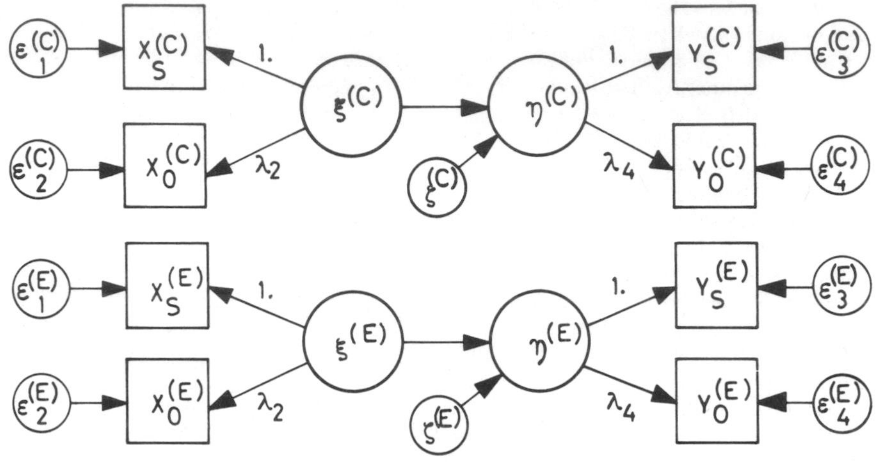

FIGURE 1
The initial model for the two groups in the example.

if this parameter is made into a free parameter, we expect the function F in (17) to decrease substantially. A study of the derivatives for the model in Figure 1 suggests that there exists a covariance for the error of the O-test between the two occasions in the E-group. A covariance of this kind can be interpreted in several ways. For example, the O-test may contain a measure of some ability not contained in the S-test, and as the same tests were used on the two occasions, it seems natural that there is some remaining correlation between the test scores for the O-test after the influence of the true scores has been removed [cf. Sörbom, 1975]. Thus when this covariation between the errors has been taken into consideration by the model, we assume that the verbal ability factor should have been "purer". For a larger set of data, with more variables incorporated, we could have been able to scrutinize this covariance further by adding true score variables to the model. However, for the present data alternatives of this kind are not identifiable.

The model with the covariance included yields a chi-square with 5 degrees of freedom equal to 17.2. The decrease in chi-square as compared with the initial model equals 17.9 (=35.1 − 17.2). Thus the hypothesis of zero covariance between the errors for the O-test in the E-group is rejected by a chi-square with 1 degree of freedom equal to 17.9. Still, the overall fit of the model is not acceptable in a strict interpretation of the chi-square measure, and an inspection of the above-mentioned derivatives for the modified model shows that there might be a covariance of the O-test errors also in the C-group. Allowing this covariance to also be a free parameter, we obtain a model with an overall fit measure equal to 2.8 with 4 degrees of freedom. As for the E-group, we can also conclude that there is a covariance for errors in the C-group. This time we can reject the hypothesis of no covariance by virtue of a chi-square with 1 degree of freedom equal to 14.4 (= 17.2 − 2.8). The overall fit of the model is now very good, the probability level for a chi-square equal to 2.8 with 4 degrees of freedom is approximately 0.59, and the sample variance-covariance matrix and the sample means are reproduced by the model parameter estimates to at least two significant digits.

By introducing covariances among the errors as done above, the identification of the model might be questioned. However, by starting with an identified model and freeing one fixed parameter with a non-zero derivative we are guaranteed to arrive at an identified model. This is so because at the minimum of F in (17) all derivatives of the free parameters are of course zero, but the derivative is also zero for each fixed parameter that would be non-identified if we made it a free parameter. This in turn follows from the fact that whatever value we give to a presumptive non-identified free parameter, we can always find values of the other free parameters that give exactly the same value of the function as the minimum arrived at for the identified model. Thus, non-identifiability implies a zero derivative and conversely we can conclude that a non-zero derivative for a fixed parameter implies that it will be identified if we make it a free parameter.

To make inferences about differences among the groups in the development of verbal ability as measured by the two tests, we can use the model with $\sigma_{\epsilon_2\epsilon_4}$ free for the two groups. Our main interest is to compare the estimated structural equations (20). An inspection of these reveals that the γ-parameters, i.e., the slopes in the regressions of true criterion score on true covariate score, are almost equal, 0.947 for the C-group and 0.854 for the E-group. The hypothesis that these slopes are equal can be tested by estimating the model once more but with the restriction $\gamma^{(C)} = \gamma^{(E)}$ added. This results in a model with an overall chi-square measure equal to 4.0 with 5 degrees of freedom. Thus, the hypothesis of equal slopes cannot be rejected, since it leads to a chi-square with 1 degree of freedom equal to 1.2, and this is not significant. The different models and their associated chi-square values are summarized in Table 2. The maximum likelihood estimates of the parameters in the final model with equal criterion-covariate slopes, are listed in Table 3.

TABLE 2.

Measures of Overall Fit for the Models

Analyzed with Olsson's Data.

Model	chi-square	d.f.	Probability level
Initial	35.056	6	0.000
$\sigma^{(E)}_{\epsilon_2\epsilon_4}$ free	17.187	5	0.004
$\sigma^{(E)}_{\epsilon_2\epsilon_4}$ and $\sigma^{(C)}_{\epsilon_2\epsilon_4}$ free	2.799	4	0.592
$\sigma^{(E)}_{\epsilon_2\epsilon_4}$ and $\sigma^{(C)}_{\epsilon_2\epsilon_4}$ free, $\gamma^{(E)} = \gamma^{(C)}$	3.989	5	0.551
$\sigma^{(E)}_{\epsilon_2\epsilon_4}$ and $\sigma^{(C)}_{\epsilon_2\epsilon_4}$ free, $\gamma^{(E)} = \gamma^{(C)}$ and $\alpha^{(E)} = \alpha^{(C)}$	55.190	6	0.000

TABLE 3

Maximum Likelihood Estimates of the Parameters for the Final Model with Olsson's Data.

(Fixed parameters are denoted by an asterisk, *)

Parameter	Control Group		Experiment Group
λ_2		0.878	
λ_4		0.907	
σ_2	29.794		47.334
γ		0.895	
σ^2_ζ	1.032		8.823
$\sigma^2_{\varepsilon_1}$	9.584		2.547
$\sigma^2_{\varepsilon_2}$	12.030		12.359
$\sigma^2_{\varepsilon_3}$	5.836		7.451
$\sigma^2_{\varepsilon_4}$	12.500		17.209
$\sigma_{\varepsilon_2\varepsilon_4}$	6.391		7.304
$E(\xi)$	0.000*		1.875
$E(\eta)$	0.000*		5.306
μ_1		18.619	
μ_2		19.910	
μ_3		20.383	
μ_4		21.203	

From the parameter estimates in Table 3, we can compute the intercept α in the structural equation (20) for the experimental group as

$$\hat{\alpha} = \hat{E}(\eta) - \hat{\gamma}\hat{E}(\xi) = 5.306 - (0.895)(1.875) = 3.628.$$

To test whether there is a significant difference in intercepts, we can reanalyze the model with the restriction $\alpha^{(C)} = \alpha^{(E)}$. This results in a model with 6 degrees of freedom and a chi-square equal to 55.2. Thus, we can conclude that the treatment had an effect by virtue of a chi-square with 1 degree of freedom equal to 55.2 − 4.0 = 51.2.

Lord's Numerical Example

In Lord [1960], a large sample test of treatment effects for the case of a fallible covariate is derived. In an example, it was shown that by accounting for the measurement errors in the covariate a significant treatment effect was demonstrated, whereas an ordinary analysis of covariance would not have detected such an effect. In this section, the example will be reanalyzed to illustrate the specification of the general model in that case, and to show how a test similar to Lord's is obtained.

The data for the example are given in Table 4. They have been taken from a study by Frederiksen and Schrader [1951]. The criterion variable, y, is the Freshman average grade and there are two covariates, x_1 and x_2. The covariate x_2 is not actually observed, since no duplicate covariate was available. It is computed to be a measure parallel to x_1. There were two groups, Male and Female, with 119 and 93 observations, respectively, and x_2 was constructed in such a way that the reliability of x_1 and x_2 was 0.8 in the Male group and 0.72738 in the Female group.

The model for the data is the same as that given by (2)-(4), except that, because of the construction of the numerical data, $\lambda_1 = \lambda_2 = 1$ in (2) and $\sigma^2_{\epsilon_1(g)} = \sigma^2_{\epsilon_2(g)} = \sigma^2_{\epsilon(g)}$, say, for $g = 1,2$. It is then seen from (5) that we can estimate σ^2_ξ by $s_{x_1x_2}$, γ by $s_{yx_1}/s_{x_1x_2}$, σ^2_ϵ by $s_{x_1x_1} - s_{x_1x_2}$, and σ^2_ζ by $s_{yy} - s^2_{yx_1}/s_{x_1x_2}$ in both groups to obtain estimates that perfectly fit the data. The estimates of γ are quite similar in the two groups; γ equals 1.1212/4.232 = 0.2649 in the Male group and 0.55278/2.8229 = 0.1958 in the Female group. To test whether the γs are equal, we estimate the model with the constraint $\gamma^{(1)} = \gamma^{(2)}$. This gives a model with a chi-square with 1 degree of freedom equal to 2.304, which means that we cannot reject the hypothesis at the 10 per cent significance level. The maximum likelihood estimates of the parameters are listed in Table 5. The next step in the procedure is to test whether there is a difference among the groups in Freshman average grade when the differences in the covariate have been taken into account, that is to test whether α in Table 5 is equal to zero. This can be done by estimating the model with the constraint $\alpha^{(1)} = \alpha^{(2)}$ added. By comparing the chi-square for this model with the chi-square for the previous one we obtain a chi-square test of the hypothesis with 1 degree of freedom. The estimation of the model results in a chi-square with 2 degrees of freedom equal to 5.221. Thus, at the 10 per cent level we can reject the hypothesis of no effect by a chi-square with 1 degree of freedom equal to 5.221 − 2.304 = 2.917, corresponding to a value of a one-sided t-statistic equal to −1.36. This value of the test statistic should be compared with the corresponding t-value that is obtained from an ordinary analysis of covariance. As reported by Lord [1960], such an analysis with the duplicate measurement x_2 used as a second predictor variable results in a t-value which is not less than −0.9. For the test proposed by Lord [1960] a comparable t-value of −1.69 was computed.

Summary and Conclusions

The examples discussed in the previous two sections were deliberately chosen because of their simplicity in order to clarify the use of the proposed methodology. They illustrate

TABLE 4.

Sample Variance-covariance Matrices and Sample Means for the Data in Lord's Numerical Example.

Male Group (N=119)

Variable	S			Means
$x_1^{(1)}$	5.29000			4.07000
$x_2^{(1)}$	4.23200	5.29000		4.07000
$y^{(1)}$	1.12120	1.12120	0.56250	1.40000

Female Group (N=93)

Variable	S			Means
$x_1^{(2)}$	3.88090			5.34000
$x_2^{(2)}$	2.82290	3.88090		5.34000
$y^{(2)}$	0.55278	0.55278	0.37210	1.57000

in what ways the method is more general than the procedure of Lord [1969] and/or the usual analysis of covariance.

To sum up:

1. There is no restriction on the number of treatment groups that can be involved. Nor is there any restriction on the number of covariates or the number of criterion variables.
2. The fallible covariates can be parallel, tau-equivalent, or congeneric measures [see Jöreskog, 1971b], or can conform to a factor analysis model.

TABLE 5.

Maximum Likelihood Estimates of the Parameters in the Model for Lord's Numerical Example.

(Fixed parameters are denoted by an asterisk, *)

Parameter	Male Group		Female Group
λ_1		1.000*	
λ_2		1.000*	
σ_ξ^2	4.318		2.691
γ		0.242	
σ^2	0.273		0.259
σ_ε^2	1.041		1.096
α	0.000*		-0.137
μ_1		4.070	
μ_2		4.070	
μ_3		1.400	

3. The method can handle fallible criterion variables as well as fallible covariates.
4. There is a provision for a wide variety of different tests regarding the parameters of the model.

REFERENCE NOTES

1. Bergman, L. R. *Change as the dependent variable* (Res. Rep. Supplement 14). Stockholm, Sweden: The University of Stockholm, Psychological Laboratories, 1972.
2. Cronbach, L. J., Rogosa, D., Price, G. & Floden, R. *Analysis of covariance, temptress and deluder? Or angel of salvation?* (Mimeograph). Palo Alto: Stanford University, Stanford Evaluation Consortium, 1976.
3. Gruvaeus, G. T. & Jöreskog, K. G. *A computer program for minimizing a function of several variables* (ETS RM 70-14). Princeton, N.J.: Educational Testing Service, 1970.
4. Jöreskog, K. G. & Sörbom, D. *Some regression estimates useful in the measurement of change* (Res. Rep. 74-2). Uppsala, Sweden: University of Uppsala, Department of Statistics, 1974.

5. Keesling, J. W. & Wiley, D. E. *Measurement error and the analysis of quasi-experimental data (Version III)* (Mimeograph). Chicago: University of Chicago, 1975.
6. Olsson, S. [*An experimental study of the effects of training on test scores and factor structure.*] Uppsala, Sweden: University of Uppsala, Department of Education, 1973.
7. Porter, A. C. *How errors of measurement affect ANOVA, regression analysis, ANCOVA and factor analysis.* Paper presented at the AERA convention, New York, 1971.

REFERENCES

Bock, R. D. *Multivariate statistical methods in behavioral research.* New York: McGraw-Hill, 1975.

Campbell, D. T. From description to experimentation: Interpreting trends as quasi-experiments. In Chester W. Harris (Ed.), *Problems in measuring change.* Madison: The University of Wisconsin Press, 1963, pp. 212–242.

Cochran, W. G. Errors of measurement in statistics. *Technometrics,* 1968, *10,* 637–666.

Elashoff, J. D. Analysis of covariance: A delicate instrument. *American Educational Research Journal,* 1969, *6,* 383–402.

Fletcher, R. & Powell, M. J. D. A rapidly convergent descent method for minimization. *Computer Journal,* 1963, *6,* 163–168.

Frederiksen, N. & Schrader, W. B. *Adjustment to college.* Princeton, New Jersey: Educational Testing Service, 1951.

Härnqvist, K. [*Manual for DBA*]. Stockholm: Skandinaviska Testförlaget, 1962.

Jöreskog, K. G. Simultaneous factor analysis in several populations. *Psychometrika,* 1971a, *36,* 409–426.

Jöreskog, K. G. Statistical analysis of sets of congeneric tests. *Psychometrika,* 1971b, *36,* 109–133.

Jöreskog, K. G. A general method for estimating a linear structural equation system. In A. S. Goldberger & O. D. Duncan (Eds.), *Structural equation models in the social sciences.* New York: Seminar Press, 1973, pp. 85–112.

Jöreskog, K. G. & Sörbom, D. Statistical models and methods for test-retest situations. In D. N. M. de Gruijter, L. J. Th. van der Kamp & H. F. Crombag (Eds.), *Advances in Psychological and Educational Measurement.* London: Wiley, 1976a.

Jöreskog, K. G. & Sörbom, D. *LISREL III—Estimation of linear structural equation systems by maximum likelihood methods: A FORTRAN IV program.* Chicago: International Educational Services, 1976b.

Jöreskog, K. G. & Sörbom, D. Statistical models and methods for analysis of longitudinal data. In D. J. Aigner & A. S. Goldberger (Eds.), *Latent variables in socioeconomic models.* Amsterdam: North Holland Publishing Company, 1977.

Lawley, D. N. & Maxwell, A. E. *Factor analysis as a statistical method.* (2nd ed.) London: Butterworth & Company, 1971.

Lord, F. M. Large sample covariance analysis when the control variable is fallible. *Journal of the American Statistical Association,* 1960, *55,* 307–321.

Lord, F. M. Elementary models for measuring change. In Chester W. Harris (Ed.), *Problems in measuring change.* Madison: The University of Wisconsin Press, 1963, pp. 21–38.

Lord, F. M. A paradox in the interpretation of group comparisons. *Psychological Bulletin,* 1967, *68,* 304–305.

Lord, F. M. & Novick, M. E. *Statistical theories of mental test scores.* Reading: Addison-Wesley Publishing Company, 1968.

Smith, H. F. Interpretation of adjusted treatment means and regressions in analysis of covariance. *Biometrics,* 1957, *13,* 282–308.

Sörbom, D. A general method for studying differences in factor means and factor structure between groups. *British Journal of Mathematical and Statistical Psychology,* 1974, *27,* 229–239.

Sörbom, D. Detection of correlated errors in longitudinal data. *British Journal of Mathematical and Statistical Psychology,* 1975, *28,* 138–151.

Reprinted from: *Psychometrika,* September 1978, Vol. 43, No. 3.

Appendix

Acquisition of Computer Programs

The computer programs EFAP (Exploratory Factor Analysis Program), COFAMM (Confirmatory Factor Analysis with Model Modification), and LISREL (Linear Structural Relations) by Karl G. Jöreskog and Dag Sörbom are distributed by

International Educational Services
1525 East 53rd Street
Suite 829
Chicago, Illinois 60615

Index